Microscopic and Macroscopic Simulation Techniques

Kharagpur Lectures

Other World Scientific titles by the author

Smooth Particle Applied Mechanics
The State of the Art
ISBN: 978-981-270-002-5

Time Reversibility, Computer Simulation, Algorithms, Chaos
Second Edition
ISBN: 978-981-4383-16-5

Simulation and Control of Chaotic Nonequilibrium Systems
With a Foreword by Julien Clinton Sprott
ISBN: 978-981-4656-82-5

Microscopic and Macroscopic Simulation Techniques
Kharagpur Lectures

William Graham Hoover

UC Davis

Carol Griswold Hoover

Ruby Valley Research Institute, Nevada, USA

World Scientific

NEW JERSEY · LONDON · SINGAPORE · BEIJING · SHANGHAI · HONG KONG · TAIPEI · CHENNAI · TOKYO

Published by

World Scientific Publishing Co. Pte. Ltd.

5 Toh Tuck Link, Singapore 596224

USA office: 27 Warren Street, Suite 401-402, Hackensack, NJ 07601

UK office: 57 Shelton Street, Covent Garden, London WC2H 9HE

British Library Cataloguing-in-Publication Data
A catalogue record for this book is available from the British Library.

MICROSCOPIC AND MACROSCOPIC SIMULATION TECHNIQUES
Kharagpur Lectures

Copyright © 2018 by World Scientific Publishing Co. Pte. Ltd.

ISBN 978-981-3232-52-5

For any available supplementary material, please visit
http://www.worldscientific.com/worldscibooks/10.1142/10777#t=suppl

We dedicate this book to those who study at the
Indian Institute of Technology at Kharagpur.

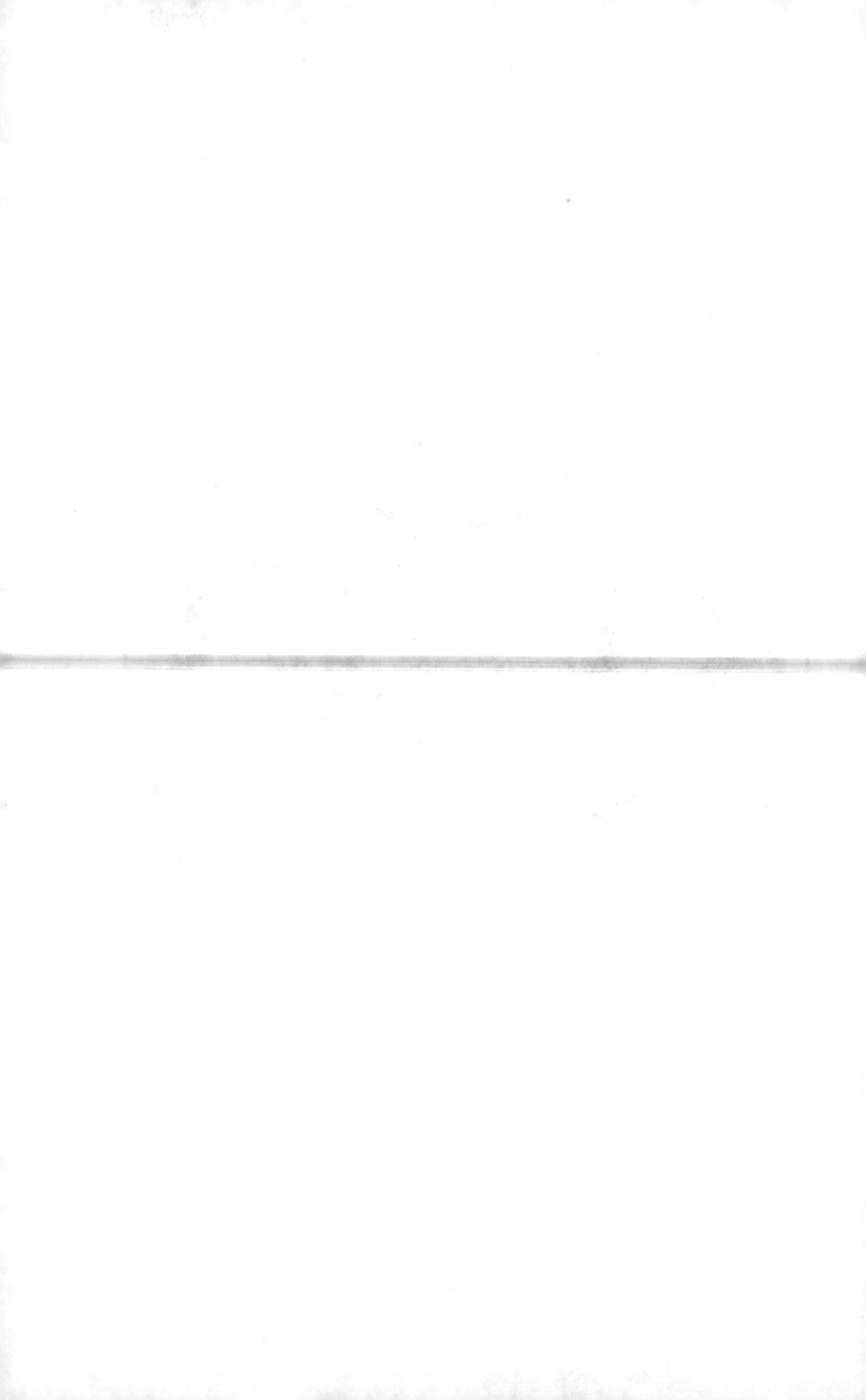

Foreword

It was nine parts Chutzpah and one part Prayer when my doctoral student Puneet Patra and I, on the last step of uploading our first ever thermostat paper in the *Journal of Chemical Physics* in December 2013, put Wm. G. Hoover's name in the required list of "suggested reviewers". Here we were, two civil engineers from the other side of the planet, unknown to him or to anyone in his community. But what did we have to lose, we reasoned, by having the world's authority in nonequilibrium molecular dynamics and the father of deterministic thermostatting read our first work, – we might as well give the great physicist a laugh as we went down in spectacular flames.

What happened three days later was the stuff of dreams. It was also the start of a beautiful friendship. Professor Hoover sent me an email praising our work. More emails were soon exchanged between the Hoovers, Puneet, and me. In less than six months, I was on my way to Ruby Valley Nevada.

Bill stresses "simplify" a lot in his conversations. And looking at the stark beauty of the desert, the clarity of the sky, the purity of the air and the snowfed streams, I realized why he and Carol chose the vast expanse east of the Ruby Mountains in northern Nevada as their home.

Two years later, Bill turned 80, and Karl Travis and Fernando Bresme decided to celebrate it at Karl's University at Sheffield, England. The pioneers, including the venerable Berni Alder, and their younger torchbearers were there. I got a ringside view of the excitement of the early days of the field, the triumphs, the advancement of theory through computer experiments, and the many open challenges in modelling and simulation of nonequilibrium thermodynamics and statistical mechanics.

By now, my Bhattacharya family had descended on Ruby Valley and the Hoovers had graciously accepted an invitation to visit Kharagpur, literally on the other side of the planet, and home of the Indian Institute of

Technology (IIT). Thus came about the two week short course on Nonequilibrium Statistical Mechanics and Molecular Dynamics (NESM/MD) under the Global Initiative of Academic Networks (GIAN) at IIT Kharagpur in December 2016. Carol and Bill taught the bulk of the course. Professor Abhishek Dhar of the International Centre for Theoretical Sciences, Bangalore, and I taught smaller parts. The course was attended by students and researchers from all over India.

The NESM/MD course was a treat in terms of the depth and range of subjects covered. I was humbled by the vast amount of new material the Hoovers created just for teaching this course. I am delighted to know that the Lectures at IIT Kharagpur were the genesis of this book in your hands.

Largely, this book amounts to a compendium of Bill and Carol's work. It starts with classical particle mechanics, the resulting differential equations, how their solutions can relate to macroscopic quantities and the meaning of equilibrium. A brief introduction is given to Newtonian, Lagrangian and Hamiltonian mechanics, and the different ways of solving the differential equations of equilibrium numerically. The book adopts "simplification". This is aptly demonstrated in Lecture 1 through the generous use of the pedagogical pendulum problem. Molecular dynamics is introduced as an extension of Newtonian and Hamiltonian mechanics. A brief history of constant-temperature mechanics follows the discussion on molecular dynamics. The last few sections of the Lecture introduce other pedagogical models for understanding material behaviour under mechanical and thermal loads – one-dimensional chains and multi-particle systems interacting through smooth pair forces. A description linking small-scale molecular dynamics and continuum scale hydrodynamics ends the Lecture.

Lecture 2 goes deeply into numerical integration of the equations of motion. It should be on the reading list of every graduate student working on coding MD algorithms. A comprehensive list of integration algorithms is discussed here. There is something for everyone – be they undergraduates, graduates, or advanced researchers – in this Lecture. Every numerical algorithm is followed by a FORTRAN code snippet. That will make the book very appealing to first-time coders.

Lectures 3 and 4 begin with thermometry and temperature, delve into the molecular dynamics of thermally constrained systems in depth, and discuss various algorithms for temperature control. The Hoovers immense knowledge and their significant contributions over the years make these Lectures a must read for anyone interested in understanding the fundamentals of thermometry. Through a simple five-particle atomistic chain rotating about its center of mass with constant angular velocity, Bill and Carol demonstrate the difficulties in interpreting the meaning of temperature using traditional means. The differences between constant isoenergetic molecular dynamics simulations and Gibbsian statistical mechanics is followed by a discussion of the classical numerical experiments performed by Fermi, Pasta, and Ulam, including the near-recurrences of initial states in their "FPU" chains. The Hamiltonian formalism of constant-temperature dynamics – Nosé, and Nosé-Hoover – presents a beautiful description of defining Hamiltonians for non-Hamiltonian dynamics by suitably augmenting the phase-space. Lecture 4 introduces additional ways of thermostatting a system applied to the prototypical "Lucy Fluid". As is a repeated theme throughout this book – "simplify" – this Lecture makes a good effort to explain the different and difficult concepts of mechanics in simple terms.

Lecture 5 examines the holy grail of ergodicity for algorithms controlling temperature. Applications to small systems are discussed in Lecture 7. Evidently, (dis)proving ergodicity is challenging, both theoretically and numerically. The Hoovers and their many coworkers have developed interesting numerical techniques that can be employed to understand the ergodicity in small scale systems. These techniques provide tests for identifying "useful algorithms". In thermostatted systems ergodicity is linked to the spectrum of Lyapunov exponents. Lectures 9 and 10 provide complementary reading to Lectures 5 and 7. A strong connection exists between dynamical systems, (ir)reversibility, and statistical mechanics. Lectures 5 through 10 are a must read for everyone interested in exploring these ideas.

Lecture 8 addresses the age-old question of how to reconcile the (nearly ubiquitous) reversibility of atomic motion with the (nearly universal) *irreversibility* of macroscopic processes. The book ends (Lecture 11) with a summary of smooth particle applied mechanics (SPAM) that is an advanced version of Lecture 6 which presents in brief the ideas of nonequilibrium hydrodynamics and its connections to equilibrium thermodynamics.

I am privileged to know the Hoovers – their graciousness to strangers, their generosity of spirit, and their humility, a side-effect of vast knowledge. In two short weeks, the students of the GIAN course at IIT Kharagpur in December 2016 experienced that same simplicity and that same spirit of giving, in addition to all the algorithms, equations, computer programs, colour figures and quizzes. This book reminds me of that wonderful time, and I am honoured to have been a part of it.

Baidurya Bhattacharya, Kharagpur, West Bengal, India, March 2018

Preface

While Puneet Kumar Patra was a doctoral student at the Indian Institute of Technology at Kharagpur, West Bengal, India, he and his Professor, Baidurya Bhattacharya, submitted a thought-provoking manuscript to *The Journal of Chemical Physics*, "A Deterministic Thermostat for Controlling Temperature Using All Degrees of Freedom". The journal sent it to me for review. I submitted an enthusiastic report and began an email correspondence with Puneet and Baidurya. This led to joint research work with Puneet and my wife Carol, and to two delightful visits of Professor Bhattacharya to our home in Ruby Valley Nevada.

The second visit, with his wife Sangeeta and their children Annapurna and Amal, led to Baidurya's suggestion that Carol and I visit India in 2016 to give lectures at a "Summer School" devoted to computer simulation of nonequilibrium systems, both microscopic and macroscopic. Weather dictated December 2016 as the time most propitious for the evolving "Summer School" project. We accepted the invitation. These Lectures are the result.

Along with the lectures came residence in the Institute's Kharagpur Guest House. Meals were delicious and spicy, augmenting familiar fare like hard-boiled eggs and bananas with many unfamiliar dishes, mostly vegetable curries augmented on special order with chicken or fish. About half the Kharagpur attendees were vegetarian. Their hunger for physics made our Lectures a real pleasure. Their enthusiasm was contagious. They were eager to learn and we to teach, a perfect combination.

With eleven lectures behind us, Professor Bhattacharya took us to his family home in Calcutta where we shared a meal with his mother Kalpana and his brother's family from England. Next came a week of sightseeing in Calcutta, Delhi, and Agra, at last seeing the wonderous Taj Mahal, the Red Fort across the river, with its troupe of monkeys, returning all too soon to Calcutta and Dubai for the sixteen hour flight to San Francisco. We returned on Christmas Eve for a reunion with our California Family.

It is no surprise that many topics covered in our recent books recur here. *Simulation and Control of Chaotic Nonequilibrium Systems* was sent to World Scientific Publishers on New Year's Eve, the last day of 2014. During 2015 we carried on an active collaboration with Patra, Bhattacharya, joined by Clint Sprott, Professor Emeritus at the University of Wisconsin-Madison. Besides terrific energy Clint has a gift for graphic renderings capturing the essence of the mathematical physics breakthroughs which foster improved understanding. As a result of these collaborations the five of us were able to settle some relatively longstanding questions dating back to Shuichi Nosé's 1984 extension of Hamiltonian mechanics to a dynamics corresponding to Gibbs' canonical ensemble. In Nosé's mechanics temperature, rather than energy, was the given independent variable.

After nearly sixty years of research work in statistical mechanics, the year 2018 looks like a viable retirement year for us. Carol and I decided on this *Kharagpur Lectures* book as the best means for passing on our current ideas to our successors. Understanding how things work through analyses of simple models is our passion. We trust that the models explored here and shared with our Indian colleagues will serve others in their own pursuits of truth through the exploration and study of new ideas.

The audience for our Lectures was varied, from advanced undergraduates to seasoned faculty. We endeavored to include ideas challenging the understanding of all of them without leaving anyone behind. In the Lectures we introduce pedagogical model systems like the Galton Board and the Baker Maps at an undergraduate level. They then recur at a more advanced level, at the state of the art of the research frontier. We provide comprehensive references for readers wishing more detail than could be included in the 22-hour Lecture format. We hope that the seeds planted at Kharagpur will produce new and interesting research around the World and foster international collaborations.

In our view it is essential that a useful understanding of mechanics be based on numerical methods, delicately supplemented with mathematical analysis. We have tried to strike that balance here and are grateful to our Indian colleagues for providing the intellectual stimulation and the physical support making this work possible. We are also grateful to our Editor, Lakshmi Narayanan for her continued encouragement.

William Graham Hoover and Carol Griswold Hoover, Ruby Valley Nevada

Contents

Chapter 1

Mechanics, Molecular Dynamics, and Gibbs' Statistical Mechanics

/ Classical Mechanics / Numerical Methods for Differential Equations / Størmer-Verlet / Runge-Kutta / Software / Equilibrium from Gibbs' Entropy / Newtonian Molecular Dynamics / Corresponding States / van der Waals' Equation / Gauss' and Nosé-Hoover Mechanics / Stiff Equations and Adaptive Integration / Lattice Dynamics / A Manybody Problem / Connecting Mechanics to Hydrodynamics / Virial Theorem / Pressure Tensor Symmetry / Heat Theorem / Summary of Lecture 1 /

1.1 Introduction

Much of computational physics includes the modelling of systems using "ordinary" differential equations which specify the time derivatives of "dependent variables", those that change with time. Time is the "independent variable". Particle mechanics, our own special interest, was formulated in terms of such ordinary differential equations by Newton (1642-1726). The subject evolved with further innovations thanks to the talents of Joseph-Louis Lagrange (1736-1813), Johann Carl Friedrich Gauss (1777-1855), William Rowan Hamilton (1805-1865), and Shuichi Nosé (1951-2005).

Typical mechanical models of physical systems (hard spheres or disks, harmonic crystals, Lennard-Jones fluids, ...) give the time evolution of the "degrees of freedom" (coordinate-momentum pairs of variables) needed to describe the state of the system and to evolve it in time. Although we happily take advantage of simple systems with analytic solutions of their motion equations (such as a particle in a constant gravitational field or a harmonic oscillator) we mainly seek numerical solutions. The numerical work required to evolve a model system in time typically requires a step-by-step integration algorithm, evolving the system of interest from time zero

1

to a time dt, and then to time $2dt$, and so on. In this Lecture we introduce the Størmer-Verlet and Runge-Kutta integration algorithms. These tools can be used to solve Newton's, Lagrange's, Gauss', Hamilton's, and Nosé's equations of motion. We will solve motion equations here with both algorithm types. Students and readers will be enabled to formulate and solve problems in all five categories of mechanics. Numerical methods have a long and sophisticated history. The more intricate details will be discussed and compared by Carol in **Lecture 2**.

This introductory Lecture is focussed on the time-dependent description of atomic and molecular systems using mechanics and numerical integration as computational tools. Numerical solutions so obtained can be correlated with laboratory experiments and with physical descriptions of macroscopic behavior. The Virial Theorem for the pressure tensor and the Heat Theorem for the heat flux vector provide essential links relating microscopic particulate descriptions to observed macroscopic behavior. Gibbs' statistical mechanics is another essential tool. Statistical mechanics provides links between microscopic averages and macroscopic behavior. At equilibrium the statistical Monte-Carlo method is another useful tool, providing an alternative source for equation of state information which can be correlated with the results of dynamical simulations.

1.2   Formulating Classical Mechanics

In writing ordinary differential equations we consistently indicate the time derivative following the motion by a superior dot : the time derivatives of the horizontal and vertical coordinates x, y are respectively $\dot{x}$ and $\dot{y}$. Newton's second law of motion gives the vertical acceleration (the second time derivative of the coordinate y is $\ddot{y}$) as the force F_y divided by the mass m .

To illustrate the analytic solution of an ordinary differential motion equation consider the vertical acceleration of a mass m due to the gravitational field g. Newton's "equation of motion" $F = ma$ becomes :
$$F = m\ddot{y} = -mg \longrightarrow \dot{y} = -gt \longrightarrow y = -g(t^2/2) \ .$$
For this model system, with velocity proportional to the time t and with the coordinate change quadratic in time, numerical methods are unnecesssary. We know precisely where the falling mass is at any time as well as its speed, from the analytic solution of Newton's second law of motion.

Lagrange's formulation of this same problem begins with the Lagrangian function $\mathcal{L}(y, \dot{y})$, the "action", which is minimized by his "Least-Action" equation of motion :

$$\mathcal{L}(y, \dot{y}) = K - \Phi \; ; \; (d/dt)(\partial \mathcal{L}/\partial \dot{y}) = (\partial \mathcal{L}/\partial y) \; .$$

Here K is the kinetic energy $(m\dot{y}^2/2)$ and Φ is the potential energy, mgy . The result is the same as Newton's

$$(d/dt)m\dot{y} = m\ddot{y} = -mg \longrightarrow \ddot{y} = -g \; .$$

Lagrange's formulation is well-suited to the description of constrained systems, where the constraints can control microscopic geometry or macroscopic thermodynamic variables.

Hamilton's formulation of mechanics is based on the Hamiltonian function $\mathcal{H}(y, p)$ where $\mathcal{H}$ is the sum (the total energy) rather than the difference (the action) of K and Φ . Hamilton's motion equations, $\dot{q} = +(\partial \mathcal{H}/\partial p)$ and $\dot{p} = -(\partial \mathcal{H}/\partial q)$, treat the coordinates q and the momenta $p = (\partial \mathcal{L}/\partial \dot{q})$ as independent variables :

$$\mathcal{H}(y, p) = K + \Phi = (p^2/2m) + mgy \; ;$$

$$\dot{y} = +(\partial \mathcal{H}/\partial p) = +(p/m) \; ; \; \dot{p} = -(\partial \mathcal{H}/\partial y) = -mg \; .$$

The result is once again identical to Newton's :

$$\{ \, \dot{y} = (p/m) \; ; \; \dot{p} = -mg \, \} \longrightarrow \ddot{y} = -g \; .$$

Hamilton's motion equations are useful in the treatment of macroscopic stress and are essential to the formulation of the dynamics of quantum systems. Evidently we can solve problems in Newtonian, Lagrangian, and Hamiltonian mechanics once we learn how to deal with "second-order" ordinary differential equations [those involving two time derivatives].

The formulations we applied to a falling particle in a gravitational field can be applied unchanged to the harmonic oscillator model. An oscillating force exerted by a linear Hooke's-Law spring is proportional to displacement $q - q_{eq}$. If we choose the mass m and force constant κ of the spring both equal to unity the second-order differential equation of motion is the simplest possible linear equation. It has sinusoidal solutions. If we start out at time $t = 0$ with the oscillator motionless and with a displacement equal to unity the trajectory is described by the analytic solution $q = \cos(t)$:

$$m\ddot{q} = -\kappa q \rightarrow \ddot{q} = -q \longrightarrow \{ \, \dot{q} = -\sin(t) \; ; \; q = \cos(t) \, \} \; .$$

In this case we have chosen the constants of integration (the velocity at the initial time 0 and the coordinate at the initial time) equal to 0 and 1, for simplicity.

> **Exercise :** Show that the oscillator problem has a sinusoidal solution
> no matter what the initial choices for q and $\dot{q}$.

Here again we have avoided the need for numerical integration because
we have an exact expression for the solution for all times. In fact the
two examples we have just considered, the falling body and the harmonic
oscillator, are unusual. Let us up the difficulty so that numerical methods
are required. We will choose a slightly harder problem next.

1.3   Numerical Methods for Differential Equations

The next step up in complexity from the falling particle and the harmonic
oscillator includes "nonlinearity" and can easily fall outside the scope of
analytic methods. Two examples are the simple pendulum problem, $\ddot{\theta} =
-\sin(\theta)$, and the Mexican Hat problem, $\ddot{q} = +q - q^3$. In these cases, where
the solutions are relatively complicated, the accuracy of numerical solutions
can be checked by analyzing reversibility and energy conservation. To see
how, let us begin with the pendulum problem :

$$\ddot{\theta} = -\sin(\theta) \ .$$

Here we have chosen the pendulum mass and length, as well as the gravitational field strength all equal to unity. How can we find $\theta(t)$? Solving
Newton's motion equations in Cartesian coordinates is most simply formulated as a centered-difference problem with a small time increment dt . We
describe the simplest such algorithm next. It is one of several developed
by the Norwegian mathematician Carl Størmer (1874-1957) and was also
used by Newton. Richard Feynman used it in his undergraduate *Lectures
in Physics* series. Loup Verlet[1] (1931-) applied it to "molecular dynamics" [meaning the microscopic manybody problem]. Let us apply this
"Størmer-Verlet" method to the motion of the pendulum.

> **Exercise :** Show that the tangential component of gravitational ac-
> celeration g of a mass constrained to lie on the unit circle is $-g\sin(\theta)$
> where θ varies from zero at the bottom to π at the top.

1.3.1 *The Størmer-Verlet Centered-Difference Algorithm*

To solve the pendulum problem $\ddot{\theta} = -\sin(\theta)$ it is natural to consider series expansions. Summing the Taylor's series expansions for θ_{t-dt} and θ_{t+dt} gives an expansion in even powers of dt :

$$\theta_{t-dt} + \theta_{t+dt} = 2\theta_t + dt^2\ddot{\theta}_t + \dots .$$

The dots indicate $(dt^4/12)\,\ddddot{\theta}_t$ and higher order terms. For small dt, we can justify ignoring the dt^4 and higher terms altogether. Rearranging the truncated expansion for $\ddot{\theta}$ gives the Størmer-Verlet algorithm :

$$\theta_{t+dt} = 2\theta_t - \theta_{t-dt} + dt^2\ddot{\theta}_t \; [\; \text{Størmer–Verlet} \;] \; .$$

A FORTRAN program implementing this algorithm for ten timesteps would begin by setting the timestep DT and the time as well as the current and previous values of the angle, THETA0 and THETAM. The time-stepping loop would then look like this :

```
DO IT = 1,10
  THETAP = 2*THETA0 - THETAM - DSIN(THETA0)*DT*DT
  TIME = TIME + DT   ! [ PRINT OUT TIME AND THETAP IF DESIRED ]
  THETAM = THETA0
  THETA0 = THETAP
ENDDO
```

Twin advantages of the Størmer-Verlet algorithm are simplicity and time-reversibility. With two successive coordinate values one can iterate forward or backward with equal ease. A disadvantage of the algorithm is the absence of the velocity $\dot{\theta}$. A simple definition is

$$\dot{\theta} \equiv (\theta_{t+dt} - \theta_{t-dt})/(2dt) \; .$$

If we choose the initial value of the pendulum at a turning point (where all of the energy is potential and the velocity vanishes) we can solve for the previous and subsequent values (which symmetry shows are equal). Let us choose the horizontal case, $\theta_0 = (\pi/2)$, as our initial condition :

$$\theta_{+dt} + \theta_{-dt} = 2\theta_0 + dt^2 a_0 = \pi - dt^2 \; .$$

At a time of 9 an accurate (Runge-Kutta) integration of the motion equation gives :

$$\theta_9 = 0.3777\;5740\;4173 \; ; \; \dot{\theta}_9 = -1.3634\;4719\;3213 \; .$$

Størmer-Verlet integration gives the following coordinate values for $dt = 0.001$, $dt = 0.002$, $dt = 0.003$, $dt = 0.004$:

$\{$ 0.3777 5740, 0.3777 5739, 0.3777 5737, 0.3777 5735 $\}$.

The errors vary close to quadratically in dt . **Figure 1** shows the variation of the errors in the coordinate θ over the range from $dt = 0.001$ to 0.006.

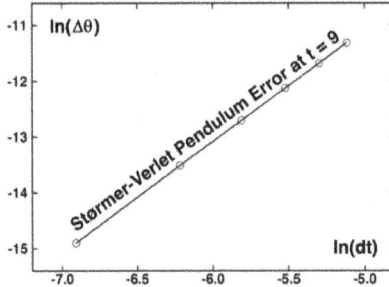

Fig. 1.1:   We see the numerical coordinate errors incurred by the Størmer-Verlet algorithm applied to the pendulum motion equation $\ddot{\theta} = -\sin(\theta)$. The pendulum is originally horizontal and its motion is followed for a time of 9. Numerical solutions from 1500 to 9000 timesteps show a quadratic dependence of the error, $|\theta_{SV} - \theta_{RK}|$, where the "exact" angle θ_{RK} is accurate to 13 significant figures. The timestep range shown here is from 0.001 to 0.006. The Størmer-Verlet algorithm is said to be "second-order accurate". The errors in the co-ordinate, velocity, and energy all vary as dt^2 if the simulation is carried out to a fixed time t. Here that fixed final time t is 9.

Exercises : Show that the single-timestep "local" error in the Størmer-Verlet algorithm is quartic in dt . Explain why this is consistent with the quadratic variation of the global error illustrated in **Figure 1.1**. The velocity in the Störmer algorithm could be estimated from either one of the two approximations :

$$\dot{\theta} \simeq (\theta_{t+dt} - \theta_{t-dt})/(2dt) \text{ or } \dot{\theta} \simeq (\theta_{t+2dt} - \theta_{t-2dt})/(4dt) .$$

Discuss the merits of a weighted average of these two approximations. Would it be worthwhile to add $(\theta_{t+3dt} - \theta_{t-3dt})/(6dt)$ to the mix ?

1.3.2 *Runge-Kutta Integration*

The Størmer algorithm just illustrated is well-suited to Newtonian mechanics in Cartesian coordinates. First-order differential equations require a different more general approach. Over a century ago Carl Runge and Wilhelm Kutta developed a family of new methods for first-order equations. The most useful of these combines four estimates of the derivative $\dot{y}$ in the neighborhood of $y(t)$ so as to eliminate error terms of order dt, dt^2, dt^3, and dt^4. The fourth-order method I give below is arguably the most useful. It is easy to program and has a single-timestep "local" error of order $(dt^5/5!)$. Here is the FORTRAN version of this integrator, which we name RK4 :

```
      SUBROUTINE RK4(YNOW,YDOT,DT)
      IMPLICIT DOUBLE PRECISION(A-H,O-Z)
      PARAMETER(NEQ=2)
      DIMENSION YNOW(NEQ),YDOT(NEQ),YNEW(NEQ)
      DIMENSION DOT1(NEQ),DOT2(NEQ),DOT3(NEQ),DOT4(NEQ)
      CALL RHS(YNOW,YDOT)
      DO 1 I = 1,NEQ
    1 DOT1(I) = YDOT(I)
      DO 2 I = 1,NEQ
    2 YNEW(I) = YNOW(I) + DT*DOT1(I)/2
      CALL RHS(YNEW,YDOT)
      DO 3 I = 1,NEQ
    3 DOT2(I) = YDOT(I)
      DO 4 I = 1,NEQ
    4 YNEW(I) = YNOW(I) + DT*DOT2(I)/2
      CALL RHS(YNEW,YDOT)
      DO 5 I = 1,NEQ
    5 DOT3(I) = YDOT(I)
      DO 6 I = 1,NEQ
    6 YNEW(I) = YNOW(I) + DT*DOT3(I)
      CALL RHS(YNEW,YDOT)
      DO 7 I = 1,NEQ
    7 DOT4(I) = YDOT(I)
      DO 8 I = 1,NEQ
    8 YNOW(I)=YNOW(I)+DT*(DOT1(I)+2*(DOT2(I)+DOT3(I))+DOT4(I))/6
      RETURN
      END
```

Here the number of equations NEQ is two. Typically a Newtonian system with N degrees of freedom will entail $2N$ first-order differential equations, NEQ=2*N. The RHS(YNOW,YDOT) subroutine furnishes the NEQ right-hand sides of the first-order differential equations. For the pendulum, with just two first-order equations, for $\dot\theta$ and $\ddot\theta$, the RHS subroutine would have the form :

```
SUBROUTINE RHS(YNOW,YDOT)
IMPLICIT DOUBLE PRECISION(A-H,O-Z)
DIMENSION YNOW(2),YDOT(2)
YDOT(1) = YY(2)
YDOT(2) = - DSIN(YY(1))
RETURN
END
```

There are *many* other Runge-Kutta integrators. A fifth-order integrator is described and applied in **Lecture 2**. It requires six force evaluations rather than four, for a loss of efficiency relative to RK4 by a factor of $(6/4) = (3/2)$. The single-step errors with RK5 vary as $(dt^6/6!)$. For the harmonic oscillator and pendulum problems the single-step coordinate, velocity, and energy errors are all of order dt^6. The difference between the two algorithms provides a good estimate of the less-accurate fourth-order method's error.

Comparing two different integration methods can yield an *adaptive integrator* useful for solving "stiff" differential equations in which the magnitudes of the righthand sides vary wildly, over many powers of 10. An adaptive integrator can also be developed by comparing the result at $t+dt$ from a single RK4 timestep to the corresponding result from two successive RK4 timesteps using half the timestep, $(dt/2)$. If the error is too large (or too small) the timestep is reduced (or increased). We will come back to the use of adaptive integrators when discussing the Nosé oscillator problem, a challenging "stiff" set of four coupled ordinary differential equations.[2]

Because the fourth-order integrator, applied to the oscillator, decays, a good error estimate at time t can be obtained by reversing the velocity and integrating back to the initial time. Comparing the errors for RK4 and RK5 shows that the energy error from RK4 is larger by roughly a factor of ten, more than compensating for the extra force evaluations using RK5. Nevertheless, considering the complexity of RK5 it is no surprise that RK4 is the preferred algorithm for solving run-of-the-mill ordinary differential equations. Let us turn now from integrators to the display of results.

1.3.3 Compiler and Graphics Software

The best approach for students seeking to master numerical algorithms is to solve example problems and to display the results. The program segments provided in this book are written in FORTRAN 77 and can be regarded as pseudo programming for those students preferring other languages or more advanced versions of FORTRAN. The FORTRAN 77 syntax is designed to resemble the structure of mathematical algorithms. This makes it specially useful as a guide for writing in other languages with more "stringent" – some would say "cumbersome" or "idiosyncratic" – syntactical rules. The gfortran compiler is a free FORTRAN compiler available in binary form for most computer systems. gfortran is well documented and conforms to the FORTRAN standards up to the year 2000 including the original FORTRAN 77. Use Google search for gfortran downloads for a free copy of the software.

Good visualization tools are essential aids in all fields of computational physics. We have found that the gnuplot software has exactly the capabilities we need for publications, books, and other research projects. These graphics capabilities include line plots, point plots, contour plots, color, labels, and the rendering of multiple plots per page. More advanced capabilities include three-dimensional hidden surfaces, transparency, and coloring of surfaces. gnuplot is well documented with a large collection of demonstration problems. Use Google's "search" feature to find it. **Figure 1.2**, which I constructed to demonstrate the quadratic dependence of the oscillator error on dt, was generated with gnuplot. The exercise below involves replicating the plot and is a necessary step for the student who seeks a working knowledge of computational mechanics. The ability to create and analyze graphic representations of computational results is absolutely necessary to success in computational physics and dynamical systems.

Exercise : Write a Störmer-Verlet program to solve the harmonic oscillator problem with unit mass, force constant, and frequency. Begin at $(q, p) = (1, 0)$ and compute the local errors, $\delta q = |q - \cos(t)|$ and $\delta p = |p + \sin(t)|$ and their maxima Δq and Δp for two periods. Use five timesteps varying twofold down from the largest, $dt = (2\pi/50)$. Use a plotting package to compare a ln-ln graph of your data to **Figure 1.2**. Explain why the slope is close to 2.

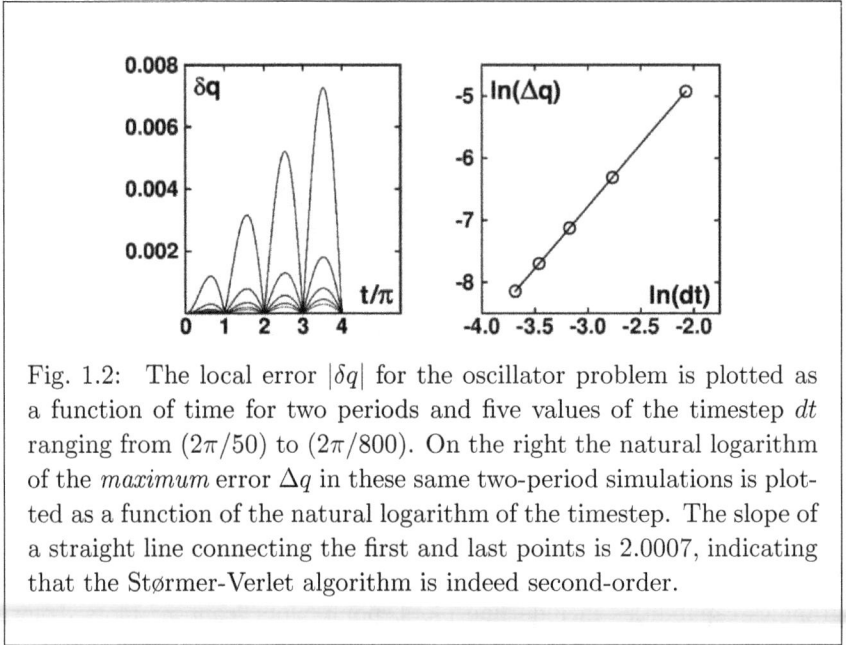

Fig. 1.2:   The local error $|\delta q|$ for the oscillator problem is plotted as a function of time for two periods and five values of the timestep dt ranging from $(2\pi/50)$ to $(2\pi/800)$. On the right the natural logarithm of the *maximum* error Δq in these same two-period simulations is plotted as a function of the natural logarithm of the timestep. The slope of a straight line connecting the first and last points is 2.0007, indicating that the Størmer-Verlet algorithm is indeed second-order.

1.4   Thermomechanical Equilibrium from Gibbs' Entropy

Josiah Willard Gibbs (1839-1903) was able to calculate the entropy for an ideal gas directly from his "sum over states" phase-volume definition :

$$S(E,V) = k\ln(\Omega) = k \prod \int dq \int dp \text{ where } E < \mathcal{H} < E + dE .$$

Gibbs showed that this statistical entropy, an integral over the phase-space volume Ω corresponding to the energy E, shares the characteristic extensive, state function, and non-decreasing qualities of the thermodynamic entropy $S(E,V)$.

Let us sketch such a calculation for a two-dimensional ideal gas confined to a volume V with total energy E. Evidently the coordinate integration gives $\int \cdots \int dq_1 \ldots dq_{2N} = V^N$. From the formal standpoint the momentum integration is more complicated because the momenta are all coupled, $\sum p_i^2 = 2mE$, where E is the (kinetic) energy. It is also good to restrict the summed-up momentum to vanish : $\sum v_x = \sum v_y \equiv 0$. For large N the resulting $(2N - 3)$-dimensional integral approaches the volume of a $(2N)$-dimensional sphere, proportional to $(2mE)^N$. It is an interesting exercise to work out the exact integrals for small systems, say $N = 2$ and

3 and 4. For two two-dimensional particles with $v_1^2 + v_2^2 \equiv 4$ the require-ment that $v_2 = -v_1$ reduces the four-dimensional velocity integral to a two-dimensional one with $v_x^2 + v_y^2 = 2$. Switching to polar coordinates gives 2π for the angular integration and $\sqrt{2}$ for the radial.

For an ideal-gas thermometer with each of its many degrees of freedom giving on average $\langle\, mv^2\, \rangle = kT$ we need only the large-N limit of the volume of a DN-dimensional velocity hypersphere. That result gives us the volume and energy dependence of Gibbs' phase integral for an ideal gas, allowing us to compute its thermodynamic derivatives of the entropy :

$$S = k \ln[\, \smallint \cdots \smallint \textstyle\prod dq dp = k \ln \Omega\,] \propto (V E^{D/2})^N \longrightarrow$$
$$(\partial S/\partial E)_V = (NDk/2E) = (1/T) \; ; \; (\partial S/\partial V)_E = (Nk/V) = (P/T) \, .$$

Gibbs supposed that at thermomechanical equilibrium *any* fluid, coupled to both a piston and an ideal-gas thermometer maximizes the overall phase volume, $\Omega_{ideal} \times \Omega_{fluid}$. Equilibrium will therefore satisfy both conditions, thermal and mechanical, for maximum entropy :

$$(\partial S_{ideal}/\partial E)_V = (\partial S_{fluid}/\partial E)_V \; ; \; (\partial S_{ideal}/\partial V)_E = (\partial S_{fluid}/\partial V)_E \, .$$

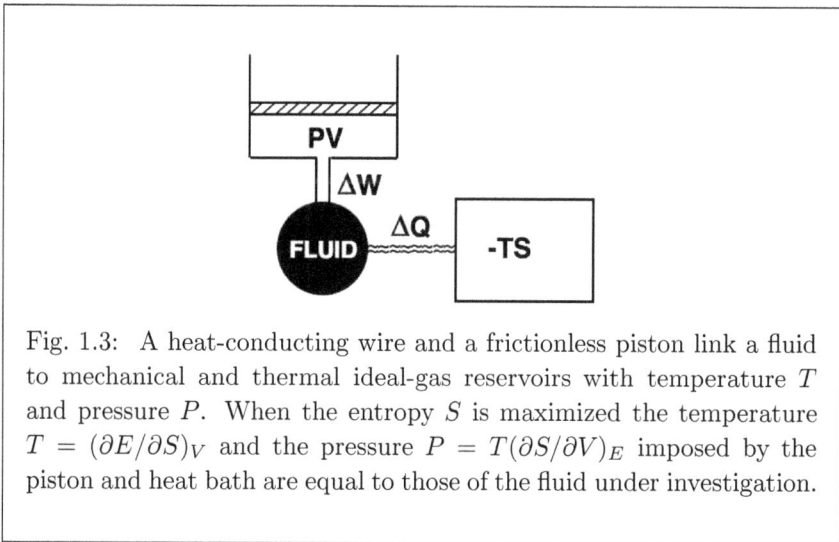

Fig. 1.3:   A heat-conducting wire and a frictionless piston link a fluid to mechanical and thermal ideal-gas reservoirs with temperature T and pressure P. When the entropy S is maximized the temperature $T = (\partial E/\partial S)_V$ and the pressure $P = T(\partial S/\partial V)_E$ imposed by the piston and heat bath are equal to those of the fluid under investigation.

Figure 1.3 is a sketch of the geometry corresponding to simultaneous mechanical and thermal equilibrium. The mechanical and thermal reser-voirs should both be envisioned as ideal gases so that we know how their en-tropy depends upon pressure and temperature. The equilibrium conditions then guarantee that (P/T) and T for the equilibrated "fluid" match those of the barostat and thermostat with which the fluid is linked.

The conclusion is that "equilibrium" corresponds to the simultaneous mechanical condition of pressure equilibrium along with the energetic condition of thermal equilibrium. These observations, physical in nature but also consequences of Gibbs' analysis, led to the thermostatted simulation techniques we employ today. Before detailing these "recent" (1984 onwards) developments we briefly review the progress that preceded present practice.

1.5   Development of Newtonian Molecular Dynamics

The early molecular dynamics simulations modelled $N < 1000$ particles interacting with pair forces inside a fixed computational box, usually periodic so as to minimize edge effects. In the late 1950s Berni Alder and Tom Wainwright studied "square-well" particles as well as hard disks and spheres.[3, 4] Disks and spheres have a high-density solid phase and a low-density fluid phase. The dynamics is simple. When two of these hard particles collide they exchange their line-of-centers relative velocity, $(v_{ij} \cdot r_{ij})(r_{ij}/\sigma^2)$, where σ is the particles' diameter. The square-well potential introduces an attractive well extending from σ to $\sigma + \Delta\sigma$ with constant depth ϵ.

The additional attractive forces induced by the potential well add a liquid phase, triple point, and critical point to the simpler solid-fluid phase diagram. A typical phase diagram including all three phases is shown in **Figure 1.4**. The "triple point" is the pressure-temperature point at which all three phases, gas, liquid, and solid can coexist. The "critical point" is the pressure-temperature point above which there is no gas-liquid equilibrium.

At densities up to two thirds of close packing hard spheres behave as fluids, with a pressure that can be understood as a power series in the density [the virial series] due to the interactions of two, three, ... particles at a time. At densities greater than three fourths of close packing spheres seek out a crystalline structure and behave like an elastic solid with a resistance to shear in addition to their bulk moduli $B_{(S \text{ or } T)} \equiv -V(\partial P/\partial V)_{(S \text{ or } T)}$.

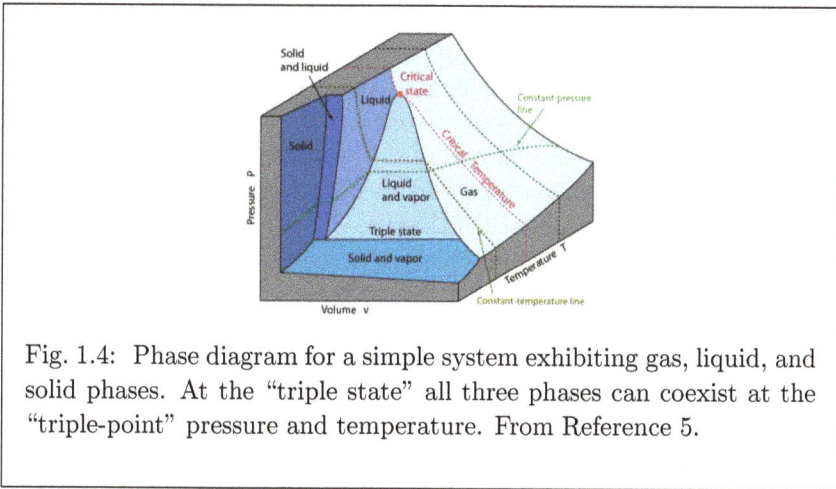

Fig. 1.4: Phase diagram for a simple system exhibiting gas, liquid, and solid phases. At the "triple state" all three phases can coexist at the "triple-point" pressure and temperature. From Reference 5.

Alder and Wainwright's impulsive forces governing the evolution of disk and sphere systems were much simpler, and therefore faster to incorporate and simulate, than were the continuous forces used by the other early practitioners of molecular dynamics.[6–8] Collisions occurring at fixed times are much more easily computed than continuous collisions with durations of dozens or hundreds of timesteps. Hard particles allow a much faster evolution than the smooth potentials used by other workers in molecular dynamics.

Alder and Wainwright's computer programs jumped or "stepped" from one collision to the next without the need for a multi-timestep tracing of the free-flight portions of the trajectories.[9] Such a dynamics is relatively straightforward to program by computing the times to each particle's next collision, sorting these times, computing the velocity changes, and then the new times required as the chain of collisions evolves.

Exercise : Consider two hard disks of unit diameter at the moment of collision, one at the origin with $\dot{x}_1 = 3$ and the other at $(x_2 = 1, y_2 = 0)$ with $\dot{x}_2 = 2$. Do the $\dot{y}$ change? What are the post-collision velocities of the two disks ? Show that the total momentum and kinetic energy are conserved. In the general case one calculates the relative radial velocity $v_{12} \cdot r_{12}$ and computes the post-collision velocities $v_1' = v_1 - r_{12}(v_{12} \cdot r_{12})$ and $v_2' = v_2 + r_{12}(v_{12} \cdot r_{12})$.

A somewhat involved, but feasible, computational project can be built around simulating the repeated collisions of two disks in a periodic box. Once this is properly programmed it is relatively easy to add additional particles. The two-particle program can form the basis of a molecular dynamics simulation code for hard-disk dynamics. What are the three types of collisions for square-well particles? How would you compute the post-collision velocities for each of the three?

Alder and Wainwright found freezing transitions for hard disks (near three-fourths of close packing) and hard spheres (near two-thirds). These outcomes were surprising and controversial at the time, but have by now been elaborately confirmed by dozens of researchers.[10]

Even earlier Fermi, Pasta, and Ulam at Los Alamos[6] and a little later George Vineyard and his many coworkers at Brookhaven[7] used the Størmer-Verlet algorithm :

$$\{ r_+ = 2r_0 - r_- + (dt)^2 F_0 \} \ [\ \text{Størmer} - \text{Verlet} \ ],$$

with smooth potentials, to study nonlinear effects in small systems and radiation damage in larger ones. This algorithm evolves coordinates. It neither requires nor produces velocities. If velocities are desired (to calculate the energy, temperature, and pressure, for example) they can be approximated in either of two ways :

$$p_0 \equiv (4/3)(q_+ - q_-)/(2dt) - (1/3)(q_{++} - q_{--})/(4dt) \ \text{or}$$

$$p_0 \equiv (q_+ - q_-)/(2dt) \ .$$

A six-time version is also possible, but differs only slightly from its four-time cousin. Carol will discuss this in **Lecture 2**.

Aneesur Rahman simulated argon at Argonne.[8] He used a centered-difference "predictor-corrector" algorithm in his simulations. The predictor makes use of the most recent momenta, $\{ p_0 \}$ as well as the "old" coordinates, $\{ q_- \}$:

$$\{ q_+ = q_- + 2dtp_0 \} \rightarrow \{ F_+ \} \ .$$

This "predictor" is just a leapfrog timestep with timestep $2dt$. It is followed by one or more applications of the "corrector", which improves the accuracy by spanning just a single timestep rather than two of them :

$$\{ p_+ = p_0 + (dt/2)[\ F_+ + F_0 \ ] \ ; \ q_+ = q_0 + (dt/2)[\ p_+ + p_0 \ ] \} \ [\ \text{Rahman} \ ] \ .$$

These last two velocity and coordinate "corrector" steps can be iterated to improve the accuracy at the expense of more computer time. This less-efficient and less-accurate method has an interesting aspect. The converged corrector solution conserves energy for the harmonic oscillator problem. But it is so inefficient as to be impractical, as **Figure 2.7** in **Lecture 2** shows.

While Fermi and Alder and Rahman were developing molecular dynamics researchers in stellar dynamics worked on their own N-body problems. They used gravitational inverse-square forces, and soon recognized the accuracy limitations imposed by Lyapunov instability. There is a large literature on their numerical simulations going back to early work by Aarseth and Miller in the 1960s. Well ahead of the computer revolution Holmberg carried out interesting "lightbulb experiments" simulating astrophysical problems in 1941![11]

Researchers simulating fluids and solids typically used smooth potentials, often with crude cutoffs. More "realistic" potentials were developed to model metals and polyatomic molecules. Biological problems including models for water are at the state of the art today. The early work sought to simulate the simplest atomistic systems, the rare gases and some simple metals, copper and sodium. In the rare-gas case, where pair potentials are adequate, averages of the potential energy, Φ , and the potential contributions to the pressure, $P(\Phi)$, can be expressed in terms of the pair distribution function $g(r)$:

$$\langle\, [\, \Phi, P(\Phi)V \,]\, \rangle = [N(N-1)/2V] \int_0^\infty 2\pi r g(r)[\, \phi, (r \cdot F/2)\,]dr \,\,[\, 2D \,]\,\,;$$

$$\langle\, [\, \Phi, P(\Phi)V \,]\, \rangle = [N(N-1)/2V] \int_0^\infty 4\pi r^2 g(r)[\, \phi, (r \cdot F/3)\,]dr \,\,[\, 3D \,]\,\,.$$

Here the $\{\, F \,\}$ are the forces derived from the pair potential $\phi(r)$. $g(r)$ is the ratio of the number of particles found a distance r from a central particle to that which would be found there in an ideal gas of the same density. $g(r)$ is therefore unity for a low-density gas, exhibits a few smooth peaks for a liquid, and can be well approximated by sets of delta functions, one for each coordination cell, for a cold solid. Because neutrons or xrays can be used to measure the pair distribution function in fluids, much of the early simulation work was devoted to calculating histograms of $g(r)$ and comparing the results with experimental data.

1.6   "Corresponding States" At and Not At Equilibrium

To a good approximation the phase diagrams of the rare gases (other than Helium) obey "corresponding states" or "scaling" principles. According to this approximation the dimensionless mechanical (P, V, E) and thermal equations of state (T, V, E) depend upon a characteristic energy ϵ and length σ. Two materials in "corresponding states" with each other exhibit identical functions of "reduced" (dimensionless) energies $(E/N\epsilon)$ and volumes $(V/N\sigma^3)$. The characteristic parameters ϵ and σ can be chosen as the microscopic energy and length scales of a pair potential, such as Lennard-Jones :

$$\phi = 4\epsilon[\ (\sigma/r)^{12} - (\sigma/r)^6\ ]\ .$$

They could equally well be chosen as the macroscopic specific energy and the (cube root of the) specific volume at a distinguished point in their phase diagram, such as the critical point or the triple point of the phase diagram shown in **Figure 1.4**. These Principles of Corresponding States provide welcome links between measured phase diagrams and the empirical interatomic potentials used in microscopic simulations.

1.6.1   *van der Waals' Equation and Corresponding States*

Johannes Diderik van der Waals (1837-1923) was awarded the 1910 Nobel Prize in Physics for his equation of state work, with many examples of the Principle of Corresponding States. For simplicity we will consider his work in the context of "simple fluids" such as the rare gases. Starting with the ideal-gas equation of state, $(PV/NkT) = 1$; $(E/NkT) = (3/2)$, van der Waals added two corrections. The first, van der Waals' a, reflects the reduction in pressure due to attractive forces. The second, van der Waals' b, reflects the increase in pressure due to repulsive forces, pictured as an excluded volume. These ideas are reminiscent of the square-well fluid studied by Alder and Wainwright, where the attractive energy ϵ and the "hard core" volume, $\sqrt{(1/2)}\sigma^3$, provide a microscopic interpretation of both of van der Waals' constants a and b.

van der Waals' approach to a universal equation of state is most easily motivated through Gibbs' microcanonical partition function. The ideal gas integral over coordinates, V^N, can be approximated by $(V - Nb)^N$, taking into account the excluded volume occupied by the repulsive atomic cores. If the attractive energy is $-(N^2a/V)$, so that an attraction felt by each atom is proportional to the number density (N/V) of nearby atoms the momentum integration becomes the volume of a hypersphere of radius $\sqrt{2K/D} = \sqrt{[\ E + (N^2a/V)\ ]}$ for a D-dimensional fluid :

$$\Omega(N, E, V) = e^{(S/k)} = (V - Nb)^N \times [\ E + (N^2a/V)\ ]^{(DN/2)}\ .$$

The thermodynamic relation for entropy, $TdS = dE + PdV$, links entropy to temperature and pressure. Differentiating Gibbs' expression with respect to the energy and to the volume gives the energy and pressure for a two-dimensional van der Waals' fluid :

$$(\partial S/\partial E)_V = (1/kT) = N/[\, E + (N^2a/V)\,] \rightarrow$$

$$E = (NDkT/2) - (N^2a/V) \,[\text{ van der Waals' energy }].$$

$$(P/kT) = (\partial S/\partial V)_E = [\, N/(V - Nb)\,] - (N^2a/V^2)/[E + (N^2a/V)] \rightarrow$$

$$(PV/NkT) = [\, V/(V - Nb)\,] - (Na/VkT) \,[\text{ van der Waals' pressure }].$$

Although it is not apparent from the form of the mechanical equation of state the pressure exhibits an instability at low temperature with the isothermal bulk modulus,

$$B_T = -V(\partial P/\partial V)_T = [\, NkTV/(V - Nb)^2\,] - 2(N^2a/V^2)\,,$$

becoming negative for temperatures lower than $kT = 2Na(V - Nb)^2/V^3$. At such temperatures the maximum entropy is obtained by taking a linear combination of states with two different volumes, one near Nb and the other at a relatively low density. van der Waals interpreted these as the volumes of two coexisting fluids [one the gas and one the liquid] below the "critical temperature". That critical temperature is determined by two conditions :

$$(\partial P/\partial V)_T = 0\,;\ (\partial^2 P/\partial V^2)_T = 0\,.$$

The solution, the "critical point", can be expressed in terms of van der Waals' a and b :

$$V_c = 3Nb\,;\ kT_c = (8a/27b)\,;\ P_c = (a/27b^2)\,,$$

from which the dimensionless compressibility factor (PV/NkT) at the critical point is $(3/8)$, a bit higher than the rare gas values which cluster around 0.3.

It is noteworthy that van der Waals' mechanical equation of state generates a "virial" [density] expansion of the pressure as does also Gibbs' statistical mechanics. In van der Waals' case the nth virial coefficient is b^{n-1} for n greater than two while the second virial coefficient, $b - (a/kT)$, is alone in reflecting the attractive forces of neighboring particles.

Exercise : Show that the critical point conditions $(\partial P/\partial V)_T = 0$ and $(\partial^2 P/\partial V^2)_T = 0$, when applied to the three-term virial equation of state $P = \rho kT[\,1 + B_2\rho + B_3\rho^2\,]$ provide a better estimate of the critical compressibility factor (PV/NkT) than does van der Waals' complete equation of state.

Similar corresponding-states observations can be made for nonequilibrium systems too. We will see that Green and Kubo's linear-response theory, a generalization of Gibbs' ensemble theory, can lead to evaluations of the transport coefficients $\{ D,\ \eta,\ \eta_v,\ \kappa \}$ (the diffusion coefficient, the shear and bulk viscosities, and the thermal conductivity). Thus there are also nonequilibrium principles of corresponding states for the four transport coefficients.

The simplest of the transport coefficients is the self-diffusion coefficient D which is a thoroughly equilibrium property. D quantifies the mean-squared distance moved by an equilibrium particle in a sufficiently long, but not "too long" time t . The diffusion coefficient D describes the mean-squared distance travelled in t by a typical gas or liquid particle. In three dimensions D has the form :

$$ D = \frac{\langle r^2 \rangle}{(6t)} = \int_0^t \int_0^t \langle\, v(t') \cdot v(t'')\, \rangle dt' dt''/(6t) = \int_{-\infty}^{+\infty} \langle\, v(0) \cdot v(t)\, \rangle dt/6 \ . $$

It is a good exercise to work through the steps necessary to convert the mean squared displacement at time t to an integral of the velocity autocorrelation function.

An illuminating "free path theory", alternative to the generation and study of correlation functions like $\langle\, v(0) \cdot v(t)\, \rangle$, comes from kinetic theory. Consider the vector displacement at time t to be the vector sum of Γt uncorrelated displacements of length $|dr|$:

$$ \langle\, (r^2/6t)\, \rangle = (1/6t)\langle\, (\sum^{\Gamma t} dr)^2\, \rangle\ = \Gamma |dr|^2/6 \ . $$

This relates the collision rate Γ and jump length $|dr|$ (which could be generalized to a distribution) to the decay of the velocity autocorrelation function. Similar results can be obtained by considering perturbations of Gibbs' distribution resulting from imposed velocities or temperatures or electric fields. The first two of these entail conductive contributions from interparticle forces, not just convection, and so are more complicated than simple diffusion. An approach we favor for its simplicity and visual appeal is the development of explicitly nonequilibrium algorithms. Such an approach requires nonHamiltonian mechanics. I will deal with this in detail in **Lectures 9 and 10**. I'll give a brief preview here.

1.7   Gauss' and Nosé-Hoover Thermal Mechanics

The mathematical genius Johann Carl Friedrich Gauss (1777-1855) extended his least-squares idea (1795) to mechanics more than 30 years later

(1829). We will apply it to isokinetic mechanics, doubtless the first means of controlling the kinetic temperature of a many-body system. Gauss' Principle of Least Constraint holds that constraint forces should be chosen such that their mean squared value is as small as possible. Keeping in mind that the ideal-gas thermometer measures the mean kinetic energy, we apply the constraint of *constant* kinetic energy :

$$K = \sum m(v^2/2) \rightarrow \dot{K} = \sum mv \cdot \dot{v} = 0 \ .$$

At the same time we add the necessary constraint forces $\{\ F_c\ \}$ to the equations of motion, including variations $\{\ \delta F_c\ \}$ which we can use to apply Gauss' Principle :

$$\{\ m\dot{v} = F + F_c + \delta F_c\ \}\ .$$

The $\{\ F\ \}$ are the usual Newtonian interparticle forces and the $\{\ F_c\ \}$ are the additional constraint forces charged with keeping K constant. Because Gauss' Principle is variational we include also an infinitesimal perturbation δF_c which we will use to minimize the squared value of F_c. We need to impose the minimum condition simultaneously with the isokinetic constraint condition $(d/dt)\sum(mv^2)/2 \equiv 0$:

$$\sum(F_c + \delta F_c)^2 = \sum F_c^2 \longrightarrow \sum F_c \cdot \delta F_c = 0 \ [\text{ minimum }]\ ;$$

$$\sum v \cdot [F + F_c + \delta F_c] = 0 \rightarrow \sum v \cdot \delta F_c = 0 \ [\text{ constraint }]\ .$$

For the two conditions to be satisfied simultaneously it must be true that

$$\sum[F_c + \zeta mv] \cdot \delta F_c = 0 \ ,$$

where the [Lagrange] multiplier ζ is an arbitrary function of time. Because the two conditions holds independently of the choice of the variation δF_c we conclude that $\sum[\ F_c + \zeta mv\ ]$ must itself vanish. Substituting the corresponding value for the multiplier into the constraint condition provides the "Gaussian isokinetic" equations of motion[12] :

$$\zeta = \sum F \cdot v / \sum mv^2 \ ; \ \{\ m\dot{v} = F - \zeta mv\ \}\ .$$

The Nosé-Hoover motion equations, which are consistent with Gibbs' canonical ensemble, include frictional forces with exactly the same form as Gauss', $m\dot{v} = F - \zeta mv$. Gauss' mechanics constrains the kinetic energy with "differential feedback" [constraining the time derivative $\dot{K} = 0$]. The Nosé-Hoover friction coefficient ζ is determined by time-averaged *integral feedback*. The time-averaged kinetic energy is controlled in such a way as to reproduce Gibbs' Gaussian distribution of velocities $\propto e^{-(mv^2/2kT)}$ and the canonical distribution of the potential, $\propto e^{-\Phi/kT}$. In the Nosé-Hoover case the evolution of the friction coefficient ζ is proportional to the difference between the present energy sum and its given long-time-averaged value :

$$\dot{\zeta}_{NH} \propto \sum mv^2 - NDkT \longrightarrow \langle \sum mv^2 \rangle = NDkT ,$$

In both Gaussian and Nosé-Hoover mechanics the friction coefficient ζ changes sign in the reversed motion so that these thermostatted motion equations are time-reversible. This means that a movie of such a motion played forward or backward satisfies exactly the same equations of motion (with the signs of both $\{\,p\,\}$ and ζ changed).

1.7.1 *Example Problem in Gaussian Isokinetic Mechanics*

The Gauss and Nosé and Nosé-Hoover motion equations allow us to fix the kinetic temperature of any selected set of degrees of freedom, an obvious requirement for maintaining steady states or boundary conditions away from equilibrium. Let us confirm the validity of our analysis by solving a periodic three-body problem with both kinds of isothermal dynamics. We imagine three point masses in a one-dimensional periodic box and describe the displacements of the particles with the coordinates $x(1)$, $x(2)$, and $x(3)$. We localize the particles with a cubic restoring force and allow each of them to interact with both of their neighbors with a Hooke's-Law force. This is a "ϕ^4" model, useful in heat-conduction simulations and named for the quartic form of the tethering potential, $(x^4/4)$:

$$\ddot{x}_1 = F_1 - \zeta\dot{x}_1 \ ; \ \ddot{x}_2 = F_2 - \zeta\dot{x}_2 \ ; \ \ddot{x}_3 = F_3 - \zeta\dot{x}_3 \ .$$

$$F_1 = x_2 + x_3 - 2x_1 - x_1^3 \ ; \ F_2 = x_3 + x_1 - 2x_2 - x_2^3 \ ; \ F_3 = x_1 + x_2 - 2x_3 - x_3^3 \ .$$

We choose initial conditions for the three particles as follows :

$$x(1) = 0 \ ; \ x(2) = 0 \ ; \ x(3) = 0 \ ; \ v(1) = 3 \ ; \ v(2) = 4 \ ; \ v(3) = -5 \ ,$$

so that the initial kinetic energy is 25.

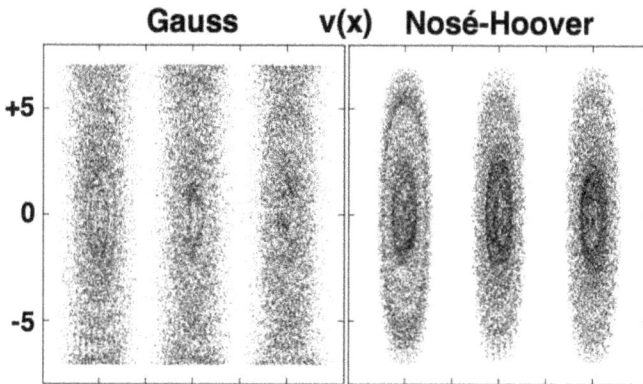

Fig. 1.5: Gaussian isokinetic and Nosé-Hoover phase-plane [$p(q)$ or $\dot{x}(x)$] trajectories for periodic chains of three ϕ^4 particles. Each particle interacts with the other two through Hooke's-Law springs and is bound to its lattice site by a quartic potential ($x^4/4$). In the Gaussian case the kinetic energy is 25, a constant of the motion (100,000 steps are shown here with $dt = 0.01$, plotting every tenth step) while in the Nosé-Hoover case the kinetic energy fluctuates but with a time-averaged value of 25. The least-energy spacing between particles has been arbitrarily chosen as 10, two tick marks on the horizontal scales.

The Gaussian mechanics requires that we solve six motion equations for the { x } and the { v }, using Gauss' recipe for ζ. The Nosé-Hoover mechanics requires a seventh motion equation for ζ. In **Figure 1.5** we plot the trajectories of the three particles for both thermostatted mechanics. 100,000 timesteps with $dt = 0.01$ are shown. The Gaussian simulation shows sharp boundaries for the velocities at $v = \pm\sqrt{50}$. The "stability" of both these sets of motion equations suggests [correctly] that these thermostatted equations of motion will be useful for manybody problems.

Stability has more than one meaning. Mathematicians sometimes use the word to indicate algorithms which stay "close" to a "true" solution of motion equations. Most interesting dynamics problems are "chaotic", meaning that a tiny perturbation in the initial conditions results in an exponentially fast departure of the perturbed solution from the "true" one. We use the term stable to mean that an algorithm produces a plausible solution, most likely chaotic, rather than numerical divergence. One of the

most interesting and simplest sets of chaotic differential equations is that
describing the Nosé-Hoover oscillator.[13, 14] Let us have a look at it next.

1.7.2 *"Stiff" Equations and the Nosé-Hoover Oscillator*

Gauss' mechanics applies a kinetic-energy constraint to control the kinetic
temperature. In 1984 Nosé conceived of a much more ambitious type of temperature control.[15, 16] He sought to reproduce the entire canonical-ensemble
distribution of velocities, $\propto e^{-mv^2/2kT}$, by introducing a "time-scaling"
variable relating the Hamiltonian momentum p to a "scaled" momentum,
(p/s). Here s ranges between 0 and 1 so that (p/s) can range from $-\infty$ to
$+\infty$. I applied Nosé's idea to the simple harmonic oscillator problem in an
effort to understand Nosé's work.

$$2\mathcal{H}_{\text{osc}} = p^2 + q^2 \to 2\mathcal{H}_{\text{Nosé}} \equiv (p/s)^2 + q^2 + \ln(s^2) + \zeta^2 .$$

Here ζ, soon to be a "friction coefficient", is the momentum conjugate to
s. Nosé's motion equations for a harmonic oscillator are as follows :

$$\dot{q} = (p/s^2) \; ; \; \dot{p} = -q \; ; \; \dot{s} = \zeta \; ; \; \dot{\zeta} = (p^2/s^3) - (1/s) .$$

Nosé's "scaling the time" simply multiplies each motion equation by s :

$$\dot{q} = (p/s) \; ; \; \dot{p} = -sq \; ; \; \dot{s} = s\zeta \; ; \; \dot{\zeta} = (p^2/s^2) - 1 .$$

Dettmann and Morriss were able to show that Gauss' isokinetic mechanics and Nosé's time-scaled isothermal mechanics can both be derived from
Hamiltonians.[17, 18] Their demonstrations rest on the assumption that the
initial conditions are chosen such that these special Hamiltonians vanish.
Here are the details of Dettmann's approach to time-scaled mechanics :

$$2\mathcal{H}_{\text{Dettmann}} \equiv 2s\mathcal{H}_{\text{Nosé}} \equiv (p^2/s) + sq^2 + s\ln(s^2) + s\zeta^2 = 0 \longrightarrow$$

$$\dot{q} = (p/s) \; ; \; \dot{p} = -sq \; ; \; \dot{s} = s\zeta \; ;$$

$$\dot{\zeta} = (1/2)[\,(p^2/s^2) - \{\,q^2 + \ln(s^2) + \zeta^2\,\}\,] - 1 \equiv (p/s)^2 - 1 .$$

Provided that the Hamiltonian vanishes, the three terms in the curly brack-
ets $\{\ldots\}$ add up to $-(p/s)^2$, providing the last equality. In that apparent
special case strictly Hamiltonian motion equations allow us to specify the
time-averaged temperature. Here it is $\langle\,(p/s)^2\,\rangle \equiv 1$. Let us compare the
trajectories for Nosé's original Hamiltonian and Dettmann's modification.
There is a detailed tutorial comparison in the American Journal of Physics[2]
which is also available on the Los Alamos arχiv.

1.7.3   *Adaptive Integration Solves Nosé's Equations*

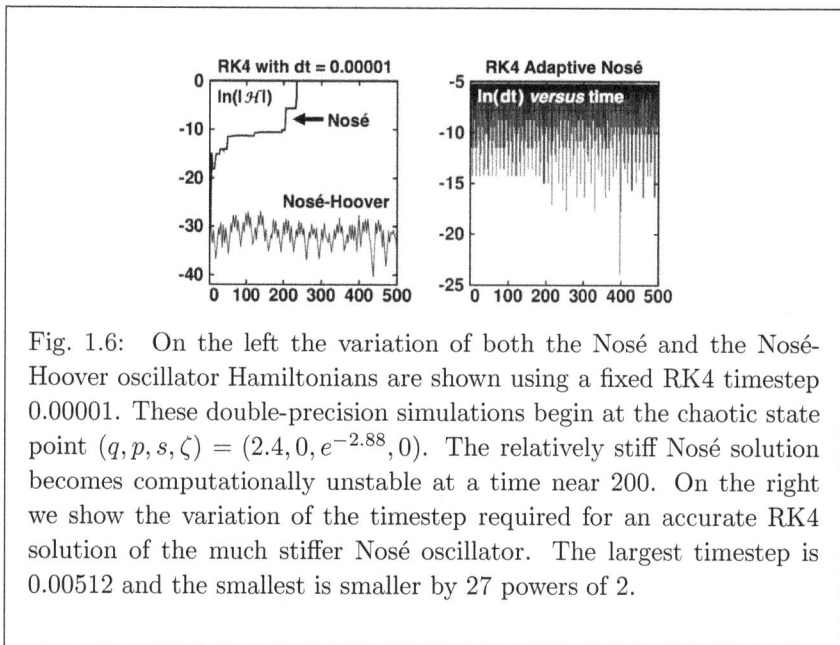

Fig. 1.6:   On the left the variation of both the Nosé and the Nosé-Hoover oscillator Hamiltonians are shown using a fixed RK4 timestep 0.00001. These double-precision simulations begin at the chaotic state point $(q, p, s, \zeta) = (2.4, 0, e^{-2.88}, 0)$. The relatively stiff Nosé solution becomes computationally unstable at a time near 200. On the right we show the variation of the timestep required for an accurate RK4 solution of the much stiffer Nosé oscillator. The largest timestep is 0.00512 and the smallest is smaller by 27 powers of 2.

In Nosé's oscillator equations the last of them, $\dot{\zeta} = (p^2/s^3) - (1/s)$ looks suspiciously "stiff", considering that there is no prohibition preventing s from approaching zero.   In fact the Nosé oscillator's stiffness is in marked contrast to the well-behaved nature of the Nosé-Hoover oscillator. **Figure 1.6** illustrates this well. For exactly the same initial condition and exactly the same (q, p, s, ζ) trajectory the two approaches are qualitatively different, with huge fluctuations in the *rate* at which the Nosé trajectory is traversed. There is simply no fixed timestep for which these "stiff" equations (denoting tremendous fluctuations in the time derivatives) generate a useful RK4 solution. A timestep less than $dt = 0.00001$ is certainly impractical for longtime studies of an oscillator problem with a characteristic frequency of order 1. With a timestep 0.00001 the Nosé-Hoover equations conserve the Hamiltonian with errors of the order of 10^{-12} for billions of timesteps. See the lefthand panel of **Figure 1.6**. The Nosé equations exhibit a computational instability at a time of 200 with the same timestep of 0.00001.

Adaptive integration provides a way out for the "stiff" Nosé integration. We take two fourth-order Runge-Kutta integration steps using the vector

XX for the dependent variables. Starting out an identical YY array with
the same initial conditions yy = xx we cover the same time increment in a
single step and compare the results, after which the less accurate YY array
is set equal to XX :

```
   CALL RK4(XX,XXP,DT/2)
   CALL RK4(XX,XXP,DT/2)
   CALL RK4(YY,YYP,DT)
   ERROR = 0
   DO 10 I = 1,4
   ERROR = ERROR + (XX(I)-YY(I))**2
10 YY(I) = XX(I)
   IF(ERROR.GT.10.0D0**(-24)) DT = DT/2
   IF(ERROR.LT.10.0D0**(-28)) DT = DT*2
```

The right panel of **Figure 1.6** shows the history of the timestep dt during an accurate adaptive integration. The timestep ranges over 28 values
with the ratio of the smallest to the largest $2^{-27} = 0.00000000745$. Because the episodes requiring tiny timesteps are relatively rare the adaptive
integrator is successful while a fixed timestep integrator fails. This example
is intended to show the power of adaptive integration and is definitely *not*
an endorsement of Nosé's original equations. In any applications the Nosé-
Hoover approach is much better behaved and requires no special numerical
techniques. Simpler centered-difference algorithms like Størmer's can be
developed and applied to Nosé-Hoover mechanics if desired.[19]

1.7.4   *Is the Restriction $\mathcal{H} = 0$ Significant?*

How serious *is* Dettmann's restriction $\mathcal{H} = 0$? Without loss of generality
we can choose our initial condition *anywhere* in the $p = 0$ plane so that the
$\mathcal{H} = 0$ condition becomes $s = e^{-(q^2/2)-(\zeta^2/2)}$. With this choice of the time
scaling variable we can still choose *any* point in the $p = 0$ plane. **Figure 1.7**
shows the chaotic sea for the oscillator, a Lyapunov unstable set of points
which includes the point(s) $(q, p, s, \zeta) = (\pm 2.4, 0, e^{-2.88}, 0)$. The "holes" in
the chaotic sea, which make up most of the measure for the oscillator, are
filled with tori surrounding stable periodic orbits. The border between the
chaotic sea and the tori is complex, with structure on all scales. This is
usual for Hamiltonian chaos.

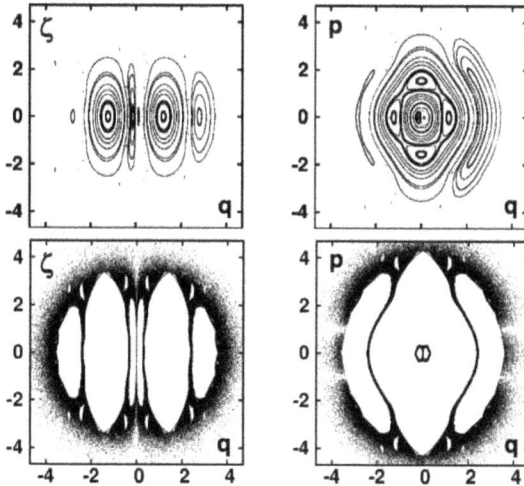

Fig. 1.7: $\zeta(q)$ and $p(q)$ sections for the Nosé-Hoover oscillator. Each chaotic section contains about half a million points generated with RK4 and a double-precision timestep of 0.001. The top-row sections are tori generated with an initial $q = 0.1$, 0.3, $0.5\ldots3.5$. The bottom-row sections are both chaotic, following from the initial condition : $(q, p, s, \zeta) = (2.4, 0, e^{-2.88}, 0)$. $\mathcal{H}$ is zero.

Let us turn next to the simpler dynamics of one-dimensional harmonic chains without any tethering potential. Because the motion equations are linear, like those of a single harmonic oscillator, they have analytic solutions. These furnish useful test cases for validating numerical integrators.

1.8 Lattice Dynamics of One-Dimensional Chains

We consider "chains" of N particles linked by nearest-neighbor "harmonic springs", $\phi(x_{ij}) = (1/2)(|x_i - x_j| - d)^2$ so that a typical particle, Particle j, experiences forces from Particles $j \pm 1$:

$$m\ddot{x}_j = \kappa[\, x_{j+1} - 2x_j + x_{j-1}\,] \, .$$

Notice that for linear springs the restlength of the spring, d, is absent from the motion equations. Exactly the same motion equations result if the $\{\, x\, \}$ represent displacements from an equally-spaced reference configuration. It is usual to choose "periodic" boundary conditions. Then Particle 1 interacts with Particles 2 and N while Particle N is linked to Particles $N-1$ and 1.

On a visit to Ann Arbor in the early 1960s Peter DeBye (1884-1966) gave Bill some good advice : to find good research problems the best approach is to look in the library for surprising or puzzling statements. These days the internet can be used. Let us consider the surprising statement of Florencio and Lee[20] that the mean squared displacement, $\langle\, x^2\,\rangle$ is *divergent* for periodic chains and but *not* for rigid chains. The divergence is a limiting case. As the chain gets longer and longer the mean squared displacement varies linearly with N .

This is a result from statistical mechanics, assuming that all states of the chain are equally likely (omitting overall motion, motion of the center of mass). The situation is quite different from the standpoint of dynamics. A little reflection reveals that this divergence property cannot and does not apply generally. Florencio and Lee had in mind phase-space averages rather than dynamical averages. The two differ for harmonic forces, where longtime averages depend on the initial conditions. To illustrate, the highest-frequency mode of a long chain has the odd atoms and even atoms moving in opposite directions so that the motion equation for any one of them is

$$m\ddot{x} = -4\kappa x \longrightarrow \omega = 2\sqrt{\kappa/m}\ ,$$

with twice the single-spring oscillation frequency. As a further aspect of this puzzle Bill stated on page 106 of *Computational Statistical Mechanics* that the mean squared displacement $\langle\, x^2\,\rangle$ differs by a factor of two, with the rigid boundary result larger due to its lower frequencies :

$$\langle\, x^2\,\rangle_{\text{periodic}} = (N+1)(kT/\kappa)/12 \ ; \ \langle\, x^2\,\rangle_{\text{rigid}} = (N+2)(kT/\kappa)/6 \ .$$

Exercise : Show that the squares of the five nonzero frequencies for a periodic six-spring chain are $\{\, 1,\ 3,\ 4,\ 3,\ 1,\, \}$ and $\{\, 2-\sqrt{3},\ 1,\ 2,\ 3,\ 2+\sqrt{3}\, \}$ for a *rigid* six-spring chain with five moving particles.

In the rigid case the wavelengths are twice as large so that the frequencies are lower. The frequencies given here, in units of $\sqrt{\kappa/m}$, are consistent with the general expressions given in Bill's online *Computational Statistical Mechanics* book. The simplest way to solve these problems is to find the eigenvalues of the dynamical matrix $\mathcal{M}$ where $\ddot{x} = \mathcal{M}x$. Here x is a shorthand notation for the vector $(x_1, x_2, x_3, x_4, x_5)$. Another more interesting method is to choose the initial values of the displacements x equal to $\sin(kx_{eq})$, where the k, one for each vibrational mode, are chosen to satisfy the boundary conditions. Then a Runge-Kutta or Størmer-Verlet solution of the motion equations will oscillate at the desired frequency. The dynamical matrix $\mathcal{M}$ for the simplest periodic five-particle case is

$$\begin{pmatrix} -2 & +1 & 0 & 0 & +1 \\ +1 & -2 & +1 & 0 & 0 \\ 0 & +1 & -2 & +1 & 0 \\ 0 & 0 & +1 & -2 & +1 \\ +1 & 0 & 0 & +1 & -2 \end{pmatrix}.$$

In St Petersburg Anton Krivtsov and Denis Tsvetkov studied the evolution of harmonic chains in which the displacements were initially zero with all of the initial energy kinetic. The results which they found were not reproducible when the initial velocities were chosen "randomly". The final values of $\langle x^2, v^2 \rangle$ typically varied by factors of two from one run to another. Evidently a "random" choice of velocities is dominated by the occasional low-frequency modes. According to Gibbs' statistical mechanics the mean energy of each vibrational mode should be kT at equilibrium. Of course there is no mechanism for a harmonic chain to attain equilibrium.

An attempt to check these solutions with lattice dynamics (choosing the initial momenta from a Gaussian distribution) reminds us that the details of the dynamics never forget the initial conditions. The ensemble averages of statistical mechanics can only be obtained by averaging over *many* sets of initial conditions. This qualitative disagreement between Gibbs' ensemble theory and molecular dynamics suggests that the harmonic chain problems are among those special cases that are interesting, good for validating integration routines, but are not specially useful in understanding the statistics of many-body systems as did Gibbs.

Exercise : Find the vibrational frequencies and dynamical matrices for the four-spring systems with fixed and periodic boundaries.

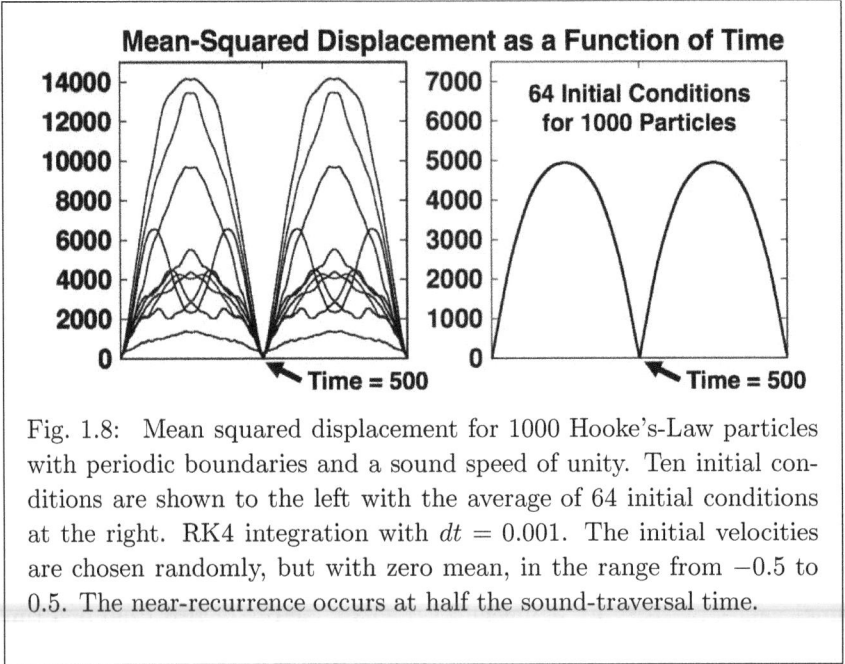

Fig. 1.8: Mean squared displacement for 1000 Hooke's-Law particles with periodic boundaries and a sound speed of unity. Ten initial conditions are shown to the left with the average of 64 initial conditions at the right. RK4 integration with $dt = 0.001$. The initial velocities are chosen randomly, but with zero mean, in the range from -0.5 to 0.5. The near-recurrence occurs at half the sound-traversal time.

The dynamics of two three-body "ϕ^4" models were illustrated in **Figure 1.5**. That model prevents low-frequency oscillations by imposing a quartic tethering potential on each particle, $\phi = (x^4/4)$, localizing the particles near their lattice sites. In the absence of tethering forces the harmonic chain can undergo unbounded low-frequency oscillations with the mean-squared amplitude proportional to the length of the chain. **Figure 1.8** shows the dependence of the mean squared displacement on time for a periodic chain of Hooke's-Law particles with initial velocities chosen randomly between -0.5 and $+0.5$. The low-frequency modes are responsible for the large amplitude with Gibbs' canonical distribution giving

$$\langle\, \delta x^2 \,\rangle = \langle\, (kT/m\omega^2) \,\rangle\,.$$

We turn next to what was a challenge prior to the advent of computers, the classical manybody problem.

1.9   A Classical ManyBody Example Problem

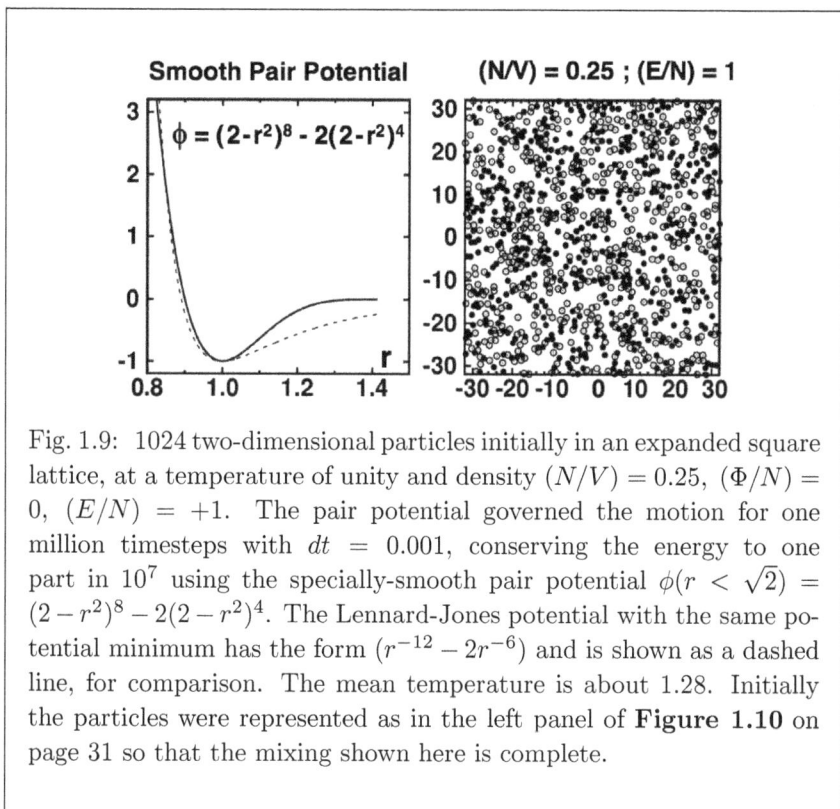

Smooth Pair Potential   **(N/V) = 0.25 ; (E/N) = 1**

$\phi = (2-r^2)^8 - 2(2-r^2)^4$

Fig. 1.9: 1024 two-dimensional particles initially in an expanded square lattice, at a temperature of unity and density $(N/V) = 0.25$, $(\Phi/N) = 0$, $(E/N) = +1$. The pair potential governed the motion for one million timesteps with $dt = 0.001$, conserving the energy to one part in 10^7 using the specially-smooth pair potential $\phi(r < \sqrt{2}) = (2 - r^2)^8 - 2(2 - r^2)^4$. The Lennard-Jones potential with the same potential minimum has the form $(r^{-12} - 2r^{-6})$ and is shown as a dashed line, for comparison. The mean temperature is about 1.28. Initially the particles were represented as in the left panel of **Figure 1.10** on page 31 so that the mixing shown here is complete.

Let us consider a prototypical chaotic problem where statistical mechanics and molecular dynamics would be expected to agree, a "manybody" problem involving 1024 particles interacting with the pair potential :

$$r < \sqrt{2} \longrightarrow \phi(r) = (2 - r^2)^8 - 2(2 - r^2)^4 .$$

In the interest of simplicity we consider the two-dimensional version. We use periodic boundary conditions, an initial energy of $N = 1024$, choosing the velocities randomly, removing the center-of-mass motion, and rescaling the velocities so that

$$\langle v_x^2 \rangle = \langle v_y^2 \rangle = 1 \ [\text{ Initial Condition }] .$$

After summing up the randomly chosen $\{ v_x(i) \}$ and $\{ v_y(i) \}$ the centering and scaling operations are as follows :

```
DO 10 I = 1,N
VX(I) = VX(I) - SUMVX/N
VY(I) = VY(I) - SUMVY/N
SUMXX = SUMXX + VX(I)*VX(I)
SUMYY + SUMYY + VY(I)*VY(I)
10 CONTINUE
DO 20 I = 1,N
VX(I) = DSQRT(N/SUMXX)*VX(I)
VY(I) = DSQRT(N/SUMYY)*VY(I)
20 CONTINUE
```

By choosing the separation of the particles greater than or equal to $\sqrt{2}$ the total energy is equal to the initial kinetic energy which we arbitrarily set equal to N. Two configurations are shown in the **Figures**, the first for a density of $(N/V) = 0.25$ and a (wholly kinetic) initial energy of $(E/N) = +1.0$; the last for a density $(N/V) = 1.0$ and initial energy of $(E/N) = -2$ (potential) $+ 1$ (kinetic) $= -1.0$. In both simulations the periodic boundary conditions are enforced at every timestep :

```
DO 18 I = 1,N
IF (X(I).GT.+ELX/2) YY(I)   = YY(I)   - ELX
IF (X(I).LT.-ELX/2) YY(I)   = YY(I)   + ELX
IF (Y(I).GT.+ELY/2) YY(I+N) = YY(I+N) - ELY
IF (Y(I).LT.-ELY/2) YY(I+N) = YY(I+N) + ELY
18 CONTINUE
```

Here `ELX` and `ELY` are the width and height of the periodic cell, 64 and 32 in the cases shown in **Figures 1.9 and 1.10**. The vector `YY` stores the $4N = 4096$ integration variables, $\{ x, y, p_x, p_y \}$ in the order indicated.

The results of such simulations are surprising, at least to those accustomed to living in a three-dimensional world. The typical arrangements of the particles look like a collection of "strings" so that one would not immediately identify gas or liquid or solid portions of the snapshots. At Sheffield Karl Travis and Amanda Bailey Hass located an approximate critical point for the 8-4 potential, $(N/V) \simeq 0.3$ and $T \simeq 0.42$ in the two-dimensional case. Evidently the simulations illustrated here, with temperatures of $T \simeq 0.71$ and 1.28, are well above the critical temperature (if there *is* a liquid phase) so that there is no observable gas-liquid equilibrium possible. At somewhat lower temperatures the solid phase is stable. Using $dt = 0.01$, ten times the proper value, with the initial conditions of **Figure 1.10** RK4 found a near-perfect triangular-lattice crystal containing three holes and no dislocations. The energy had declined from -1024 to -2470 over the million-timestep run. We can easily prevent such a decline by using Nosé-Hoover dynamics. A fringe benefit of that improvement is that the dynamics is stable for a timestep 50 times larger, $dt = 0.050$. (!)

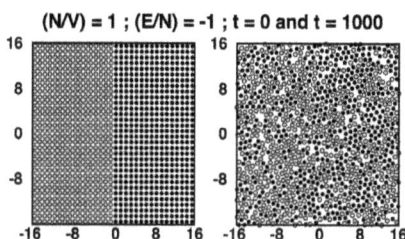

Fig. 1.10:   1024 two-dimensional particles initially at unit density in a square lattice : $(N/V) = 1$, $(\Phi/N) = -2$, $(E/N) = -1$. The pair potentials governed the motion for one million timesteps with $dt = 0.001$, conserving the energy to one part in 10^7 using the potential function of **Figure 1.9** with an initial temperature of unity. The initial and final states are shown. The mean temperature is about 0.71. Note the diffusive mixing, thorough, but considerably slower than at the four times lower density of **Figure 1.9**.

At the lower density, where the mean free path is on the order of unity, the diffusion coefficient should likewise be of order one so that a typical particle diffuses a distance on the order of 32 in the simulation time of 1000. The free path at a density of unity could be as small as 0.2 (for a collision with a line-of-centers velocity of order 1, corresponding to a collision diameter less than 0.9) giving a particle displacement on the order of 10 during the simulation shown in **Figure 1.10**. The potential chosen for these simulations has roughly the shape one expects for simple materials. Its specially-smooth structure generates forces with two continuous derivatives at the cutoff radius $\sqrt{2}$. This makes energy conservation simple. A less-expensive approach is to include a thermostat of the Nosé-Hoover type in the dynamics, avoiding any longterm drift.

Following this brief intoduction to manybody dynamics let us consider the analysis necessary to connect microscopic models to macroscopic behavior. We address the query: How does one make contact with the continuum disciplines describing matter at and away from equilibrium?

1.10 Simulating Hydrodynamic Manybody Properties

The foundations of equilibrium Thermodynamics and nonequilibrium Hydrodynamics are empirical. The equilibrium case is simpler : for a given amount of fluid at equilibrium the energy, temperature, pressure, and volume are the usual state variables. It is an empirical characteristic of fluids that an undisturbed equilibrium sample can be described by just *two* state variables. It is natural to choose energy and volume for Newtonian fluids of the type just illustrated. These independent variables, maintained constant throughout a standard "equilibrium" molecular dynamics simulation, make it possible to measure the other two, the [kinetic] temperature and pressure. Keeping the ideal-gas thermometer in mind we *define* the temperature in terms of the mean squared values of the three velocity components (just one or two in one or two dimensions) ,

$$(kT/m) \equiv \langle\, v_x^2\, \rangle = \langle\, v_y^2\, \rangle = \langle\, v_z^2\, \rangle \,.$$

Maxwell and Boltzmann and Gibbs showed that, independent of the form of the potential energy, the most likely velocity distribution is Gaussian, centered on the center-of-mass velocity and with its halfwidth given by the temperature, as detailed above. The usual formal proof rests on the idea that the phase-space volume Ω occupied at equilibrium is a maximum, so that two systems able to exchange heat with one another are at equilibrium with a common temperature T when

$$\Omega = \Omega_1 \times \Omega_2 \to (\partial \ln \Omega_1/\partial E_1)_V = (\partial \ln \Omega_2/\partial E_2)_V = (1/kT) \,.$$

In statistical mechanics a parallel derivation shows that the volume derivative of $\ln \Omega$ is $(N/V) = (P/kT)$ where P is the pressure. So long as the potential and kinetic energies are not coupled the pressure can be determined by applying the Virial Theorem and the heat-flux vector can be measured by applying the Heat Theorem. Let us elucidate !

1.10.1 *The Virial Theorem for the Pressure Tensor P*

The virial theorem relates individual particles' coordinates, velocities, and forces to the pressure tensor, P. There are several ways to derive the virial theorem. For more details see Bill's *Molecular Dynamics* book at williamhoover.info on the web. We consider here the simplest interesting situation, a fluid with a pairwise-additive potential energy, $\Phi = \sum \phi$. Differentiation of this potential yields pairwise-additive forces, $F_i = \sum_j F_{ij}$ where F_{ij} is the force on Particle i due to Particle j. To express the tensor component P_{xx} in terms of particle properties take the x component $m\ddot{x}$ of the equation of motion of Particle i, multiply it by x_i and sum over all i, symmetrizing the result :

$$\sum_i m x_i \ddot{x}_i = \sum_{\text{pairs}} (x^2 F/r)_{ij} - P_{xx} V = \sum_i m(d/dt)(x_i \dot{x}_i) - \sum_i m\dot{x}_i^2$$

where we have combined

$$x_i F_{ij}(x/r)_{ij} + x_j F_{ji}(x/r)_{ij} = x_{ij}^2 F_{ij}/r_{ij} = (Fx^2/r)_{ij} .$$

We have included also the contributions of the long-time-averaged forces on the walls, $\pm P_{xx} V$. Again, $F_{ij} = -F_{ji}$ is the force *on* Particle i due to its interaction with Particle j. Because the $x\dot{x}$ terms are all bounded the long-time average $\langle \dots \rangle$ of their time derivative is necessarily zero, giving the xx component of the virial theorem :

$$P_{xx} V = NkT + \langle \sum_{i<j} (x^2 F/r)_{ij} \rangle \ ; \ \langle \sum_i m\dot{x}_i^2 \rangle \equiv NkT.$$

This derivation is correct for both fluids and solids at equilibrium. It also applies just as well for a solid undergoing static stress or a fluid undergoing viscous shear. For example, in the case of "simple shear" where the x component of the fluid velocity varies with y :

$$P_{xy} V = \sum_i m\dot{x}\dot{y} + \sum_{i<j} (xyF/r)_{ij} = P_{yx} V .$$

An alternative derivation of the theorem can be made locally without any need for system wall-averaging. Pressure is the force per unit area exerted *by* a volume element *on* its surroundings. P_{ij} is the force per unit area exerted in the i direction onto the fluid's boundary with outward normal in the j direction. The pictorial definition of **Figure 1.11** expresses this concept well. The force exerted by the surroundings normal to the righthand face of the volume element in the figure is $-P_{xx}dy$. The force exerted *by* the element on the surrounding fluid is $+P_{xx}dy$. An alternative and fully equivalent picture results if we *define* pressure as the *flux* of momentum. This is the flow per unit time and area of momentum as measured in a small sampling box δV within a much larger system.

1.10.2   *The Pressure Tensor is Necessarily Symmetric*

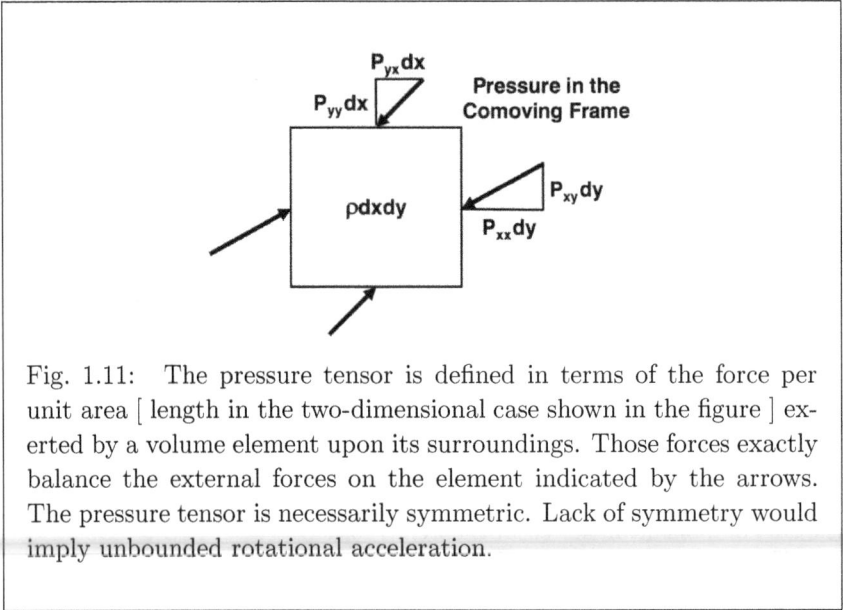

Fig. 1.11:   The pressure tensor is defined in terms of the force per unit area [length in the two-dimensional case shown in the figure] exerted by a volume element upon its surroundings. Those forces exactly balance the external forces on the element indicated by the arrows. The pressure tensor is necessarily symmetric. Lack of symmetry would imply unbounded rotational acceleration.

In two dimensions flux is the flow per unit *length* and time. Any particle within δV contributes its individual kinetic contribution to the momentum flux, $P_{xx}\delta V \supset \sum m x_i^2$. Any *pair* of particles in δV close enough to interact contributes to $P_{xx}\delta V$ through the pair-force contribution to the flux. Thus there are two kinds of tensor contributions to the pressure, [1] mvv, single-particle convection and [2] particle-pair momentum transfer $(Fr)_{ij}$. The first is convective while the second is purely configurational, independent of the particle velocities. Configurational pressure is the result of "action at a distance". The interparticle force term describes the rate at which particle i transfers momentum to particle j and is likewise a symmetric tensor Fr .

The symmetry of the pressure tensor is important. Of the nine components P_{ij} only six are independent. In two dimensions and with pairwise-additive forces the symmetry of the pressure tensor is obvious. Of the four components only three are independent as $P_{xy} \equiv P_{yx}$. Continuum mechanics requires this symmetry. Were the pressure tensor *not* symmetric the asymmetry would cause a small volume element of the fluid or solid to accelerate infinitely rapidly as the size of the element decreases. This point, on a subject where the literature is spotty, deserves its own exercise!

Exercise : Calculate the rate of energy increase in a small $L \times L$ two-dimensional fluid zone $dxdy$ subject to clockwise shear forces $P_{xy}L$ on its vertical sides and opposing counterclockwise forces $P_{yx}L$ on each of the two horizontal sides. If $P_{xy} - P_{yx} = \Delta P$ show that the work done on the element by rotating the square through a small angle θ is $L^2 \Delta P \theta$ and that the kinetic energy is $\rho L^4 \dot{\theta}^2 / 12$. Differentiation with respect to time and dividing by $\dot{\theta}$ gives $\ddot{\theta} = (6 \Delta P / M)$ where $M = \rho L^2$. This exercise shows that a fixed shear stress difference gives a divergent angular acceleration for an infinitesimal element, $M \to 0$. Carry out the analogous calculation in three dimensions.

1.10.3   The Heat Theorem for the Heat-Flux Vector Q

The flux of heat (per unit area and time) is a vector quantity. According to Jean-Baptiste Joseph Fourier's "Law" the heat flux is proportional to the temperature gradient, with the heat flowing from hot to cold.

$$Q = -\kappa \nabla T .$$

κ is the thermal conductivity. The derivation of the microscopic description of energy flow follows steps similar to those of the virial theorem. We begin by associating an energy e_i with the ith particle :

$$e_i = mv_i^2 + (1/2) \sum_j \phi_{ij} \longrightarrow Q_x \delta V \supset \sum_i \dot{x}_i e_i .$$

So long as the particles are identical this step is not controversial. The convective portion of the heat flux within δV depends upon the velocity and energy of each particle in the volume. The contribution from the pair potential ϕ_{ij} is carried along by the particles' velocity $(v_i + v_j)/2$ so that the x component of the heat-flux vector includes $x_{ij}(F_{ij} \cdot (v_i + v_j)/2$. The general expression for the flux (with pairwise forces) is simply the sum of both contributions :

$$Q \delta V = \sum_i (ve)_i + \sum_{i<j} x_{ij} [\, F_{ij} \cdot (v_i + v_j)\,]/2 .$$

We will find this expression particularly useful in describing the flow of energy in strong shockwaves.

1.11 Summary of Lecture 1

The ordinary differential equations of Newtonian, Lagrangian, and Hamiltonian mechanics can be solved using centered-difference or Runge-Kutta algorithms. Gauss' and Nosé-Hoover thermostatted mechanics greatly expand the possible applications of molecular dynamics and also allow for a substantially larger timestep due to the stabilizing effect of the thermostats. The kinetic energy provides the temperature and the virial theorem provides the pressure tensor giving all the quantities necessary for the equilibrium equation of state for solids and fluids.

The nonequilibrium situation is more interesting and requires more thought. The number of state variables has to be determined empirically. At least one of them defines the nature of the nonequilibrium state being simulated. The strain rate in viscous flows or the boundary temperatures in heat flows are good examples. Typically gradients lead to nonlinear as well as linear responses. Pressure contributions quadratic in the shear strain rate and heat flux contributions proportional to the square of the angular velocity are examples.[21, 22] Initial conditions are particularly important in nonequilibrium simulations. This can be a difficult task, particularly where solids are concerned. The field, and fields, of continuum mechanics have been undergoing rapid development so that a variety of constitutive relations are used to describe the response of materials to sources of work and heat. Much remains to be done.

For the most part we restrict ourselves to macroscopic models exhibiting "Newtonian" viscosity and Fourier heat conduction. In studying shockwaves more complicated concepts must be included such as anisotropic temperature and time delay. Time delay, with the response to a gradient coming afterward rather than instantaneously, certainly plays a part in understanding how reversible dynamics can produce irreversible behavior. We will have much more to say about these problems in the last three Lectures where we describe the Lyapunov-unstable nonequilibrium flows and introduce smooth-particle methods useful for bringing the atomistic and continuum approaches into consonance.

References

1. L. Verlet, "Computer 'Experiments' on Classical Fluids. I. Thermodynamical Properties of Lennard-Jones Molecules", Physical Review **159**, 98-103 (1967).
2. Wm. G. Hoover, J. C. Sprott, and C. G. Hoover, "Adaptive Runge-Kutta Integration for Stiff Systems: Comparing Nosé and Nosé-Hoover Dynamics for the Harmonic Oscillator", American Journal of Physics **84**, 786-794 (2016).
3. B. J. Alder and T. E. Wainwright, "Molecular Dynamics by Electronic Computers", Proceedings of the International Union of Pure and Applied Physics Symposium on the *Statistical Mechanical Theory of Transport Properties* (Brussels, 1956) (Interscience Publishers, New York, 1958).
4. B. J. Alder and T. E. Wainwright, "Studies in Molecular Dynamics. I. General Method", The Journal of Chemical Physics **31**, 459-466 (1959).
5. L. Woodcock, "Percolation Transitions and Fluid State Boundaries", Computational Methods in Science and Technology **23**, 281-294 (2017).
6. E. Fermi, J. Pasta, and S. Ulam, "Studies of Nonlinear Problems", Los Alamos Scientific Laboratory Report LA-1940 (1955).
7. J. B. Gibson, A. N. Goland, M. Milgram, and G. H. Vineyard, "Dynamics of Radiation Damage", Physical Review **120**, 1229-1253 (1960).
8. A. Rahman, "Correlations in the Motion of Atoms in Liquid Argon", Physical Review A **136**, 405-411 (1964).
9. M. A. M. Karlsen, "The Early Years of Molecular Dynamics and Computers", *Advances in the Computational Sciences*, pages 176-183 in the Proceedings of the Symposium in Honor of Dr. Berni Alder's 90th Birthday, editted by E. Schwegler, B. M. Rubenstein, and S. B. Libby (World Scientific, Singapore, 2017).
10. M. Engel, J. A. Anderson, S. C. Glotzer, M. Isobe, E. P. Bernard, and W. Krauth, "Hard-Disk Equation of State: First-Order Liquid-Hexatic Transition in Two Dimensions with Three Simulation Methods", Physical Review E **87**, 042134 (2013).
11. H. J. Rood, "The Remarkable Extragalactic Research of Erik Holmberg", Astronomical Society of the Pacific **99**, 921-951 (1987).
12. Wm. G. Hoover, A. J. C. Ladd and B. Moran, "High-Strain-Rate Plastic Flow Studied *via* Nonequilibrium Molecular Dynamics", Physical Review Letters **48**, 1818-1820 (1982).
13. H. A. Posch, Wm. G. Hoover, and F. J. Vesely, "Canonical Dynamics of the Nosé Oscillator: Stability, Order, and Chaos", Physical Review A **33**, 4253-4265 (1986).
14. Wm. G. Hoover, "Remark on 'Some Simple Chaotic Flows' ", Physical Review E **51**, 759-760 (1995).
15. S. Nosé, "A Molecular Dynamics Method for Simulations in the Canonical Ensemble", Molecular Physics **52**, 255-268 (1984).
16. S. Nosé, "A Unified Formulation of the Constant Temperature Molecular Dynamics Methods", The Journal of Chemical Physics **81**, 511-519 (1984).
17. C. P. Dettmann and G. P. Morriss, "Hamiltonian Formulation of the Gaussian Isokinetic Thermostat", Physical Review E **54**, 2495-2500 (1996).

18. C. P. Dettmann and G. P. Morriss, "Hamiltonian Reformulation and Pairing of Lyapunov Exponents for Nosé-Hoover Dynamics", Physical Review E **55**, 3693-3696 (1997).

19. B. L. Holian, A. J. De Groot, Wm. G. Hoover, and C. G. Hoover, "Time-Reversible Equilibrium and Nonequilibrium Isothermal-Isobaric Simulations with Centered-Difference Størmer Algorithms", Physical Review A **41**, 4552-4553 (1990).

20. J. Florencio, Jr, and M. H. Lee, "Exact Time Evolution of a Classical Harmonic-Oscillator Chain", Physical Review A **31**, 3231-3236 (1985).

21. Wm. G. Hoover, C. G. Hoover, and J. Petravic, "Simulation of Two- and Three-Dimensional Dense-Fluid Shear Flows *via* Nonequilibrium Molecular Dynamics. Comparison of Time-and-Space-Averaged Stresses from Homogeneous Doll's and Sllod Shear Algorithms with Those from Boundary-Driven Shear", Physical Review E **78**, 046701 (2008).

22. Wm. G. Hoover, B. Moran, R. M. More, and A. J. C. Ladd, "Heat Conduction in a Rotating Disk *via* Nonequilibrium Molecular Dynamics", Physical Review A **24**, 2109-2115 (1981).

Chapter 2

Numerical Integration and Error Analysis

/ Numerical Errors, Accuracy, Convergence / Second-Order Methods / Runge-Kutta Methods / Symplectic Integrators / Lyapunov Instability / Nonlinear Chain / Chaotic Sea / Seven Cell-Model Integrators / Which to Use ? / Predictor-Corrector Algorithms / Fourth-Order Bit-Reversible Algorithm / Relative Accuracy and Efficiency / Summary of Lecture 2 /

2.1 Introduction

This Lecture surveys numerical techniques for solving ordinary differential equations. We wish to solve motion equations in particle mechanics as well as evolution equations for models like the Lorenz attractor. These techniques can solve a broad class of particle problems in molecular dynamics. Though we concentrate on atomistic systems with pairwise-additive forces the same integrators can be applied to systems with many species of large molecules. We will illustrate smooth-particle applications to problems in continuum dynamics. We will not consider astrophysical applications though the same techniques can be applied there as well.

There are many excellent references discussing numerical methods: William Milne's 1949 classic *Numerical Calculus*[1] is useful today. See Daan Frenkel and Berend Smit's book[2] on computational chemistry, our own books on statistical mechanics, thermodynamics, nonlinear systems, and chaos,[3–6] and Clint Sprott's book(s)[7] with hundreds of worked-out examples of chaotic systems. Tamás Tél and Márton Gruiz' book[8] includes a compendium of two-dimensional dynamical systems with practical solution algorithms and graphical results. Billy Todd and Peter Daivis' recent book[9] provides an extensive theoretical background for molecular dynamics written at an advanced level. The last five chapters of their book include many new applications and results. Their rheological research into steady periodic shear deformation and elongation is novel and well worth reading.

Nearly all differential equations we encounter can be solved with explicit methods. Explicit methods formulate the solution at time $t + dt$ in terms of functions and their derivatives at one or more earlier timesteps. We will consider "stiff" equations where explicit fixed-timestep algorithms are computationally unstable unless that timestep is excessively small. Such stiff equations can often be treated with predictor-corrector or adaptive techniques. Instead of calculating an approximate solution one step ahead once and for all, predictor-corrector methods iterate, finding an "implicit" solution at time $t+dt$ which is itself a function of the variables at time $t+dt$. Aneesur Rahman used a two-timestep predictor-corrector method. Applied to our favorite test case, the harmonic oscillator, Rahman's integrator takes the form[10] :

$$\widetilde{q}_{t+dt} = q_{t-dt} + 2dtp_t$$

$$p_{t+dt} = p_t - (dt/2)(q_t + \widetilde{q}_{t+dt}) \; ; \; q_{t+dt} = q_t + (dt/2)(p_t + p_{t+dt}) \; .$$

Stiff equations can also be approximated with adaptive techniques, changing the timestep to control the error. An adaptive method was applied to the Nosé oscillator in **Lecture 1**, page 23.

Different numerical methods can be compared by applying them to a typical problem with a known solution. The harmonic oscillator is the prototypical problem we choose as a test case. The accuracy of a method is expressed as a powerlaw in the timestep. We assess the accuracy of a numerical method as the "global error", the error at a fixed time, or its maximum in a fixed interval. Nonlinear systems often exhibit a physical exponential instability, Lyapunov instability, the hallmark of chaos. That physical instability cannot be eliminated by numerical methods.

An alternative assessment of accuracy results from systematically changing the timestep and recording the solution's dependence, at a fixed time, on a series of smaller timesteps, $\{(dt/2), (dt/4), ...\}$. The mathematical concept of "consistency" is related to this idea. A solution algorithm is said to be consistent if the limiting small-timestep numerical solution agrees with the true solution.

2.2 Numerical Errors, Accuracy, and Convergence

Types of integration errors include amplitude and energy errors, perpendicular to the trajectory, and phase errors parallel to it. In the case of conservative isoenergetic systems like the oscillator, energy drift can be measured along with trajectory errors. Symplectic integrators preserve the comoving many-dimensional volume in phase space in accord with Liouville's theorem. Broken symmetries are another error type. We will consider all of these error types in this Lecture. Our primary tool is the analytic solution of the harmonic oscillator problem. Unless stated otherwise I make the following choices for the oscillator's mass, force constant, and frequency :

$$m = 1 \; ; \; \kappa = 1 \longrightarrow \omega = 1 \;.$$

I use the initial condition $\{\, q, p \,\} = \{\, 1, 0 \,\}$. The solutions of the first-order and second-order motion equations are $\{\, q, p \,\} = \{\, \cos(t), -\sin(t) \,\}$:

$$\dot{q} = p = -\sin(t) \;,\; \dot{p} = -q = -\cos(t) \;[\text{first} - \text{order}] \;;$$

$$\ddot{q} = -q = -\cos(t) \;[\text{second} - \text{order}] \;.$$

Fixed-timestep numerical integration methods approximate the solution of differential equations as a time series :

$$\{\, q_0, \; q_1\;, q_2, \; ... \; q_n \,\} = \{\, q(0), \; q(dt), \; q(2dt), \; ... \; q(ndt) \,\} \;.$$

The algorithms are generated by using a combination of Taylor's series (or equivalent polynomial interpolations and extrapolations) to estimate the solution at the next point q_{n+1} in terms of earlier points in the time series. For example, the Størmer-Verlet algorithm approximates the sum of two Taylor's series at q_{n+1} and q_{n-1} expanded around the time ndt :

$$q_{n+1} + q_{n-1} \equiv 2q_n + dt^2 \ddot{q}_n \;;\; \delta q_n \simeq \tfrac{1}{12} dt^4 \dddot{q}_n \;.$$

Here δq_n is the *local* single-timestep error. This algorithm provides the solution at time $t = (n \pm 1)dt$ with an error term proportional to the fourth power of the timestep. The odd-order terms cancel because we are using a centered fixed-timestep algorithm. Centered-difference methods like Størmer-Verlet are typically "time reversible", a desirable feature matching this aspect of the physical equations of motion. The algorithm can be followed equally well either forward or backward in time.

Let us use the harmonic-oscillator Størmer-Verlet algorithm to illustrate the *order* of an integration algorithm. Errors in the oscillator solution are calculated as the absolute values of differences between the analytical solution and the numerical solution. Thus the single-step "local" errors for the oscillator are

$$\delta q \equiv |\cos(t_n) - q_n| \;;\; \delta p \equiv |-\sin(t_n) - p_n| \;;\; \delta \mathcal{H} \equiv |\tfrac{1}{2} - \mathcal{H}_n| \;.$$

The "global" errors are the timestep-dependent errors measured at a (not too large) fixed time or the maximum error within a fixed time interval. For definiteness I choose the *maximum errors* in the first oscillator period :

$$\Delta q \equiv max\{\delta q_n\} \ ; \ \Delta p \equiv max\{\delta p_n\} \ ; \ \Delta \mathcal{H} \equiv max\{\delta \mathcal{H}_n\} \ .$$

I will use global metrics evaluated over one oscillator period to describe the order of the accuracy of all the integrators presented in this Lecture.

The underlying time series yields "local" numerical errors proportional to a power of the timestep dt : $\delta q \propto dt^n$. "Global" numerical errors measure *accumulated* errors up to a fixed time or within a fixed interval. As global errors correspond to a sum of local errors, they likewise obey a limiting powerlaw relationship for short times. **Figure 2.1** illustrates the local and global oscillator coordinate errors using the Størmer-Verlet algorithm.

The upper curves in the figure are log-log plots of the global error Δq as a function of the timestep at four times, $t = (2, 4, 8, 16)$. The slopes show that the global errors are quadratic in the timestep. Numerical methods are best characterized by the short-time global errors. Following this definition the Størmer-Verlet method is second-order. This results from the twofold integration of the local fourth-order error δq .

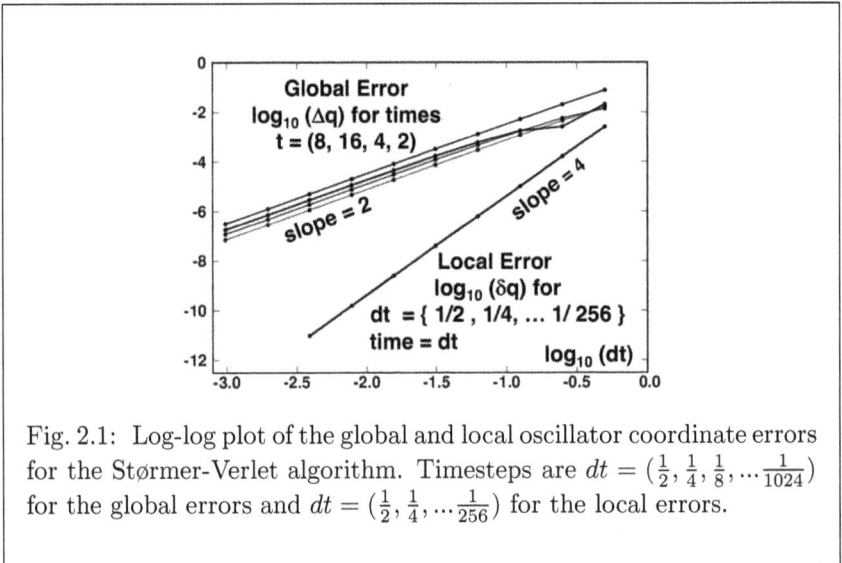

Fig. 2.1: Log-log plot of the global and local oscillator coordinate errors for the Størmer-Verlet algorithm. Timesteps are $dt = (\frac{1}{2}, \frac{1}{4}, \frac{1}{8}, ... \frac{1}{1024})$ for the global errors and $dt = (\frac{1}{2}, \frac{1}{4}, ... \frac{1}{256})$ for the local errors.

These same metrics are applied to the oscillator coordinate and energy in **Figure 2.2**. The slopes are all 2 : the global errors vary as the *square* of dt.

The logarithm of the energy errors using a two-point momentum definition (middle curve, $p_2 \equiv [\, q_{n+1} - q_{n-1} \,]/2dt$) are three times larger than the errors using a four-point momentum approximation (bottom curve). The Figure shows that the Størmer-Verlet method is second-order accurate and that this same accuracy applies to the energy and momentum.

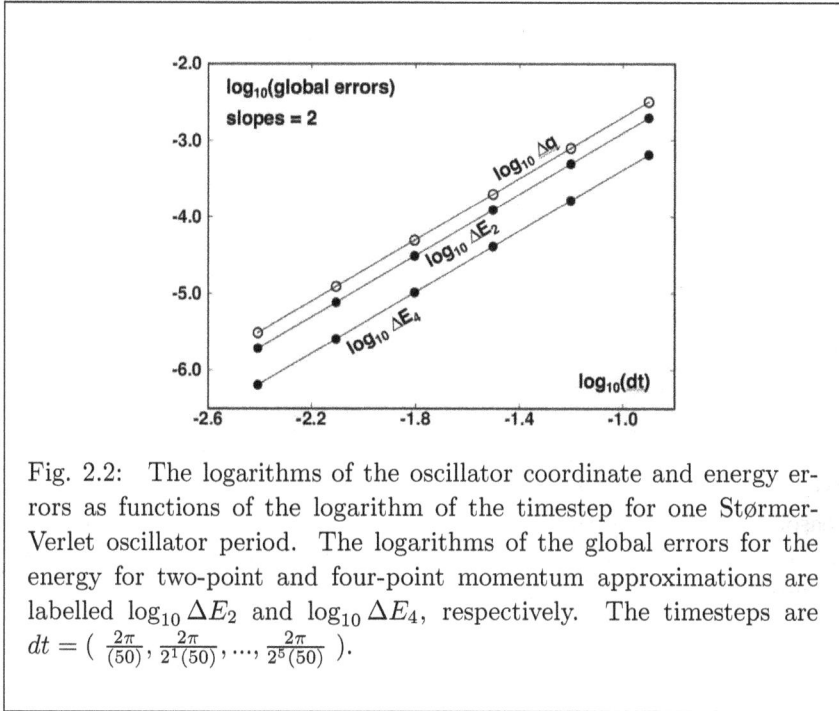

Fig. 2.2: The logarithms of the oscillator coordinate and energy errors as functions of the logarithm of the timestep for one Størmer-Verlet oscillator period. The logarithms of the global errors for the energy for two-point and four-point momentum approximations are labelled $\log_{10} \Delta E_2$ and $\log_{10} \Delta E_4$, respectively. The timesteps are $dt = (\, \frac{2\pi}{(50)}, \frac{2\pi}{2^1 (50)}, ..., \frac{2\pi}{2^5 (50)} \,)$.

The following is a simple program calculating the single-step local coordinate error for the Størmer-Verlet method for timesteps $dt = (\frac{1}{2}, \frac{1}{4}, ... \frac{1}{256})$.

```
DO 30 IT = 1,8
DT       = 1.0D0/(2.0D0**IT)
QNOW     = 1.0D0
QOLD     = 1.0D0 - 0.5D0*DT*DT
TIME     = IT*DT
QNEW     = 2.0D0*QNOW - QOLD + DT*DT*(-QNOW)
DQ       = DABS(DCOS(TIME)-QNEW)
30 WRITE(6,*) DLOG10(DT),DLOG10(DQ)
```

Exercise : Formulate and check a time-centered four-point momentum definition based on $q_{n\pm1}$ and $q_{n\pm2}$.

Most practical problems we will encounter are nonlinear and subject to a physical Lyapunov instability. This exponential growth of the Lyapunov instability soon dominates the powerlaw error growth. I demonstrate this in **Section 2.6.1**, pages 67-68 with a four-particle ϕ^4 model, a nonlinear Hooke's Law harmonic chain with an added quartic tethering potential.

What about the convergence of the numerical approximation? We can demonstrate a converging solution by examining the global error as the timestep is reduced. For example, consider oscillator coordinate errors for a series of 22 timesteps $dt = \{\frac{2\pi}{50}, \frac{2\pi}{2^1(50)}, \frac{2\pi}{2^2(50)}, \dots \frac{2\pi}{2^{21}(50)}\}$. The resulting log-log plot of the errors is linear. The last calculation results in an error of $\Delta q = 7.2 \times 10^{-16}$ for a timestep $dt = 0.0000000599$. To minimize round-off errors I used quadruple precision. This is clearly a convergent solution.

Most interesting problems we solve today, such as the motion of atoms in a biomolecule, are not amenable to an analytical solution. For these problems we use a technique similar to that demonstrated in the figures. We set up a series of short calculations with the timestep varying from a reasonable starting estimate ($dt = 0.01$ or 0.001 for molecular dynamics) and reduced twofold for each succeeding run. The error can be estimated by taking the difference between the solution at a given timestep and the estimated true solution limit. For Lyapunov unstable systems the true-solution concept needs further clarification. We return to it in later Lectures.

Exercise : Write a program to implement the Størmer-Verlet method for a single period of the standard oscillator with $\{q_0, p_0\} = (1, 0)$. Calculate the global error Δq for $dt = (\frac{2\pi}{50}, \frac{2\pi}{100}, \frac{2\pi}{200})$. Plot the logarithm of the global error as a function of the logarithm of the timestep.

2.3 Second-Order Methods

The Størmer-Verlet algorithm, which goes back to Newton, is arguably the most popular method used for Hamiltonian molecular dynamics. It was ideal when computer time was expensive and storage was limited. Today the method is still useful for massively-parallel calculations. Only three coordinates are used for each timestep. It is a single-step method calculating the force only once each timestep. These two facts minimize interprocessor communication when the problem is distributed among processors, making the problem more nearly scalable (computer time varying inversely with the number of processors) for parallel computers. In this Section I describe Størmer-Verlet and three algorithms for solving two first-order differential equations. Two of them produce coordinates equivalent to the Størmer-Verlet algorithm.

2.3.1 The Størmer-Verlet Method

We begin with the Størmer-Verlet algorithm to solve for the coordinate of the second-order differential equation for the oscillator

$$q_{n+1} = 2q_n - q_{n-1} + dt^2 \ddot{q}_n \; ; \; \ddot{q}_n = F = -q_n \; ; \; \delta q \simeq \tfrac{1}{12} dt^4 \; \ddddot{q}_n \; .$$

The local error is fourth-order and as shown in **Figure 2.2** the global error is second-order in dt.

Writing a program to solve for the motion of the harmonic oscillator involves three steps. Follow the steps below to develop and test your program for the Størmer-Verlet approximation to the harmonic oscillator motion.

[1] Specify the initial condition. Notice that for the Størmer-Verlet approximation we need the solution at the initial condition and also at one previous timestep. Most algorithms need more than one value of the coordinate or its derivatives at times earlier than the initial condition. Such algorithms are not "self starting". For the oscillator I choose the initial condition at one of the turning points, $\{q_0, p_0\} = (1, 0)$. A Taylor's series for q_{-1} around the point q_0 gives $q_{-1} = q_0 - \tfrac{1}{2} dt^2 q_0$. We can use the cosine solution as an alternative for the starting condition, $q_{-1} = \cos(-dt)$.

[2] Add diagnostics to your program. Energy conservation is the best diagnostic. For the Størmer-Verlet method it is necessary to estimate the momentum in order to define an energy. Two-point, four-point, and six-point centered-difference approximations for the momentum can be found on page 99 of Milne's book[1] :

$$p2_n = \frac{(q_{n+1} - q_{n-1})}{2dt} \; ; \; \delta p2 \simeq -\frac{dt^2}{6} \dddot{q}_n \; ;$$

$$p4_n = \frac{4}{3} \frac{(q_{n+1} - q_{n-1})}{2dt} - \frac{1}{3} \frac{(q_{n+2} - q_{n-2})}{4dt} \; ; \; \delta p4 \simeq +\frac{dt^4}{30} \ddddot{q}_n \; ;$$

$$p6_n = \frac{3}{2} \frac{(q_{n+1} - q_{n-1})}{2dt} - \frac{3}{5} \frac{(q_{n+2} - q_{n-2})}{4dt} + \frac{1}{10} \frac{(q_{n+3} - q_{n-3})}{6dt} \; ; \; \delta p6 \simeq \frac{-dt^6}{140} \dddddd{q}_n \; .$$

Add the calculation of the local and global errors, the logarithm of the global error and the logarithm of the timestep. Add output statements in your program to print and plot the trajectory values $q(t)$ and $p(t)$, the total energy $E(t) = (q^2 + p^2)/2$, the kinetic energy $K(t)$, and the potential energy $\phi(t)$. Plot the orbit in (q, p) "phase space" for the oscillator.

[3] Estimate a "reasonable" timestep and carry out a few runs with timesteps both smaller and larger than this. Plot the energy and coordinate error. Select a timestep that gives errors that are second-order accurate as shown in **Figures 2.1** and **2.2**.

A program segment for the Størmer-Verlet algorithm is shown here. Add output statements for the error diagnostics and the energy values.

```
ITMAX = 50
TWOPI = 2.0D0*3.141592653589793D0
DT = TWOPI/ITMAX
QNOW = 1.0D0
QOLD = QNOW - 0.5D0*DT*DT*QNOW
DO IT = 1,ITMAX
TIME = DT*IT
QNEW = 2.0D0*QNOW - QOLD - DT*DT*QNOW
PNOW = (QNEW - QOLD)/(2.0D0*DT)
ENOW = 0.5D0*(QNOW*QNOW+PNOW*PNOW)
QOLD  = QNOW
QNOW  = QNEW
END DO
```

Exercise : Use a solution of the form $e^{i\omega t}$ to show that the analytic solution of the Størmer-Verlet method for the oscillator is :

$$q(t_n) = \cos(\omega ndt) \text{ for } \omega = (1/dt)\arccos[\ (1 - (dt^2/2)\ ]$$

The difference between the cosine solution and the analytic solution of the algorithm is

$$\Delta q(t_n) = max\{\sqrt{[\ \cos(t_n) - \cos(\omega t_n)\ ]^2}\ \}\ .$$

Show that these analytic solutions match your numerical results.

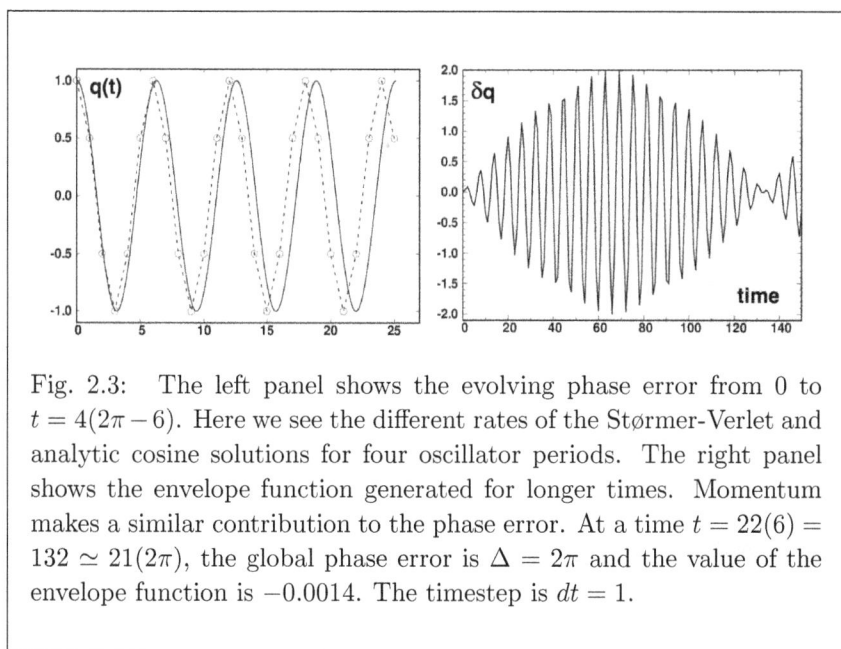

Fig. 2.3: The left panel shows the evolving phase error from 0 to $t = 4(2\pi - 6)$. Here we see the different rates of the Størmer-Verlet and analytic cosine solutions for four oscillator periods. The right panel shows the envelope function generated for longer times. Momentum makes a similar contribution to the phase error. At a time $t = 22(6) = 132 \simeq 21(2\pi)$, the global phase error is $\Delta = 2\pi$ and the value of the envelope function is -0.0014. The timestep is $dt = 1$.

A large timestep simplifies the visualization of "phase errors", where an algorithm's trajectory is faster or slower than the true one. The left panel in **Figure 2.3** shows the oscillator coordinates for the Størmer-Verlet algorithm (dashed line) and the cosine solution (solid line) with $dt = 1$. The Figure illustrates phase error. The dashed curve linking the Størmer-Verlet points, with period 6, is *faster* than the true cosine solution, with period 2π. The global phase error between the two solutions (the distance along the abscissa between the two curves) is $2\pi - 6 \simeq 0.2832$ for each oscillator period and increases as a function of time. The results shown are for a time of 25.0, a bit less than four cosine periods, $t = 4(2\pi) = 25.13$. The right panel in the Figure shows that for longer times the accumulation of the global phase error is bounded by an envelope function with maximum and minimum values of ± 2. For clarity in the plot I did not take the absolute value of the error in plotting the envelope function shown in **Figure 2.3**. Similarly **Figure 2.4** shows two envelope functions from a longer time simulation with a smaller timestep $dt = \pi/10$. The global phase error accumulates an additional 2π each time the envelope function reaches a multiple of (about) 132.

Exercises :
(a) Find the motion of a pendulum in a gravitational field. Use the motion equation $\ddot{\theta} = -\sin(\theta)$ and initial values $(\theta, \dot{\theta}) = (\pi/2, 0)$. Determine the period of the oscillation and compute the energy through one period using $p2$ and $p4$. Which is the better choice ?

(b) Consider a falling particle in a gravitational field. Use the motion equation $\ddot{y} = -1$ with initial values $(y, \dot{y}) = (0, 0)$. Write a program and calculate the Størmer-Verlet solution for a series of timesteps. Compare the results with the analytical solution. Discuss the trajectory and energy errors and the convergence of the solution. How would you implement the possibility of the particle's bouncing at $y = -1$?

Exercises : Extend your Størmer-Verlet program for the oscillator :
(a) For $dt = (\pi/10)$ calculate the maximum coordinate and energy errors with $p2$ in one period. Run five more calculations reducing the timestep by a factor of two in each run. Plot the trajectory $q = q(t)$ and the orbit $p(q)$ for dt $= \pi/10$.

(b) Rerun the problem for 6000 steps using the timestep $dt = (\pi/10)$. Write the numerical solution and the cosine solution to a file. Plot the numerical trajectory $q(ndt)$ and the cosine solution $\cos(ndt)$ for several time ranges: 100 to 120, 500 to 520, 1000 to 1020 and 1500 to 1520. What is the maximum amplitude? Notice that the phase shift increases between the numerical solution and the cosine solution but the maximum amplitude remains equal to 1.

(c) Follow the dynamics of three nearby points and compute the area of the corresponding triangle as a function of time. How closely does the result follow Liouville's Theorem ?

The left panel in **Figure 2.4** shows that the analytical solution of the Størmer-Verlet local coordinate error is useful up to a time $t = 200$ (31.8 periods) for a timestep of $(\pi/10)$. The global phase error calculated using a linear series reversion is $\delta\tau \approx (tdt^2/24)$ and is shown as the straight line in the longer-time simulation in the right panel of the figure.

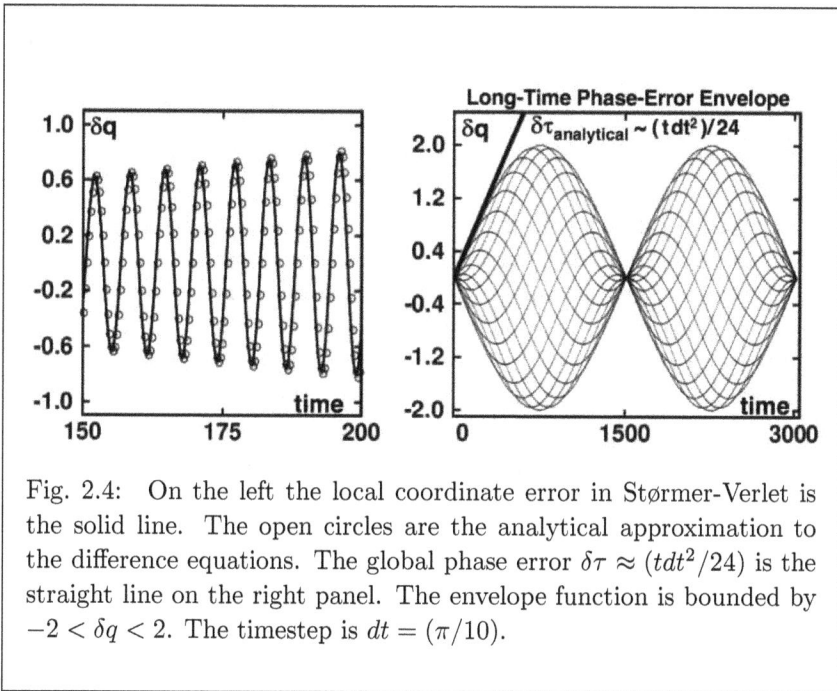

Fig. 2.4: On the left the local coordinate error in Størmer-Verlet is the solid line. The open circles are the analytical approximation to the difference equations. The global phase error $\delta\tau \approx (tdt^2/24)$ is the straight line on the right panel. The envelope function is bounded by $-2 < \delta q < 2$. The timestep is $dt = (\pi/10)$.

The right panel of **Figure 2.4** is a plot of the local phase error in the coordinate up to a time $t = 3000$ with $dt = (\pi/10)$. This corresponds to two periods of the envelope of the local phase error. We can verify this calculation analytically. Using a solution of the form $e^{i\omega t}$ the coordinate and frequency of the oscillator for the Størmer-Verlet method are

$$q_n = \cos(\omega t) \text{ for } \omega = (1/dt) \arccos[\, (1 - (dt^2/2) \,]\,.$$

The phase error in the analytical solution is $\delta q = \cos(\omega t) - \cos(t)$. A calculation of the times at which there is a minimum in the absolute value of the phase error corresponds to times when the envelope function is closest to zero and the two solutions are approximately in phase. These times and local phase error values for the first two minima in the envelope for the local error are

$$t = 1511.106 \,;\ \delta q = 5.175 \times 10^{-7} \,;\ \Delta q = 2\pi \,;$$
$$t = 3022.212 \,;\ \delta q = 2.070 \times 10^{-6} \,;\ \Delta q = 4\pi \,.$$

Notice that the times differ by a factor of 2. As time increases each period of the envelope function adds 2π to the global error. This provides us with the time for subsequent envelope minima.

The interesting pattern in the right panel of the **Figure 2.4** is a Moiré pattern. This is the interference pattern between the Størmer-Verlet solution and the cosine solution for the oscillator. I generated it by adjusting the output interval for the data and then plotting the points as small dots.

An investigation of the coordinate data up to a time of 1500 shows that the maximum amplitude is one throughout the calculation of 240 cosine periods. This is no accident ! Liouville's theorem for Hamiltonian flows proves that the comoving phase volume (here a two-dimensional area) evolves without change. The invariant corresponding to the total volume in phase space is the entropy. Symplectic numerical methods are algorithms which maintain this invariant (and others in higher-dimensional phase spaces) for Hamiltonian systems. Applying these ideas to a numerical solution of the oscillator shows that an orbit which encloses a circle initially can neither expand nor decay in time. However, the rate at which the points are reached on the circle depends on the timestep. Evidently the numerical error is solely a phase error, parallel rather than perpendicular to the trajectory.

In contrast, the fourth- and fifth-order Runge-Kutta algorithms applied to the oscillator in Section 2.4 show that the long-time orbits decay with RK4 and grow with RK5, causing energy "drift" errors. Part (**c**) in the last of the exercises in this section, page 48, demonstrates the symplectic nature of the Størmer-Verlet algorithm.

2.3.2 *Reversibility and Bit-Reversibility*

Time reversibility is a fundamental characteristic of the physical equations describing particle dynamics, and indeed nearly *all* equations describing physics. It is desirable that numerical approximations for dynamical systems maintain time reversibility for as long as possible despite the growth of numerical errors due first to roundoff errors and later to Lyapunov instability. The Størmer-Verlet algorithm is time-reversible. To see this replace $dt \rightarrow -dt$. Because the form of the equation remains unchanged the algorithm *is* time-reversible. The reversibility results from the time centering of the second derivative :

$$\text{Forward} \longrightarrow q_{n+1} = 2q_n - q_{n-1} + (+dt)^2(-q_n) \ ;$$
$$\text{Reverse} \longleftarrow q_{n-1} = 2q_n - q_{n+1} + (-dt)^2(-q_n) \ .$$

The motion displayed in a movie obeying time-reversible equations of motion can be reversed by projecting the frames in the opposite order. Despite the unchanged form of the forward and reverse equations, the time reversibility of a numerical algorithm is typically limited by numerical errors. We will show this by checking reversibility numerically for several integrators applied to a nonlinear cell model in Section 2.7, page 73. Let us consider a remarkable instance of *perfect* time reversibility next, the integer algorithm developed by Levesque and Verlet.[11]

Dominique Levesque and Loup Verlet pointed out that an *integer* integration algorithm in an integer phase space can be made precisely time-reversible by a particular choice of roundoff error. Their algorithm is an *integer* version of the Størmer-Verlet method. For the oscillator it is :

$$IQ_{n+1} - 2IQ_n + IQ_{n-1} = \text{INT}(-IQ_n dt^2) \ \rightarrow \ \text{Bit} - \text{Reversible} \ ! \ .$$

The "INT" operation drops all digits following the decimal point, $INT(\pm 2.5) = \pm 2$. The lefthandside { IQ } variables are evaluated as integers (four-byte integers in the example below). The product of the acceleration and the square of the timestep is converted to an integer to match the precision of the lefthandside. It is quite feasible to use 64-bit (8-byte) integers with the **gfortran** compiler options. **Figure 2.5** on the next page illustrates the forward and backward bit-reversible oscillator solution. Care must be taken that the righthandside of the equation is evaluated with the floating point product of the integer and the square of the timestep calculated *before* truncating the result with the INT function. The top panel of the figure illustrates that an incorrect truncation destroys the bit-reversibility. The bottom panel illustrates the correct bit-reversible trajectory with the INT function applied after the product of the integer and the square of the timestep is calculated.

$$IQ_{n+1} = \text{INT}(2IQ_n - IQ_{n-1} - IQ_n dt^2) \ ; \ [\text{ incorrect }] \ ;$$

$$IQ_{n+1} = 2IQ_n - IQ_{n-1} + \text{INT}(-IQ_n dt^2) \ \rightarrow \ [\text{ bit} - \text{reversible }] \ .$$

The bit-reversible algorithm is of interest pedagogically as well as in practice when a truly reproducible trajectory is needed in both the forward and reverse directions. A correctly implemented bit-reversed trajectory reproduces all the bits in the integer coordinates in the forward direction. An example described in Section 2.10 on page 89 illustrates the use of a second-order bit-reversible method. In that example two balls each composed of 400 smooth particles undergo an inelastic (but perfectly Hamiltonian) collision, merging to form an 800-particle ball. Particles colored black in the Figures make above-average contributions to the local Lyapunov instability. Completely different particles are the important ones in the reverse motion. The simulation shows that although the dynamics is reversible, stability is

not. We will see that this irreversibility is closely related to the Second
Law of Thermodynamics. A precisely-reversed trajectory is required for
the two-ball problem. Without a bit-reversible algorithm for this problem,
it would be necessary to write 1600 coordinates every timestep for 100,000
steps in the forward trajectory to a file. The bit-reversible method eliminates the need for storing the trajectory and more importantly calculates
an identical backward trajectory, exact to every bit.

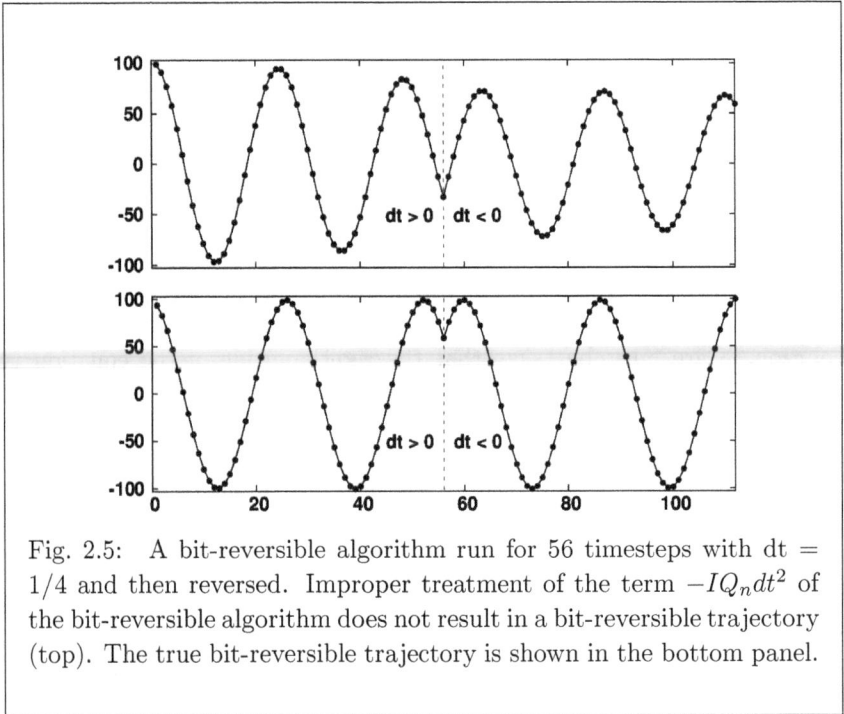

Fig. 2.5: A bit-reversible algorithm run for 56 timesteps with dt =
$1/4$ and then reversed. Improper treatment of the term $-IQ_n dt^2$ of
the bit-reversible algorithm does not result in a bit-reversible trajectory
(top). The true bit-reversible trajectory is shown in the bottom panel.

An excerpt from the program generating **Figure 2.5** is shown below
for 56 timesteps in each direction. Notice the order of the coordinates
is reversed to start the backward solution making it unnecessary to set
$dt = -dt$.

```
C PROGRAM TO GENERATE A BIT REVERSIBLE TRAJECTORY
      INTEGER IQNEW, IQNOW, IQOLD, IFORCE
      IQNOW = 100
      IQOLD = 100
      DT = 0.25D0
      DO 10 IT = 1,56
```

```
C GET THE INTEGER FORCE
      IFORCE = -DT*DT*IQNOW
      IQNEW   = 2*IQNOW - IQOLD + IFORCE
      TIME = IT*DT
      WRITE(6,*)IT,TIME,IQNOW
      IQOLD = IQNOW
  10  IQNOW = IQNEW
C REVERSE THE TRAJECTORY
      II = IQNOW
      JJ = IQOLD
      IQOLD = II
      IQNOW = JJ
      DO 30 IT = 57,112
C REPEAT THE SAME LOOP AS IN THE FORWARD DIRECTION
  30  CONTINUE
```

2.3.3 *Rahman, Symplectic-Leapfrog, Velocity-Verlet*

Let us take up three variants of the Størmer-Verlet method for solving two first-order differential equations : [1] Rahman's predictor-corrector algorithm; [2] A Symplectic-Leapfrog algorithm; and [3] the Velocity-Verlet algorithm. In the early days of molecular dynamics calculations Rahman used a "predictor-corrector" technique for simulating the motion and correlations of 864 liquid argon atoms in a periodic cube.[10]

Rahman's algorithm predicts a value of the coordinate at the new timestep. This prediction is then used in an implicit corrector step, which can be repeated. The corrector step can be iterated until the change it produces is sufficiently small. The two-step Rahman method is :

$$\mathbf{P} : q_{n+1} = q_{n-1} + 2dtp_n \ ; \ \delta q \simeq (1/3)dt^3 \ \dddot{q}_n \ [\text{The Predictor}] \ ;$$

$$\mathbf{C} : p_{n+1} = p_n + (dt/2)(\dot{p}_{n+1} + \dot{p}_n) \ ; \ q_{n+1} = q_n + (dt/2)(p_{n+1} + p_n) \ .$$

The local errors incurred in each of these steps are third-order in dt. For the oscillator we would simply use $\dot{p} = -q$ in the corrector step $\mathbf{C}$ above.

The Symplectic-Leapfrog algorithm is a "multistage" method. Symplectic methods calculate the coordinate and momentum updates in one or more stages "within" a timestep. In this method $\tilde{q}$ denotes a mid-timestep stage used to advance from time t_n to time t_{n+dt} :

$$\tilde{q} = q_n + p_n(dt/2) \ ; \ p_{n+1} = p_n + \dot{p}_n dt \ ; \ q_{n+1} = \tilde{q} + p_{n+1}(dt/2) \ .$$

For the oscillator this becomes

$$\tilde{q} = q_n + p_n(dt/2) \; ; \; p_{n+1} = p_n - \tilde{q}dt \; ; \; q_{n+1} = \tilde{q} + p_{n+1}(dt/2) \; .$$

In each stage the new coordinate or momentum is based on the momentum or coordinate calculated just previously. This is a characteristic of multistage methods. In our later discussion of symplectic methods we will observe that the weights for each stage [here $(1/2)$ and $(1/2)$] in a timestep sum to one. This is a requirement for achieving a consistent solution, one which agrees with the true solution as the timestep approaches zero.

The third variant of the Størmer-Verlet algorithm is the Velocity-Verlet algorithm. The coordinate and momentum updates are

$$q_{n+1} = q_n + p_n dt + \ddot{q}_n \tfrac{dt^2}{2} \; ; \; \delta q \simeq \tfrac{dt^3}{6} \dddot{q}_n$$

$$p_{n+1} = p_n + (\ddot{q}_n + \ddot{q}_{n+1}) \tfrac{dt}{2} \; ; \; \delta p \simeq \tfrac{dt^3}{12} \dddot{p}_n \; .$$

The error terms for the coordinate and momentum are both third-order. Substitute $\ddot{q} = -q$ to get the oscillator version of the algorithm.

We often compare algorithms using a relatively-large timestep $dt = 1$ to examine trajectory details visibly, such as the amplitude and phase errors and the growth or decay of the energy. To compare these three algorithms for the harmonic oscillator we illustrate this idea with $dt = 1$. The numerical results show that the coordinate values calculated for the Symplectic-Leapfrog and the Velocity-Verlet methods are identical :

$$\{q\} = (1.0, +0.5, -0.5, -1.0, -0.5, +0.5, 1.0) \; ,$$

but with different approximations for the momentum. These coordinate values also agree with the coordinates calculated with the Størmer-Verlet algorithm. We can demonstrate that the coordinates for Rahman's method agree with the three here, but with a different timestep, "scaling the time".

Let us consider the analytic solution of Rahman's difference equations and find the timestep Δ which produces the same coordinate set as the other methods with $dt = 1$. Substituting Rahman's momentum approximation into the coordinate update gives a quadratic equation in Δ :

$$\Delta^2 - \frac{4p_n \Delta}{(q_{n+1} + q_n)} + \frac{4(q_{n+1} - q_n)}{(q_{n+1} + q_n)} = 0.$$

For n=0 the initial values q_0, p_0 and the value $q_1 = +0.5$ at $t = 1.0$ substituted into this expression give the timestep $\Delta = \sqrt{4/3}$ for the Rahman algorithm. Using this timestep the Rahman coordinates reproduce the coordinates for the other variants of the Størmer-Verlet algorithm.

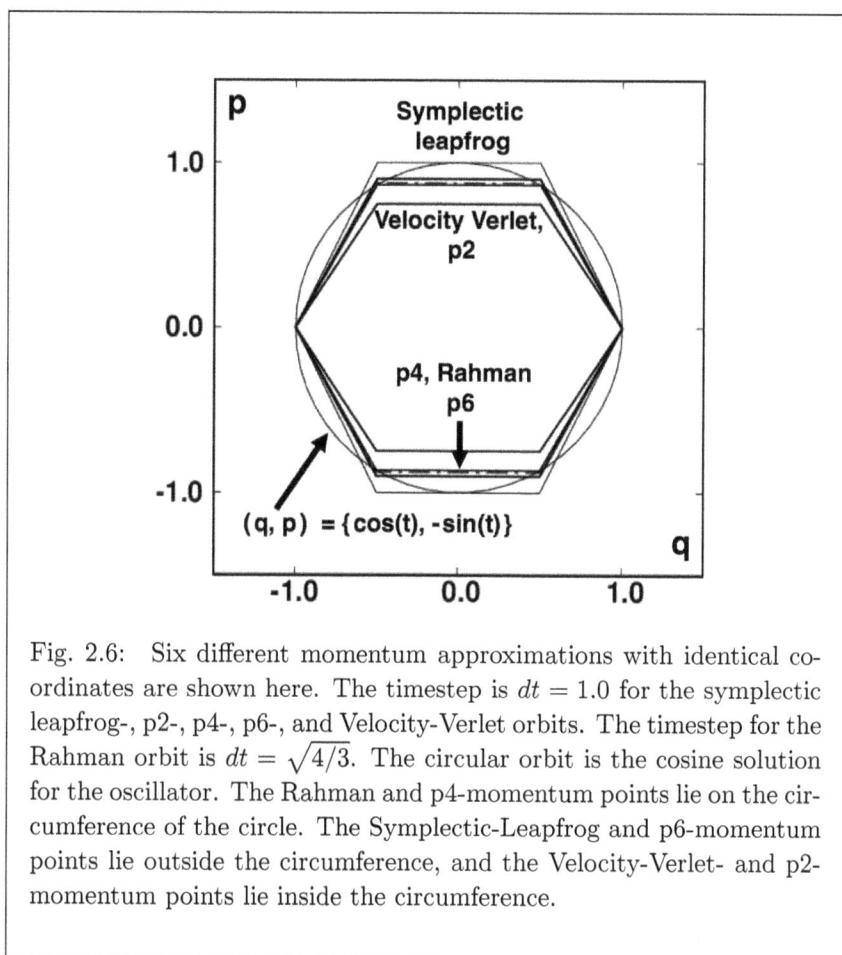

Fig. 2.6: Six different momentum approximations with identical coordinates are shown here. The timestep is $dt = 1.0$ for the symplectic leapfrog-, p2-, p4-, p6-, and Velocity-Verlet orbits. The timestep for the Rahman orbit is $dt = \sqrt{4/3}$. The circular orbit is the cosine solution for the oscillator. The Rahman and p4-momentum points lie on the circumference of the circle. The Symplectic-Leapfrog and p6-momentum points lie outside the circumference, and the Velocity-Verlet- and p2-momentum points lie inside the circumference.

In **Figure 2.6** six orbits are compared using Velocity-Verlet, Symplectic-Leapfrog, Rahman, p2-, p4-, and p6-momentum approximations. The coordinates are exactly the same for all six orbits. The timestep is $dt = 1.0$ for the first five and $dt = \sqrt{4/3}$ for the Rahman orbit. We can solve for the momentum in terms of the coordinates. The nonzero values of the six momenta are listed below with their defining equations. These can be used to check and correct programs.

$$p_{n+1} = \frac{(q_{n+1}-q_n)}{dt} - \tfrac{1}{4}(q_n + q_{n+1})dt \; ; \; [\text{ Rahman Corrector }] \; ;$$

$$p_{n+1} = (q_{n+1} - q_n - \tfrac{1}{2}q_{n+1}dt^2)/dt \; [\text{ Velocity} - \text{Verlet }] \; ;$$

$$p_n = [\, q_{n+1} - q_n (1 - \tfrac{dt^2}{2}) \,] \, / \, [\, 1 - \tfrac{dt^3}{4} \,] \; [\text{ Symplectic }] \; ;$$

$$p2_n = \tfrac{(q_{n+1} - q_{n-1})}{2dt} \; [\text{ p2 momentum }] \; ;$$

$$p4_n = \tfrac{4}{3} \tfrac{(q_{n+1} - q_{n-1})}{2dt} - \tfrac{1}{3} \tfrac{(q_{n+2} - q_{n-2})}{4dt} \; [\text{ p4 momentum }] \; ;$$

$$p6_n = \tfrac{3}{2} \tfrac{(q_{n+1} - q_{n-1})}{2dt} - \tfrac{3}{5} \tfrac{(q_{n+2} - q_{n-2})}{4dt} + \tfrac{1}{10} \tfrac{(q_{n+3} - q_{n-3})}{6dt} \; [\text{ p6 momentum }] \; .$$

For $dt = 1$ the nonzero momenta are ± 1 (Leapfrog) ; ± 0.900 (p6 momentum) ; ± 0.875 (p4 momentum) ; ± 0.866 (Rahman) ; and ± 0.750 (Velocity-Verlet and p2 momentum) .

Figures 2.7 and 2.8 compare the accuracies of the second-order methods discussed so far. The most significant difference between Rahman and the other algorithms is that the Rahman energy error is zero! The energy errors in **Figure 2.8** for the Størmer-Verlet algorithm with $p2$, Velocity-Verlet, and Symplectic-Leapfrog algorithms are too similar to distinguish. The energy errors for these approximations with $dt = (2\pi/50)$ are listed here :

$$\Delta E = 7.9946 \times 10^{-3} \, [\text{Leapfrog}] \; ; \; \Delta E = 7.8684 \times 10^{-3} \, [\text{Velocity} - \text{Verlet}] .$$

$$\Delta E = 7.8684 \times 10^{-3} \, [\text{Størmer} - \text{Verlet} + \text{p2}] .$$

The Størmer-Verlet algorithm with the four-point momentum $p4$ results in an energy error three times smaller than the same algorithm with the two-point approximation :

$$\Delta E = 2.5115 \times 10^{-3} \, [\text{Størmer} - \text{Verlet} + \text{p4}] .$$

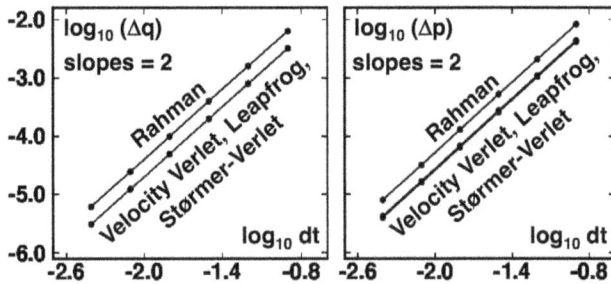

Fig. 2.7: Trajectory errors for Rahman, Velocity-Verlet, Størmer-Verlet, and Symplectic methods with $dt = (\, \frac{2\pi}{(50)}, \frac{2\pi}{2^1 (50)}, \ldots, \frac{2\pi}{2^5 (50)} \,)$.

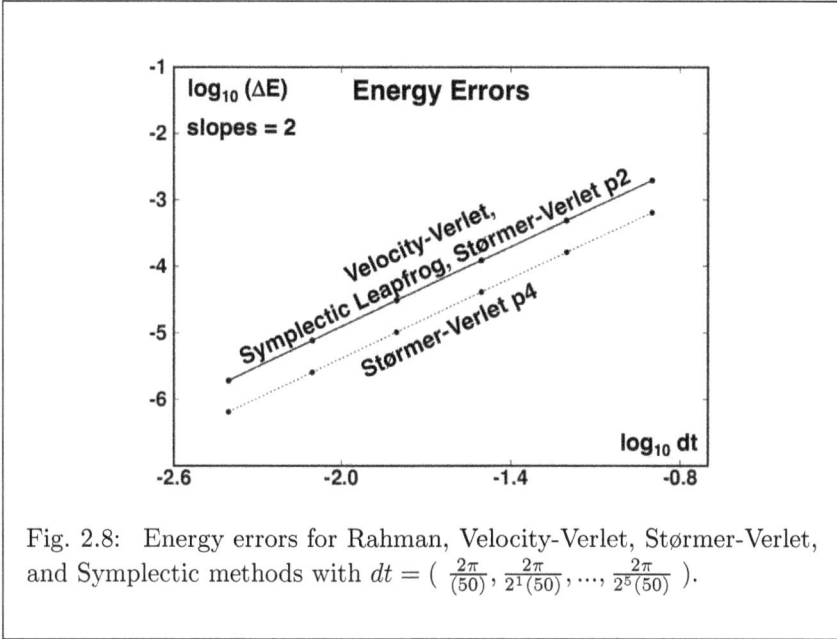

Fig. 2.8: Energy errors for Rahman, Velocity-Verlet, Størmer-Verlet, and Symplectic methods with $dt = (\frac{2\pi}{(50)}, \frac{2\pi}{2^1(50)}, \dots, \frac{2\pi}{2^5(50)})$.

Bill, Oyeon Kum, and Nancy Owens[12] developed a second-order Monte-Carlo method with very accurate trajectory and energy values. The method is symplectic. They used a Monte-Carlo search for six coefficients subject to two constraints in addition to time reversibility : [1] Invariant phase-space volume (symplectic) and [2] Intermediate coefficients in the steps that sum to one (a consistency condition).

$$q = q + 0.005904pdt \; ; \; p = p + 0.171669(F/m)dt \; ; \; \text{for } (F/m) = \ddot{q} \; ;$$

$$q = q + 0.515669pdt \; ; \; p = p - 0.516595(F/m)dt \; ;$$

$$q = q - 0.021573pdt \; ; \; p = p + 1.689852(F/m)dt \; ; \; q = q - 0.021573pdt \; ;$$

$$p = p - 0.516595(F/m)dt \; ; \; q = q + 0.515669pdt \; ;$$

$$p = p + 0.171669(F/m)dt \; ; \; q = q + 0.005904pdt \; .$$

The global errors produced with this Monte Carlo algorithm are shown in **Figure 2.9**. For the timestep $dt = (\pi/100)$ the Monte Carlo trajectory errors are almost five orders of magnitude smaller than the trajectory errors for the Velocity-Verlet method.

$$\log_{10} \Delta q_{VV} - \log_{10} \Delta q_{MC} = 4.813 \; ; \; \log_{10} \Delta p_{VV} - \log_{10} \Delta p_{MC} = 4.752 \; .$$

The Monte Carlo energy error is four orders of magnitude smaller than the energy error for the Velocity-Verlet method.

$$\log_{10} \Delta H_{VV} - \log_{10} \Delta H_{MC} = 4.149 \ .$$

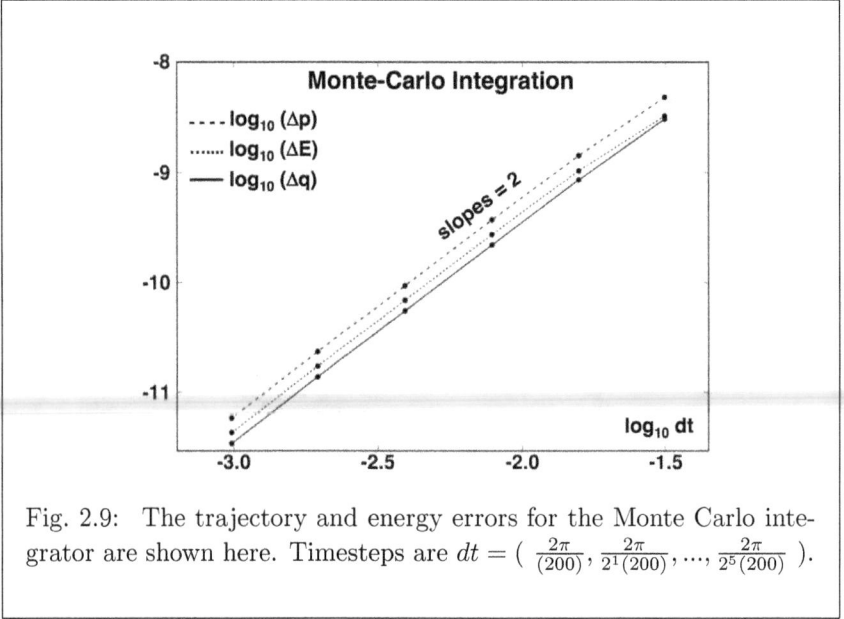

Fig. 2.9: The trajectory and energy errors for the Monte Carlo integrator are shown here. Timesteps are $dt = (\frac{2\pi}{(200)}, \frac{2\pi}{2^1(200)}, ..., \frac{2\pi}{2^5(200)})$.

The Monte-Carlo algorithm has been tested for the classical many-body problem. An advantage of the algorithm is that it is not necessary to store the intermediate results. For the oscillator the error in the algorithm behaves irregularly for timesteps $dt > 0.03$.

Exercise : Show that the Rahman, Symplectic-Leapfrog, and Velocity-Verlet algorithms are all time-reversible.

Exercise : Write programs implementing the algorithms in this section. Find the timestep that reduces the global energy error to 1%.

2.4 Runge-Kutta Methods

There are many Runge-Kutta integrators for first-order differential equations. All of them are multistage methods. The weights for the stages sum to one for each phase-space variable. The most useful Runge-Kutta algorithms for our work are the fourth-order (RK4) and fifth-order (RK5) integrators. These two integrators can be used to investigate the errors and the growth of errors in the dynamics of chaotic systems. We have recently used the two integrators to develop adaptive integration methods. An example, integrating the $(\dot{q}, \dot{p}, \dot{s}, \dot{\zeta})$ very stiff Nosé oscillator equations,[13] is described in Section 1.73 on page 23.

The RK4 update combines derivatives, $\dot{q}$ or $\dot{p}$, in the neighborhood of $q(t)$ or $p(t)$ so as to eliminate error terms of order dt, dt^2, dt^3, and dt^4. It has a single timestep error of order $dt^5/5!$. The algorithm computes the values of q_{n+1} and p_{n+1} by averaging four values of the derivatives $\{p_1, p_2, p_3, p_4\}$ and $\{F_1, F_2, F_3, F_4\}$ at times $\{t_n, t_{n+1/2}, t_{n+1/2}, t_{n+1}\}$. Each of the four stages for evaluating the derivative involves a force evaluation. This is significant as the majority of the computation time in molecular dynamics is spent calculating the forces. The four stages are :

$$\text{Stage 1} \ : \ p_1 = \dot{q}_n \ ; \ F_1(q_n) = \dot{p}_n \ ;$$

$$q_{n+(1/2)} = q_n + \tfrac{dt}{2}p_1 \ ; \ p_{n+(1/2)} = p_n + \tfrac{dt}{2}F_1 \ ;$$

$$\text{Stage 2} \ : \ p_2 = \dot{q}_{n+(1/2)} \ ; \ F_2(q_{n+(1/2)}) = \dot{p}_{n+(1/2)} \ ;$$

$$q_{n+(1/2)} = q_n + \tfrac{dt}{2}p_2 \ ; \ p_{n+(1/2)} = p_n + \tfrac{dt}{2}F_2 \ [\text{ New Values }] \ ;$$

$$\text{Stage 3} \ : \ p_3 = \dot{q}_{n+(1/2)} \ ; \ F_3(q_{n+(1/2)}) = \dot{p}_{n+(1/2)} \ ;$$

$$q_{n+1} = q_n + dt\,p_3 \ ; \ p_{n+1} = p_n + dt\,F_3 \ ;$$

$$\text{Stage 4} \ : \ p_4 = \dot{q}_{n+1} \ ; \ F_4(q_{n+1}) = \dot{p}_{n+1} \ .$$

The solutions are computed in the final step as a weighted sum of the derivatives from each of the four stages.

$$q_{n+1} = q_n + \tfrac{dt}{6}(p_1 + 2p_2 + 2p_3 + p_4) \ ; \ p_{n+1} = p_n + \tfrac{dt}{6}(F_1 + 2F_2 + 2F_3 + F_4) \ ;$$

$$\delta q \simeq -(1/5!)dt^5 \ \dddot{q} \ ; \ \delta p \simeq +(1/5!)dt^5 \ \dddot{p} \ .$$

The local error is fifth-order in both the coordinates and the momenta.

The RK5 integrator approximates the derivatives in six relatively complex stages at five equally-spaced times, ending with a weighted average of these derivatives at a final update to time t_{n+1}. There are six force evaluations for this algorithm.

The single-step errors for the fourth- and fifth-order Runge-Kutta integrators are decay ($\simeq -dt^5/5!$) and growth ($\simeq +dt^6/6!$), respectively as shown in **Figure 2.10**. This also illustrates the Runge-Kutta energy drifts, monotonically decreasing for RK4 and monotonically increasing for RK5.

A program implementing the RK4 algorithm is presented in Section 1.3. A FORTRAN version of RK5 is shown here. Notice that the weights for the five derivative approximations add up to unity. The six times for evaluating the derivative with the subroutine RHS are commented in the program.

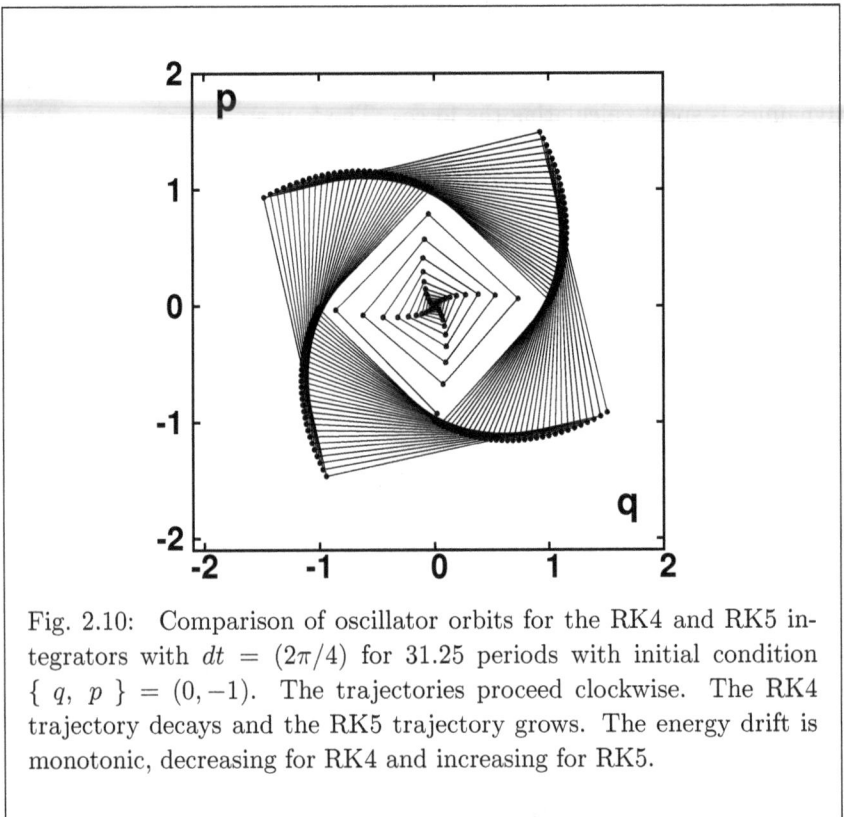

Fig. 2.10: Comparison of oscillator orbits for the RK4 and RK5 integrators with $dt = (2\pi/4)$ for 31.25 periods with initial condition $\{ q, p \} = (0, -1)$. The trajectories proceed clockwise. The RK4 trajectory decays and the RK5 trajectory grows. The energy drift is monotonic, decreasing for RK4 and increasing for RK5.

```
 SUBROUTINE RK5(YY,YYP,DT)
c Insert storage allocations and PARAMETER statement
 CALL RHS(YY,YYP)    ! The solution and derivative at t = ndt
 DO I = 1,NEQ
 Y1(I) = YYP(I)
 END DO
 DO I = 1,NEQ
 YNEW(I) = YY(I) + Y1(I)*DT/2
 END DO
 CALL RHS(YNEW,YYP) ! The solution and derivative at t = (n+1/2)dt
 DO I = 1,NEQ
 Y2(I) = YYP(I)
 END DO
 DO I = 1,NEQ
 YNEW(I) = YY(I) + (3*Y1(I)+Y2(I))*DT/16
 END DO
 CALL RHS(YNEW,YYP)   ! The solution and derivative at t = (n+1/4)dt
 DO I = 1,NEQ
 Y3(I) = YYP(I)
 END DO
 DO I = 1,NEQ
 YNEW(I) = YY(I) + Y3(I)*DT/2
 END DO
 CALL RHS(YNEW,YYP) ! The solution and derivative at t = (n+1/2)dt
 DO I = 1,NEQ
 Y4(I) = YYP(I)
 END DO
 DO I = 1,NEQ
 YNEW(I) = YY(I) + (-3*Y2(I)+6*Y3(I)+9*Y4(I))*DT/16
 CALL RHS(YNEW,YYP) ! The solution and derivative at t = (n+3/4)dt
 DO I = 1,NEQ
 Y5(I) = YYP(I)
 ENDDO
 DO I = 1,NEQ
 YNEW(I) = YY(I) +
& (Y1(I)+4*Y2(I)+6*Y3(I)-12*Y4(I)+8*Y5(I))*DT/7
 END DO
 CALL RHS(YNEW,YYP) ! The solution and derivative at t = (n+1)dt
 DO 20 I = 1,NEQ
```

```
Y6(I) = YYP(I)
END DO
DO I = 1,NEQ
YYP(I) =
& (7*Y1(I)+32*Y3(I)+12*Y4(I)+32*Y5(I)+7*Y6(I))/90
END DO
DO I = 1,NEQ
YY(I) = YY(I)+DT*YYP(I)
CONTINUE
RETURN
END
```

Exercise : Develop programs for the "springy" pendulum. The pendulum supports a unit mass from the origin using a Hooke's law spring. The gravitational field strength is unity and the spring constant is 4. The restlength of the pendulum is unity. Using polar coordinates the pendulum's y-coordinate is $y = -r\cos(\theta)$ where θ is measured from the downward direction $(x, y) = (0, -1)$. Write the Hamiltonian for this system in both Cartesian and polar coordinates. Write and debug programs using RK4 and RK5 to solve the problem in both coordinate systems with initial conditions $\{x, y, p_x, p_y\} = (1, 0, 0, 0)$. Check your results using the energy diagnostic and the plotted orbits. Compare also the four programs with initial conditions $\{x, y, p_x, p_y\} = (2^{-1/2}, 2^{-1/2}, 0, 0)$, which exhibits chaos.

Exercise : Show that the fourth-order Runge-Kutta approximation for the oscillator agrees with the analytic solution of the difference equations : $q_n = [\, 1 - (dt^6/72) + (dt^8/576) \,]^{n/2} \cos(ndt\lambda)$. λ is given by the relation : $\tan(dt\lambda) = [\, dt - (dt^3/6) \,]/[\, 1 - (dt^2/2) + (dt^4/24) \,]$.

2.5 Symplectic Integrators

Symplectic integrators are designed to mimic a property of Hamiltonian systems. Such integrators preserve the comoving phase-space volume in accord with Liouville's Theorem. The logarithm of this conserved phase-space volume represents the entropy of the system. There are additional conserved quantities, the Poincaré invariants. Meiss[14] develops the mathematical definition of the invariants. The Poincaré invariants define a symplectic structure. The phase-space invariance can be interpreted as allowing volume-preserving shear deformations in the phase-space flow but not allowing compressive or dilatational deformations.

There are some interesting unexpected consequences of Liouville's Theorem. Yoshida[15] has pointed out that a symplectic algorithm can be designed to reproduce either the energy or the frequency but not both. Another rather surprising result is that the constant density established by Liouville's theorem is not affected by external forces $F(t)$. Let me demonstrate with an example from Bill's *Computational Statistical Mechanics*.

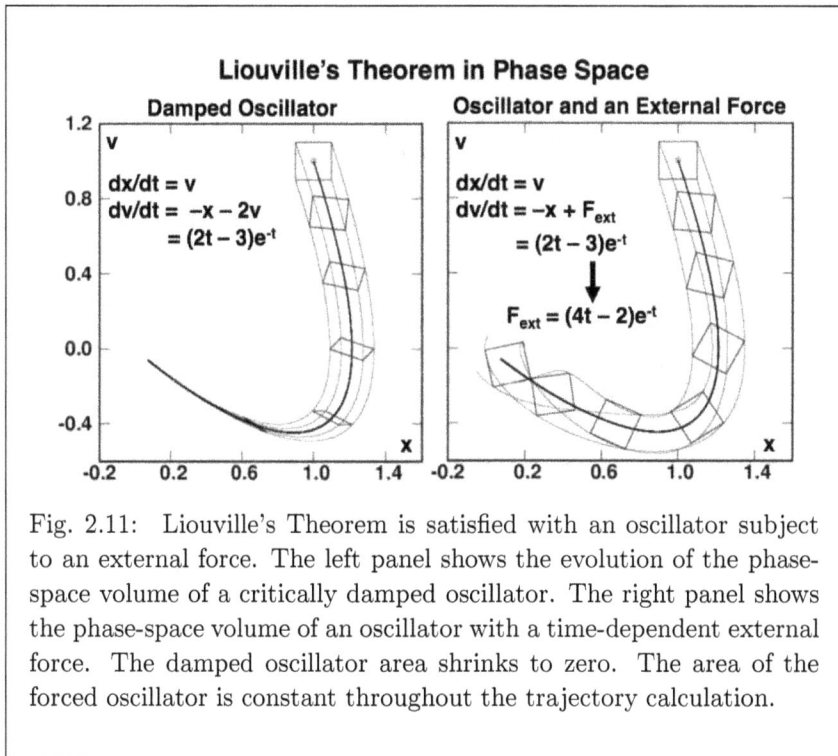

Fig. 2.11: Liouville's Theorem is satisfied with an oscillator subject to an external force. The left panel shows the evolution of the phase-space volume of a critically damped oscillator. The right panel shows the phase-space volume of an oscillator with a time-dependent external force. The damped oscillator area shrinks to zero. The area of the forced oscillator is constant throughout the trajectory calculation.

Bill[3] compared the motions in phase-space of two harmonic oscillators, one critically damped and the other subject to a time-dependent force. He showed that though the trajectories of the two are identical, in the damped case the phase-space density diverges as e^{+2t} whereas the undamped but forced oscillator follows exactly the same trajectory but with a constant comoving phase-space density.

The equations of motion for the critically damped oscillator are

$$\dot{x} = v \; ; \; a \equiv \dot{v} = -x - 2v \; .$$

With (x, v, t) initially $(1, 1, 0)$ the solution and its two time derivatives are

$$x = (2t + 1)e^{-t} \; ; \; \dot{x} = v = (1 - 2t)e^{-t} \; ; \; a = (2t - 3)e^{-t} \; .$$

The equations of motion for the forced oscillator with the external force chosen to make the trajectories $x(t), v(t)$ identical are

$$\dot{v} = -x + F(t) \text{ with } F(t) = (4t - 2)e^{-t} \; .$$

Figure 2.11 shows the trajectories traced out by a square of initial conditions with the square's center at $\{q_0, p_0\} = (1, 1)$ and with sidelength 0.2. The right panel shows that the comoving square maintains its area despite the external force $F(t)$. The square at the left panel quickly deforms and collapses to a point at the (x, v) origin. In both cases the initial point $(1, 1)$ follows the same trajectory. The equations of motion for the forced oscillator obey Liouville's Theorem :

$$(\partial \dot{x} / \partial x) + (\partial \dot{v} / \partial v) = 0 \; .$$

The damped oscillator equations do not.

Yoshida's[15] 1993 paper reports the motivation and theoretical development for symplectic algorithms along with applications to planetary motion. This interesting work points out one of the reasons symplectic algorithms became popular in the late 1980s in the astronomy community. Symplectic methods by definition are not subject to the energy drift present in nonsymplectic methods. Yoshida compares symplectic and nonsymplectic methods for the Kepler problem and the motion of the outer planets. He makes a good case for symplectic methods. Gray, Noid, and Sumpter[16] compare explicit symplectic methods useful for molecular dynamics simulations. The second-order Størmer-Verlet method can be cast into a symplectic form. Here we include two more symplectic algorithms, Candy-Rozmus' fourth-order and Yoshida's sixth-order algorithms.

The Candy-Rozmus fourth-order method[17] consists of four coordinate updates and only three momentum updates with three force evaluations.

$$q = q + 0.6756036pdt \; ; \; p = p + 1.3512072(F/m)dt \; ;$$

$$q = q - 0.1756036pdt \; ; \; p = p - 1.7024144(F/m)dt \; ;$$

$$q = q - 0.1756036pdt \ ; \ p = p + 1.3512072(F/m)dt \ ;$$

$$q = q + 0.6756036pdt \ .$$

The Yoshida 6th-order method[18] consists of eight coordinate updates and seven momentum updates with seven force evaluations :

$$q = q + 0.39225680523878pdt \ ; \ p = p + 0.78451361047756(F/m)dt \ ;$$

$$q = q + 0.51004341191846pdt \ ; \ p = p + 0.23557321335936(F/m)dt \ ;$$

$$q = q - 0.47105338540976pdt \ ; \ p = p - 1.1776799841789(F/m)dt \ ;$$

$$q = q + 0.06875316825252pdt \ ; \ p = p + 1.3151863206839(F/m)dt \ ;$$

$$q = q + 0.06875316825252pdt \ ; \ p = p - 1.1776799841789(F/m)dt \ ;$$

$$q = q - 0.47105338540976pdt \ ; \ p = p + 0.23557321335936(F/m)dt \ ;$$

$$q = q + 0.51004341191846pdt \ ; \ p = p + 0.78451361047756(F/m)dt \ ;$$

$$q = q + 0.39225680523878pdt \ .$$

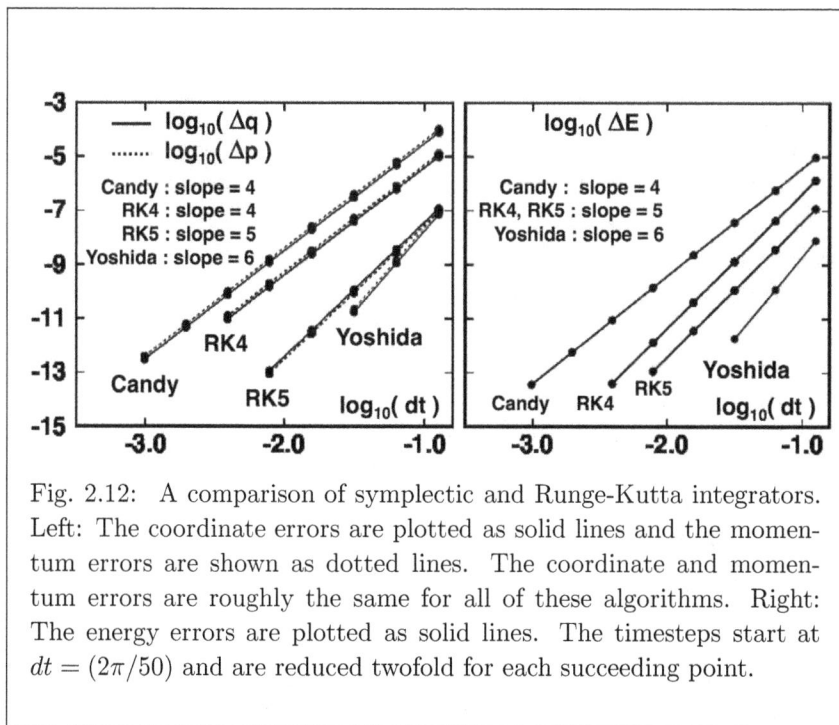

Fig. 2.12: A comparison of symplectic and Runge-Kutta integrators. Left: The coordinate errors are plotted as solid lines and the momentum errors are shown as dotted lines. The coordinate and momentum errors are roughly the same for all of these algorithms. Right: The energy errors are plotted as solid lines. The timesteps start at $dt = (2\pi/50)$ and are reduced twofold for each succeeding point.

We compare the oscillator errors for these two symplectic integrators and the fourth- and fifth-order Runge-Kutta methods in **Figure 2.12**. Notice that the trajectory for the fourth-order Candy-Rozmus integrator is considerably less accurate than the RK4 trajectory. It is also noteworthy that the energy error for RK4 is fifth-order accurate. The trajectory and energy errors for RK5 are both fifth-order accurate. The trajectory and energy are sixth-order accurate for Yoshida's algorithm.

2.6   Lyapunov Instability

Most of the problems we have considered so far have addressed solutions for linear problems such as the linear harmonic chain or two-dimensional systems of particles interacting with Hooke's law forces. In the natural world around us most motions are nonlinear and are not amenable to analytical techniques for their solution. Examples include continuum fluid or solid mechanics applications with nonlinear stress-strain relationships and dissipative constitutive relations. Examples in molecular dynamics include systems with nonlinear forces (including singular cutoffs) and with velocity and temperature gradients imposed through boundary interactions. As we seek numerical solutions to nonlinear dynamics problems it is important to be aware that physical instabilities may occur.

More than a hundred years ago a host of talented theoretical physicists applied statistical mechanics to an understanding of the microscopic basis for macroscopic behavior. The best known of these men were Boltzmann, Gauss, Hamilton, Lagrange, Lyapunov, Maxwell, and Poincaré. In 1892 Alexander Lyapunov published a book, "The General Problem of Stability of Motion". It received little notice for half a century. But in the 1950s his ideas were perceived to be useful in analyzing the stability and control of aerospace guidance systems. These systems produce strong nonlinearities which cannot be treated analytically. Today there are millions if not billions of systems exhibiting Lyapunov instability.[7] Researchers recognizing the frequent occurrence of Lyapunov instability have been in large part responsible for the rapid growth of chaos studies.

Lyapunov instability is a state-space instability. Lyapunov exponents measure the exponential direction-dependent rates at which neighboring trajectory points in phase-space separate in time ,

$$\dot{\delta} \propto \delta \longrightarrow \delta(t) \simeq \delta_0 e^{\lambda t} \ .$$

Here δ_0 represents the initial separation and $\dot{\delta}$ represents the rate at which the separation grows, decays, or oscillates. For a Hamiltonian system of

N particles in three dimensions there are $3N$ (q,p) degrees of freedom and $6N$ phase-space dimensions. The Lyapunov spectrum, the set of $6N$ time-averaged Lyapunov exponents, measures the rate of change of the separation of nearby trajectories. At equilibrium an N-particle Hamiltonian system has a long-time-averaged *symmetric* spectrum with each positive exponent paired with a negative exponent of the same magnitude. The two change roles if time is reversed. Hamiltonian systems are stable systems with no net growth or decay in the comoving phase-space volume. Let us consider the details for a very simple system, a four-particle chain which exhibits chaos, Lyapunov instability.

2.6.1 *The Growth of Errors in a Nonlinear Chain*

The four-particle nonlinear chain is a simple problem exhibiting the exponential growth of numerical errors due to Lyapunov instability. We consider a periodic Hooke's-Law chain with an additional quartic tethering potential, ϕ^4, which directs the particles to their equally-spaced lattice sites. Aoki and Kusnezov[19] developed and studied this lattice model as a way better to understand Fourier heat conduction in nonequilibrium steady states. The Hamiltonian for a periodic four-particle ϕ^4 chain is a sum of twelve terms :

$$\mathcal{H} = \sum_{i=1}^{4}[\ (p_i^2/2) + (q_i^4/4)\ ] + \sum_{4}^{springs} (q_{ij}^2/2)\ .$$

Here the displacement of a particle from its lattice site is q_i and the Hooke's Law displacements are $\{q_{ij}\}$. The periodic boundary condition includes the tethering forces as well as the Hooke's law forces linking particles 1 and 4.

$$\ddot{q}_1 = q_2 + q_4 - 2q_1 - q_1^3\ ;\ \ddot{q}_4 = q_3 + q_1 - 2q_4 - q_4^3\ .$$

We begin our study of the chain with two simultaneous calculations of its motion: the first uses RK4 integration and the second uses RK5 integration. We use an initial condition which leads to chaos : $\{q_{i0}\} = (0,0,0,0)$ and $\{p_{i0}\} = (2,2,2,-2)$. A suitable timestep with this very smooth Hamiltonian is $dt = 0.01$. We carry out the pair of simulations using 1.2 million timesteps with each integrator. The RK5 integration is more-nearly-accurate. We estimate the error in the RK4 integration as the root-mean-square error between the RK4 and RK5 trajectories.

Exercise : Estimate the time-averaged ratio of the RK5 to the RK4 integrator error and check your estimate for the harmonic oscillator.

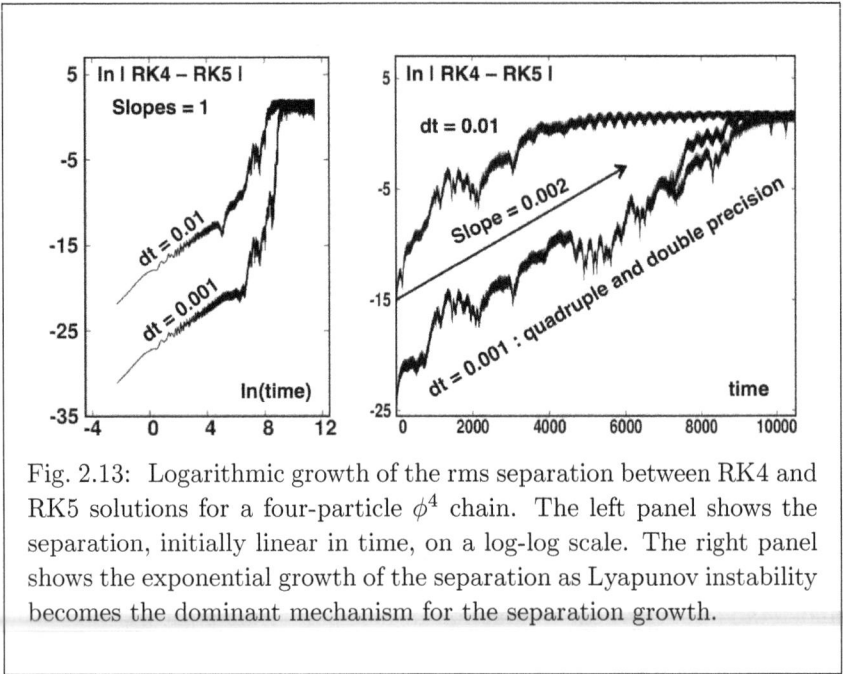

Fig. 2.13: Logarithmic growth of the rms separation between RK4 and RK5 solutions for a four-particle ϕ^4 chain. The left panel shows the separation, initially linear in time, on a log-log scale. The right panel shows the exponential growth of the separation as Lyapunov instability becomes the dominant mechanism for the separation growth.

Figure 2.13 shows the results of the ϕ^4 calculations. The panel on the left is a plot of the logarithm of the root-mean-square separation between RK4 and RK5 as a function of the logarithm of the time. At early times the separation between the two integration methods varies linearly in the time. At later times the variation becomes exponential in the time, reflecting Lyapunov instability. The difference in the errors for timesteps $dt = (0.01, 0.001)$ is a factor of 10^4 during the linear growth phase, representing the global compounding of local errors. The panel on the right of the Figure is a plot of the logarithm of the separation as a function of the time. In this case the slope of the curve is the "local" Lyapunov exponent $\lambda(t)$. The approximate value of the local Lyapunov exponent is 0.002 .

Detailed studies (See **Lectures 9 and 10**) show that the local Lyapunov exponent varies wildly as the chaotic trajectory evolves. Much longer simulations with as many as a billion timesteps are needed to calculate the time-averaged maximum Lyapunov exponent which characterize the behavior of a chaotic system. Although the Lyapunov growth appears saturated in this problem, a longer run[20] for the same initial condition using methods described in later Lectures shows that this four-particle nonlinear chain has a time-averaged Lyapunov spectrum with a maximum exponent of 0.7 .

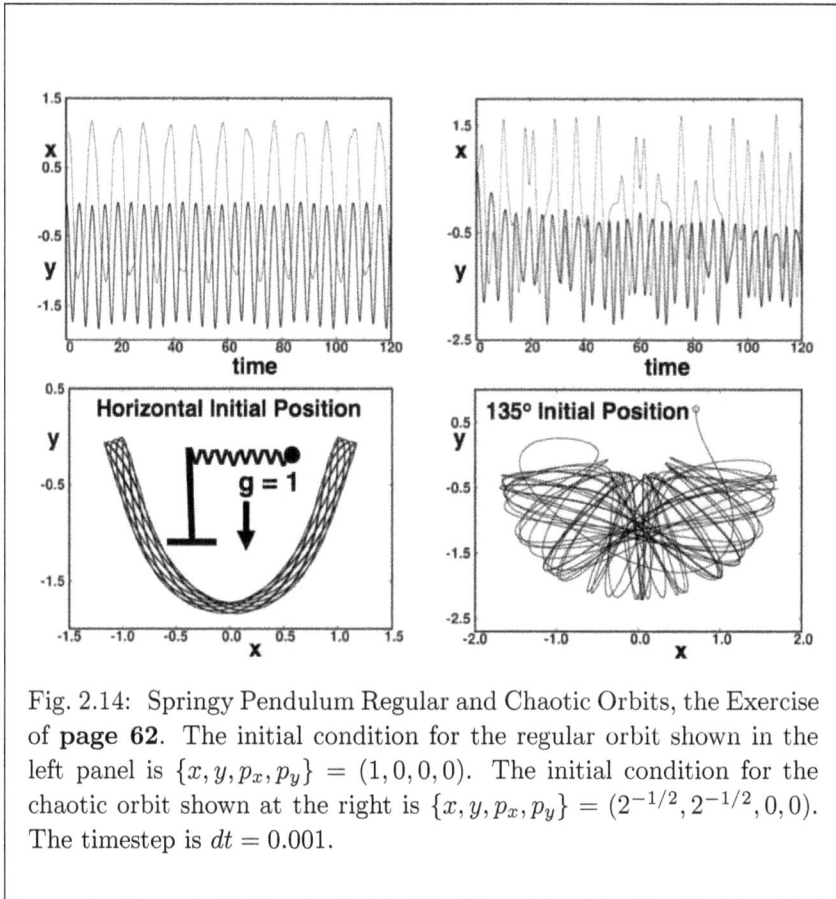

Fig. 2.14: Springy Pendulum Regular and Chaotic Orbits, the Exercise of **page 62**. The initial condition for the regular orbit shown in the left panel is $\{x, y, p_x, p_y\} = (1, 0, 0, 0)$. The initial condition for the chaotic orbit shown at the right is $\{x, y, p_x, p_y\} = (2^{-1/2}, 2^{-1/2}, 0, 0)$. The timestep is $dt = 0.001$.

The "springy pendulum" of **Figure 2.14** has two qualitatively different solution types. The geometry is shown in the lower left panel. A horizontal initial condition $\{x, y, p_x, p_y\} = (1, 0, 0, 0)$ at an angle $\theta = (\pi/2)$ results in the orbit shown in the left panels of **Figure 2.14**. This is a periodic "regular" orbit of the kind we observe in a clock pendulum. The orbit is quite different if we start the pendulum at $\theta = (3\pi/4)$ using the initial condition $\{x, y, p_x, p_y\} = (2^{-1/2}, 2^{-1/2}, 0, 0)$. This is the chaotic orbit shown in the right panels of the Figure. The long-time-averaged maximum Lyapunov exponent for the chaotic orbit is 0.122 . This means that if we follow the separation between this orbit and a nearby orbit with a constrained separation $\delta = 0.000001$, the averaged tendency of the separation δ to grow is proportional to $e^{+0.122t}$. In the next section we will develop a method for calculating the maximum Lyapunov exponent.

2.6.2 The Lyapunov Exponent in the Chaotic Sea

In this section I describe a method for calculating the maximum Lyapunov exponent using a reference-and-satellite-trajectory calculation.[21, 22] There is also a more elegant method replacing the rescaling of the offset reference-to-satellite vectors with Lagrange multipliers.[23, 24] I use the thermostatted Nosé-Hoover oscillator trajectory in a three-dimensional phase-space as an example. The equations of motion are :

$$\dot{q} = p \; ; \; \dot{p} = -q - \zeta p \; ; \; \dot{\zeta} = p^2 - 1 \; [\text{ Nosé} - \text{Hoover Oscillator }] \; .$$

Two simultaneous trajectories, the "reference" and the "satellite" trajectory, are calculated. Runge-Kutta RK4 or RK5 is a good choice for this problem. The chaotic tendency toward separation can be measured at each timestep by comparing the separation at $t + dt$ to that at t :

$$\dot{\delta} \propto \lambda\delta \longrightarrow \lambda(t) = \ln \left[\frac{|(x_s - x_r)_{t+dt}|}{|(x_s - x_r)_t|} \right] / dt \; ,$$

where x represents the trajectory $\{q, p, \zeta\}$. At the end of each timestep the separation between the two trajectories is rescaled to the original separation δ_0 :

$$x_s = x_r + [\; \delta_0/|x_s - x_r| \;](x_s - x_r) \; .$$

The maximum Lyapunov exponent is the time average of the local Lyapunov exponent $\lambda(t)$:

$$< \lambda > = \int_0^\tau (\lambda(t')dt') / \int_0^\tau dt' \longrightarrow \sum (\lambda(t)dt)/\tau \; .$$

where τ is the sufficiently long total time for the numerical calculation.

The following program segment uses fourth-order Runge-Kutta for the integration of the trajectories. There are three equations solved for each trajectory. The initial separation is $\delta_0 = 0.000001$ and the chaotic initial condition is $\{q, p, \zeta\} = (2.4, 0, 0)$. The timestep is $dt = 0.001$. The problem is run for two billion timesteps. This number of steps is limited by the size, four bytes (32 bits), of the default maximum integer on our computers. To run a larger number of steps I process the problem in batches of a size less than or equal to two billion steps each.

```
DT = 0.001D00
DELTA = 0.000001D00
ITMAX = 2 000 000 000
QR =   2.40D00       ! REFERENCE TRAJECTORY
PR =   0.0D00
ZR =   0.0D00
XX(1) = QR
XX(2) = PR
XX(3) = ZR
YY(1) = QR + DELTA ! SATELLITE TRAJECTORY
YY(2) = PR
YY(3) = ZR
TIME = 0.0D00
SUMYDT = 0.0D00

DO IT = 1,ITMAX
CALL RK4(XX,XXP,DT)
CALL RK4(YY,YYP,DT)
TIME = TIME + DT
RR = (XX(1)-YY(1))**2 + (XX(2)-YY(2))**2 +  (XX(3)-YY(3))**2
R  = DSQRT(RR)
FACTOR = DELTA/R
YAP1 = -DLOG(FACTOR)/DT
SUMYDT = SUMYDT + YAP1
ENDIF
YY(1) = XX(1) + FACTOR*(QS - QR)
YY(2) = XX(2) + FACTOR*(PS - PR)
YY(3) = XX(3) + FACTOR*(ZS - PR)
ENDDO
```

> **Exercise :**   Calculate the maximum Lyapunov exponent for the springy pendulum using a chaotic initial condition.

Figure 2.15 shows the convergence of the maximum Lyapunov exponent for the nonlinear Nosé-Hoover oscillator with the initial separation of the satellite and reference trajectories of $\delta = 0.000001$. The trajectories were integrated using RK4 with a timestep of 0.001 for 20 billion steps. Rescaling the separation of the trajectories after each time step provides consistent estimates of the largest Lyapunov exponent, $\lambda_1 = 0.014$.

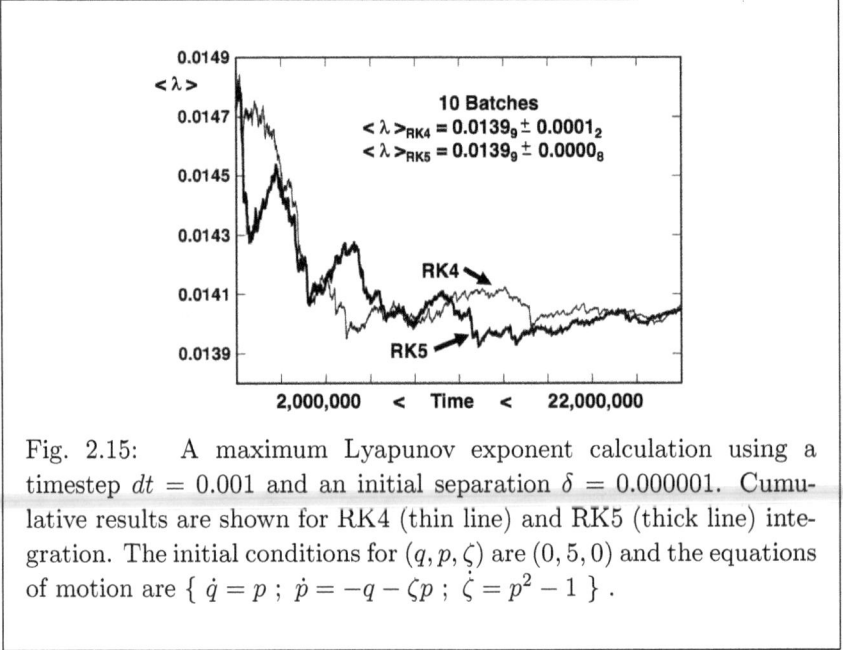

Fig. 2.15: A maximum Lyapunov exponent calculation using a timestep $dt = 0.001$ and an initial separation $\delta = 0.000001$. Cumulative results are shown for RK4 (thin line) and RK5 (thick line) integration. The initial conditions for (q, p, ζ) are $(0, 5, 0)$ and the equations of motion are $\{ \dot{q} = p \; ; \; \dot{p} = -q - \zeta p \; ; \; \dot{\zeta} = p^2 - 1 \}$.

Exercise : Write a program to calculate the maximum Lyapunov exponent $< \lambda >$ for the Nosé-Hoover oscillator. Use the initial condition $\{q_0, p_0, \zeta_0\} = (2.4, 0, 0)$ and an initial separation $\delta = 0.000001$.

Exercise : Consider the equations of motion

$$\{ \dot{q} = (p/s) \; ; \; \dot{p} = -qs \; ; \; \dot{s} = s\zeta \; ; \; \dot{\zeta} = [\, (p^2/s^2) - 1 \,] \, .$$

Write a program to calculate the maximum Lyapunov exponent using the method described in this section. Notice that the substitution $(p/s) \rightarrow p$ in the above equations gives the Nosé-Hoover equations described in the previous exercise.

2.7 Comparison of Seven Integrators Using a Cell Model

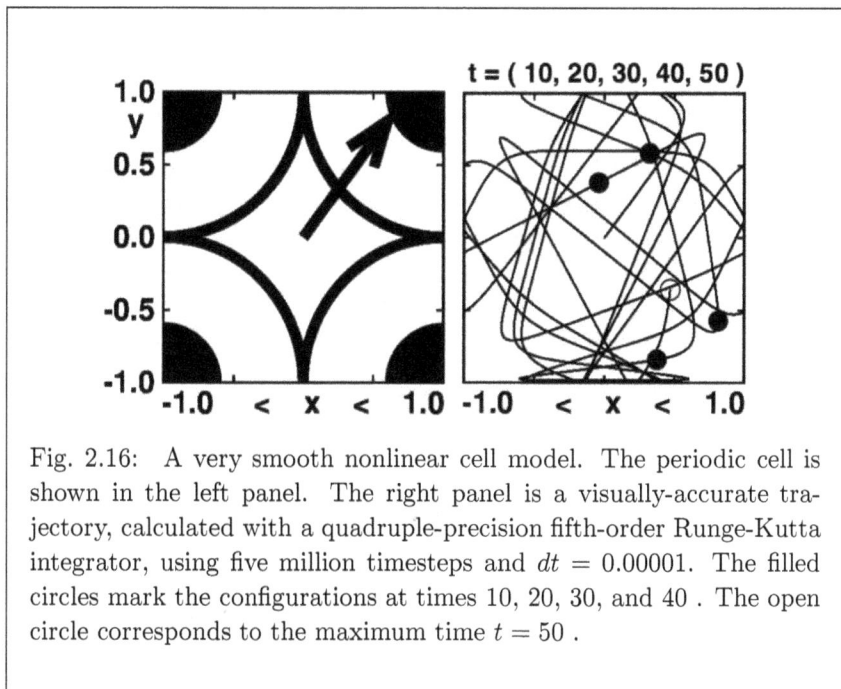

Fig. 2.16: A very smooth nonlinear cell model. The periodic cell is shown in the left panel. The right panel is a visually-accurate trajectory, calculated with a quadruple-precision fifth-order Runge-Kutta integrator, using five million timesteps and $dt = 0.00001$. The filled circles mark the configurations at times 10, 20, 30, and 40 . The open circle corresponds to the maximum time $t = 50$.

Here we study a very smooth nonlinear cell model and compare results for seven integrators. We use the two Runge-Kutta Integrators and five symplectic integrators to follow the motion of a point mass, a "wanderer particle", in a periodic two-dimensional square lattice[25] with a nearest-neighbor distance of two. The wandering particle interacts with fixed lattice scatterers at the four corners of the cell. The scatterers repel the moving wanderer point with forces which act whenever one of the scatterers' distance is less than unity :

$$\phi(r) = (1 - r^2)^4 \rightarrow F_x = 8x(1 - r^2)^3 \; ; \; F_y = 8y(1 - r^2)^3 \; .$$

The left panel of **Figure 2.16** illustrates this interaction. The black regions, where the potential energy is greater than one half, are inaccessible to the wanderer particle. Outside the central diamond-shaped region the scatterers at the cell corners exert repulsive forces on the wanderer particle. The initial condition chosen for the problem is

$$\{ x, y, p_x, p_y \} = (0, 0, 0.6, 0.8) \; .$$

Periodic boundary conditions are applied whenever the point mass escapes
the boundaries, where $|x|$ or $|y|$ or both exceed unity :

$$x < -1 \rightarrow x = x + 2 \; ; \; x > 1 \rightarrow x = x - 2 \; ;$$

$$y < -1 \rightarrow y = y + 2 \; ; \; y > 1 \rightarrow y = y - 2 \; .$$

Ernst Teloy,[26] Christoph Schlier, and Ansgar Seiter developed an *eighth-order* symplectic integrator which we use as an accurate reference trajectory
for comparing the lower-order methods. Coupled with quadruple precision
(128 bits $\approx$ 35 decimal digits) this integrator can provide precise trajectories provided that the time is not too long. When the moving particle is
initially at the cell center with velocity (0.6,0.8), the accuracy of the trajectory is lost at the rate of one binary bit in unit time where the moving
mass is equal to unity. At a time of 50 the loss is $2^{50} = 10^{15}$. The loss of 15
digits allows an accurate benchmark of the moving particle's coordinates.
For this integrator the timestep dependence of the coordinates (x,y) with
quadruple precision at a time of 50 is as follows :

```
dt = 0.00000100  ⟶  (0.46961 40145, -0.35682 95856)
dt = 0.00000010  ⟶  (0.46961 40143, -0.35682 95861)
dt = 0.00000001  ⟶  (0.46961 40142, -0.35682 95862) .
```
The coefficients for the algorithm are in **Reference 25** which is readily
available online at www.arχiv.org.

This precise and accurate five-billion-timestep trajectory with $dt = 10^{-8}$
was used to evaluate the accuracy of the other six lower-order integrators.
A comparison of the abilities of all six integrators to [1] return to the
origin by reversing the time (done by changing the sign of the velocity at
time 50), [2] maintain the initial energy, and [3] stay within a distance
0.01 to the eighth-order benchmark trajectory is shown in the **Table**.

The Table's "Accurate Trajectory Time" is the maximum time for which
coordinate errors are less than 0.01 when compared to the accurate eighth-
order trajectory. "Energy Accuracy" is the difference between the energy
at time 50, $E(50)$ and $E(0)$. "Reversal Time" is the maximum time for
which a reversed trajectory returns to the origin within a distance of 0.01 .

The simple second-order Symplectic-Leapfrog integrator was able to return to the origin within a distance of 0.01 for velocity reversals up to a
time of 47, longer than the reversal time for any of the other integrators.
On the other hand, the Symplectic-Leapfrog algorithm was only *accurate*
[within a distance of 0.01 in (x, y) space] up to a time of 18. The last

column in the Table shows the number of force evaluations per timestep for each method. The force evaluation is usually the most computationally expensive part of the calculation for systems with many particles. An evenhanded comparison of any two integrators must also include the ratios of the force evaluations and a modification of the timestep to account appropriately for the ratio of the number of force evaluations.

Table–Comparison of Cell-Model Integration Methods

Method	Order	Accurate Trajectory Time	Energy Accuracy $t = 50$	Reversal Time	Force Evaluations
Leapfrog	2nd	18	10^{-7}	47	1
Monte-Carlo	2nd	31	10^{-11}	43	5
Candy-Rozmus	4th	34	10^{-12}	42	3
Runge-Kutta	4th	35	10^{-12}	42	4
Runge-Kutta	5th	34	10^{-13}	42	6
Yoshida	6th	36	10^{-13}	42	7

2.8 Which Integrator Should We Use?

The choice of integrator is arbitrary. Possible criteria are accuracy, simplicity, reversibility, energy conservation, phase-volume conservation, and æsthetics. Our own preference is for the fourth-order Runge-Kutta integrator. This choice requires four force evaluations per step with an error of order $(dt^5/5!)$. We specially value the ability to make quick and simple modifications of the integrator for debugging and analysis.

The numerical solution of the cell model using the RK4 integrator was accurate to time $t = 35$ and reversible to time $t = 42$ with a timestep of $dt = 0.001$. The Candy-Rozmus integrator was accurate to time $t = 34$ and reversible to time $t = 42$ with the same timestep and with three force evaluations per step rather than four. Evidently these two integrators are roughly equivalent for smooth Hamiltonian problems like the cell model. On the other hand the symplectic integrators are not directly applicable to more general first-order motion equations with velocity-dependent forces, equations we will consider in detail throughout these lectures.

Some choose to believe that symplectic integrators are "better" on the grounds that conserving phase-space volume (as opposed to energy or trajectory accuracy) is the most important criterion for numerical integration. Even for situations like the cell model the numerical evidence for this view is not convincing. An argument against symplectic integrators could compare the error terms to those from Runge-Kutta, $[\ (dt^n)$ rather than $(dt^n/n!)\ ]$ leading to adopting Runge-Kutta. Again, the cell-model results suggest that the main difference is æsthetic, as opposed to numeric.

Let us close this discussion by comparing the single-timestep coordinate errors for an oscillator with initial conditions $(q = p = 1)$ and using a timestep $dt = 1$:

$$(q,p) \longrightarrow (+1.381773, -0.301168)_{\text{exact}} \ ; \ (+1.375000, -0.291667)_{\text{RK4}}$$
$$(+1.382552, -0.300781)_{\text{RK5}} \ ; \ (+1.463729, -0.131066)_{\text{CR}} \ .$$

The Candy-Rozmus symplectic errors, $(\ 0.082,\ 0.170\ )_{\text{CR}}$, are at least an order of magnitude larger than those from the two Runge-Kutta integrators, $(\ 0.007,\ 0.010\ )_{\text{RK4}}$ and $(\ 0.001,\ 0.000\ )_{\text{RK5}}$. I compare the optimized symplectic errors to the Runge-Kutta integrators in Section 2.11.

As advice from our cell-model studies we conclude : [1] it is useful and informative to compare results from *different* integrators whenever the *accuracy* of a calculation is in question. [2] In most interesting calculations it is the exponential Lyapunov instability which is responsible for inaccuracies – that effect is about the same for *all* viable integrators. Lyapunov instability for the manybody problem certainly overshadows other types of numerical errors. [3] Apart from regular astronomical problems, there is no evidence that symplectic integrators have an advantage over traditional integrators. Nonequilibrium simulations typically require the solution of first-order differential equations where phase volume is not conserved.

2.9   Predictor-Corrector Algorithms

Predictor-corrector integrators solve differential equations with two distinct stages. The first stage is explicit. It is used to predict the implicit terms in the second-stage corrector step. The corrector step can be iterated more than once, even until two successive iterates are identical within the machine precision. The most general packaged-software forms of the predictor-corrector algorithms were developed for stiff equations. They rely on matrix methods and iteration techniques that are both time consuming to program correctly and difficult to modify. Adaptive techniques are simpler. We discussed a simple adaptive method using Runge-Kutta integration in Section 1.7.3, pages 23-24, and applied it to the Nosè-Hoover oscillator. The method was programmed in a few lines and is stable. In this section we find that Milne's first-order methods are unstable for that problem, even at equilibrium.

Nonequilibrium systems are typically described with first-order differential equations involving velocity-dependent forces. We generally solve such problems with Runge-Kutta integrators. Numerical results show that Milne's methods,[1] pages 132 and 135 of his book, are unstable for the equilibrium Nosé-Hoover oscillator. C. William Gear's Algorithms applied to nonequilibrium systems are stable. Below I discuss the Milne and Gear algorithms along with applications to small-scale problems. All of them can just as easily be applied to large-scale state-of-the-art simulations.

2.9.1 *Methods for First-Order Differential Equations*

Milne[1] published his book *Numerical Calculus* after World War II when numerical integration was in its infancy using "electronic calculators". He predicted at that time the fast rise in the use of electronic calculators. Although true enough, those calculators were soon superceded by mainframe computers located in government and University laboratories. Milne described two methods, one of them explicit and the other an implicit predictor-corrector method. Both of them were developed for solving first-order differential equations. Applied to the harmonic oscillator Milne's simplest explicit integrator for first-order equations has the form :

$$q_{n+1} = q_{n-1} + 2p_n dt \; ; \; p_{n+1} = p_{n-1} - 2q_n dt \; [\text{ Explicit Method }] \; .$$

The local errors are third-order in the timestep.

The oscillator form of Milne's more-nearly-accurate fourth-order predictor-corrector integrator for first-order equations is :

$$q_{n+1} = q_{n-3} + (4dt/3)(2p_n - p_{n-1} + 2p_{n-2}) \; ;$$

$$p_{n+1} = p_{n-3} - (4dt/3)(2q_n - q_{n-1} + 2q_{n-2}) \; [\text{ Explicit Predictor }] \; ;$$

$$q_{n+1} = q_{n-1} + (dt/3)(p_{n+1} + 4p_n + p_{n-1}) \; ;$$

$$p_{n+1} = p_{n-1} - (dt/3)(q_{n+1} + 4q_n + q_{n-1}) \; [\text{ Implicit Corrector }] \; .$$

Local errors for the corrector are fifth-order in the timestep :

$$\delta q \simeq \delta p \simeq (1/90)dt^5 \; \dddot{q} \simeq (1/90)dt^5 \; \dddot{p} \; .$$

The number of iterations of the corrector step is determined either by specifying an error level or by terminating the corrector phase when two successive iterations give the same result.

The simplicity and time-symmetry of this pair of algorithms is appealing. The fourth-order algorithm for the oscillator is stable for 10^9 steps with $dt = \pi/25$. Nevertheless, neither one of the two algorithms has retained much of a following. To see why that is consider the following exercise :

Exercise : Show that both Milne integrators are computationally unstable for the Nosé-Hoover oscillator. At time zero $(q, p, \zeta) = (2.4, 0, 0)$. Use RK4 to find starter values of (q, p, ζ) at $t = -dt$. Check the stability for timesteps $dt = (0.100, \; 0.010, \; 0.001)$.

In contrast, Table 11.1 on page 217 in Gear's book[27] contains a useful Table of algorithms well-suited to first-order differential equations including the Nosé-Hoover oscillator both at and away from equilibrium $(\epsilon > 0)$:

$$\{ \; \dot{q} = p \; ; \; \dot{p} = -q - \zeta p \; ; \; \dot{\zeta} = p^2 - T \text{ with } T = 1 + \epsilon \tanh(q) \; \} \; .$$

We consider two Gear algorithms, "GEAR4$_1$" based on four previous coordinate values and "GEAR5$_1$" based on five. These are useful examples. Using x to denote the vector (q, p, ζ) the GEAR4$_1$ implicit algorithm requires a single derivative evaluation at a predicted vector x_{n+1} :

$$x_{n+1} = (48x_n - 36_{n-1} + 16x_{n-2} - 3x_{n-3} + 12\dot{x}_{n+1}dt)/25 \; [\; \text{Gear4}_1 \;] \; .$$

Replace $\dot{x}_{n+1}$ by $\dot{x}_n$ in GEAR4$_1$ for a useful predictor expression. Storing an additional past vector x_{n-4} gives the higher-order GEAR5$_1$ algorithm :

$$x_{n+1} = (300x_n - 300x_{n-1} + 200x_{n-2} - 75x_{n-3} + 12x_{n-4} + 60\dot{x}_{n+1}dt)/137 \; .$$

Errors for these first-order-equation solvers are shown in **Figure 2.17**.

Exercise : Show that Gear's implicit GEAR4$_1$ integrator can give a stable solution of the nonequilibrium Nosé-Hoover oscillator problem where the temperature $T(q) = 1 + 0.1 \tanh(q)$. Use RK4 to generate "starter" values of $x(t) = (q, p, \zeta)$ at times -0.1, -0.2, and -0.3 .

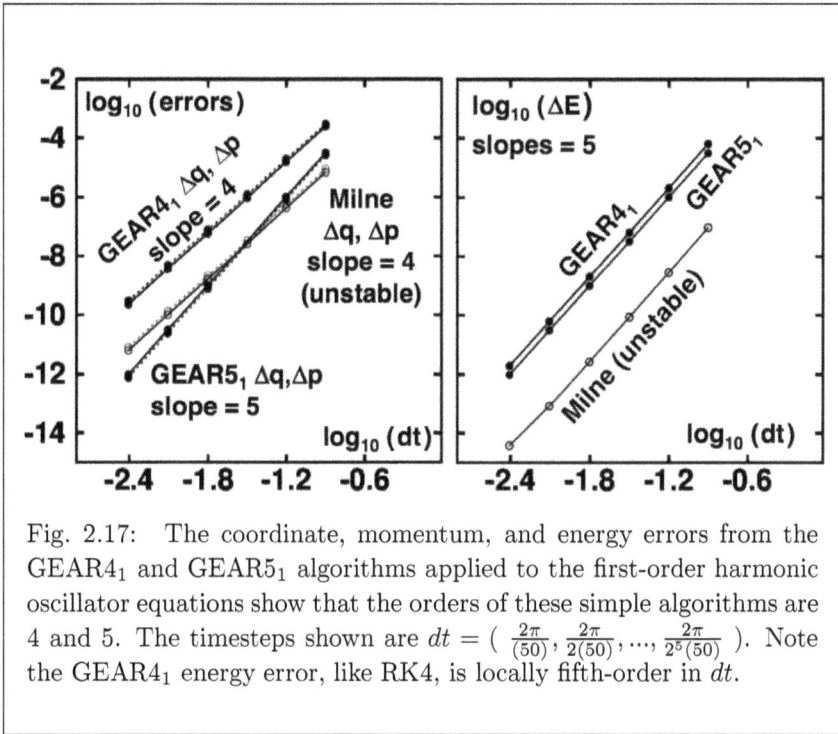

Fig. 2.17: The coordinate, momentum, and energy errors from the $GEAR4_1$ and $GEAR5_1$ algorithms applied to the first-order harmonic oscillator equations show that the orders of these simple algorithms are 4 and 5. The timesteps shown are $dt = (\frac{2\pi}{(50)}, \frac{2\pi}{2(50)}, ..., \frac{2\pi}{2^5(50)})$. Note the $GEAR4_1$ energy error, like RK4, is locally fifth-order in dt.

Developing a good predictor to start the $GEAR4_1$ and $GEAR5_1$ corrector iterations may be very challenging for some problems. A predictor with a local error comparable to the corrector error may be needed. The consequence of a poor choice for the predictor is the need for more corrector iterations to reach convergence to within a specified accuracy.

Adaptive methods using the RK4 and RK5 are much more easily programmed and implemented than the Gear algorithms. The following programming implements a fourth-order adaptive Runge-Kutta algorithm.

```
CALL RK4(X,XP,DT/2.0D0)
CALL RK4(X,XP,DT/2.0D0)
CALL RK4(Y,YP,DT/1.0D0)
ERROR = DABS(X-Y)
IF(ERROR.GT.10.D0**(-10)) DT = 0.5D0*DT
IF(ERROR.LT.10.D0**(-12)) DT = 2.0D0*DT
```

In this example two calls using $dt/2$ are made to RK4 to advance X one step and one call is made to advance Y one step. The error bounds are between 10^{-12} and 10^{-10}. The timestep is halved if the error is too large and doubled if the error is too small.

The RK4 and RK5 algorithms require four and six derivative evaluations per timestep so that Gear wins out from the standpoint of computational efficiency if only a few corrector iterations are needed. It is useful to evaluate both methods for stiff problems and important to test both the efficiency and the stability of each algorithm.

Giulia Venneri and Bill[28] studied a third-order algorithm using equations 16 and 6a from Beeman's[29] 1976 paper. We can show that when the two third-order equations are applied to the oscillator the result is a fourth-order trajectory with coordinates identical to those from Milne's accurate and stable fourth-order algorithm, found on page 140 of the Milne book.[1] The Beeman equations for the oscillator are :

$$q_{n+1} = q_n + p_n dt - (q_{n+1} + 6q_n - q_{n-1})(dt^2/12) \; ;$$

$$p_{n+1} = p_n - (5q_{n+1} + 8q_n - q_{n-1})(dt/12) \; .$$

Notice that we can solve for the oscillator coordinate by calculating

$$(q_{n+1} - q_n) - (q_n - q_{n-1})$$

and substituting $(p_{n+1} - p_n)$ into this expression. This eliminates the momentum and the remaining oscillator coordinate expression is the centered implicit Milne corrector approximation for a second-order differential equation !!

$$q_{n+1} = 2q_n - q_{n-1} - (q_{n+1} + 10q_n + q_{n-1})(dt^2/12) \; .$$

This fourth-order Milne coordinate equation can be solved analytically using a solution of the form $e^{i\omega t}$. Substituting the solution into the Milne corrector provides the frequency-timestep relationship and the analytical solution of the difference equation.

$$q_n = \cos \omega n dt \; ; \quad \omega = \arccos\left[\,(12 - 5dt^2)/(12 + dt^2)\,\right]/dt \; .$$

The error calculation for the Milne analytical solution to the oscillator equation provides a very accurate fourth-order oscillator coordinate in agreement with **Figure 2.17**. We will revisit this solution in the next Section.

Let us next consider a more specialized but very common case, classical Newtonian or Hamiltonian mechanics described by second-order equations of motion : $\{\, m\ddot{x} = F(x) \,\}$.

2.9.2 *Methods for Second-Order Differential Equations*

Milne's predictor-corrector combination for second-order differential equations (page 140 of his book) is the pair of updates :

$$q_{n+1} = q_n + q_{n-2} - q_{n-3} + (dt^2/4)(5\ddot{q}_n + 2\ddot{q}_{n-1} + 5\ddot{q}_{n-2}) \; [\text{ Predictor }] \; ;$$

$$q_{n+1} = 2q_n - q_{n-1} + (dt^2/12)(\ddot{q}_{n+1} + 10\ddot{q}_n + \ddot{q}_{n-1}) \; [\text{ Corrector }] \; .$$

The local error terms for this second-order-differential-equation solver are sixth-order, $\delta q \simeq (17/240)dt^6 \; \overset{......}{q}$ for the predictor and $\delta q \simeq -(1/240)dt^6 \; \overset{......}{q}$ for the corrector. We will refer to this algorithm as MILNE$_2$. This algorithm is stable.

Exercise : Evaluate the order of the errors and the accuracy of the MILNE$_2$ predictor with the two-point, four-point and six-point approximations to the momentum. Would the predictor or the corrector be useful for bit-reversible calculations? Why or why not?

Several equivalent Gear approximations can be written as Taylor's series in terms of the previous coordinate and its derivatives, only in terms of previous coordinates, or in terms of previous coordinates and selected derivatives. We will display one of each and denote them as GEAR4$_2$ and GEAR5$_2$.

In addition to the coordinate and velocity the GEAR4$_2$ algorithm solves for the first and second derivatives of the velocity $\dot{v} = a$ and $\ddot{v} = b$. Initial values must be specified for these two velocity derivatives. All four of the variables are calculated by the corrector step at least once in each timestep. The GEAR4$_2$ predictor step is a set of truncated Taylor's series for each of the four variables:

$$q^p_{n+1} = q_n + dt v_n + \tfrac{1}{2} dt^2 a_n + \tfrac{1}{6} dt^3 b_n \; ;$$

$$v^p_{n+1} = v_n + dt a_n + \tfrac{1}{2} dt^2 b_n \; ;$$

$$a^p_{n+1} = a_n + dt b_n \; ;$$

$$b^p_{n+1} = b_n \; .$$

After the predictor step a new value of the force is calculated for use in the corrector step. For the standard oscillator the acceleration using the predicted coordinate is

$$a^c_{n+1} = -q^p_{n+1} \; .$$

Next a correction term for the acceleration Δa_{n+1} is calculated as the difference between the acceleration computed with the predictor coordinate a_{n+1}^c and the predictor acceleration :

$$\Delta a_{n+1} = a_{n+1}^c - a_n^p = -q_{n+1}^p - a_{n+1}^p .$$

The GEAR4$_2$ corrector updates the new coordinate and the derivatives using the correction term for the acceleration :

$$q_{n+1}^c = q_{n+1}^p + (dt^2/12)\Delta a_{n+1} ;$$

$$v_{n+1}^c = v_{n+1}^p + (5dt/12)\Delta a_{n+1} ;$$

$$a_{n+1}^c = -q_{n+1}^p ;$$

$$b_{n+1}^c = b_{n+1}^p + (1/dt)\Delta a_{n+1} .$$

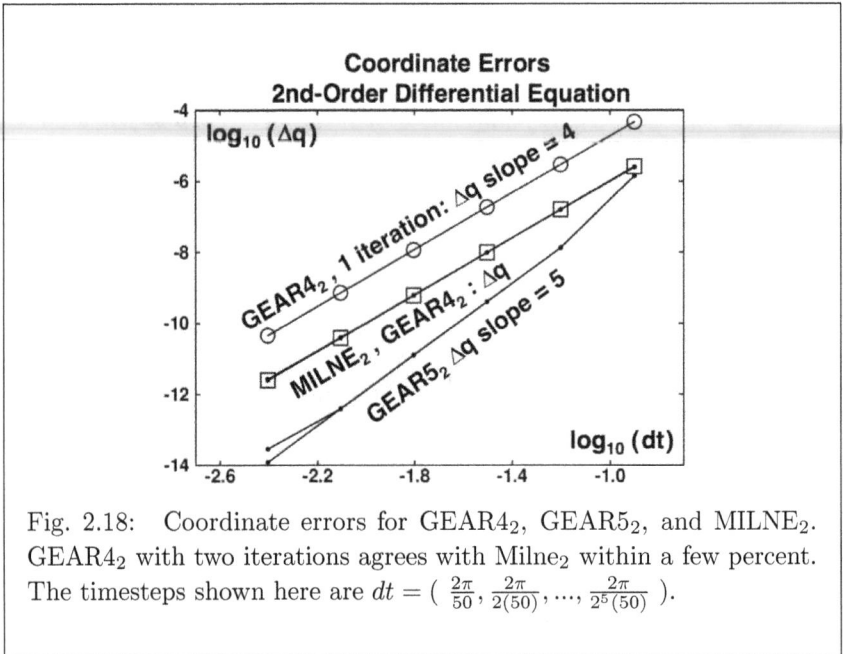

Fig. 2.18: Coordinate errors for GEAR4$_2$, GEAR5$_2$, and MILNE$_2$. GEAR4$_2$ with two iterations agrees with Milne$_2$ within a few percent. The timesteps shown here are $dt = (\frac{2\pi}{50}, \frac{2\pi}{2(50)}, ..., \frac{2\pi}{2^5(50)})$.

Figure 2.18 compares the coordinate accuracies of the MILNE$_2$, GEAR4$_2$, and GEAR5$_2$ algorithms. The GEAR4$_2$ and MILNE$_2$ errors are fourth-order. The open circles are GEAR4$_2$ with one corrector iteration. The open squares are GEAR4$_2$ with two corrector iterations. The solid points are MILNE$_2$ with one corrector iteration. The GEAR$_4$ algorithm with two iterations agrees within a few percent with the MILNE$_2$ algorithm. Notice that GEAR4$_2$ with two iterations is more than an order of magnitude more accurate than with one iteration. The GEAR5$_2$ errors are fifth-order. The first point on the GEAR5$_2$ line corresponds to a timestep that is too large. The last point on the GEAR5$_2$ line is computed in both double and quadruple precision. The double-precision result is affected by roundoff error. Quadruple precision corrects this error.

The GEAR5$_2$ predictor with $\ddot{v} = c$ is :

$$q^p_{n+1} = q_n + dtv_n + \tfrac{1}{2}dt^2 a_n + \tfrac{1}{6}dt^3 b_n + \tfrac{1}{24}dt^4 c_n \; ;$$

$$v^p_{n+1} = v_n + dta_n + \tfrac{1}{2}dt^2 b_n + \tfrac{1}{6}dt^3 c_n \; ;$$

$$a^p_{n+1} = a_n + dtb_n + \tfrac{1}{2}dt^2 c_n \; ;$$

$$b^p_{n+1} = b_n + dtc_n \; ;$$

$$c^p_{n+1} = c_n \; .$$

The correction term for the force is $\Delta a_{n+1} = a^c_{n+1} - a^p_n = -q^p_{n+1} - a^p_{n+1}$.

The GEAR5$_2$ corrector updates are

$$q^c_{n+1} = q^p_{n+1} + (19/24)dt^2 \Delta a_{n+1} \; ;$$

$$v^c_{n+1} = v^p_{n+1} + (3/8)dt\Delta a_{n+1} \; ;$$

$$a^c_{n+1} = a^p_{n+1} \; ;$$

$$b^c_{n+1} = b^p_{n+1} + (3/2dt)\Delta a_{n+1} \; ;$$

$$c^c_{n+1} = c^p_{n+1} + (1/dt^2)\Delta a_{n+1} \; .$$

Figures 2.19 and 2.20 present the errors in the GEAR4$_2$ and GEAR5$_2$ algorithms. The left panels of the Figures show the Gear errors in the momentum and energy using the momentum computed at the end of the corrector steps. The right panels of the Figures show the momentum and energy errors using the centered $p4$ and $p6$ approximations. Notice in the left panels that if the Gear momenta are used instead of $p4$ and $p6$ the coordinate and momentum errors are not symmetric and are in fact quite different. This disparity in the errors should not happen for a symmetric solution involving sines and cosines. The right panel of each figure shows that the errors in the Gear coordinates and the centered momenta have been rendered symmetric. It is equally important that they are much more nearly accurate !

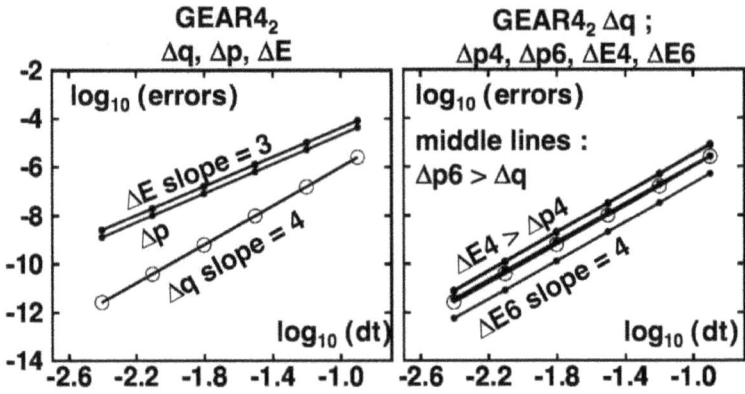

Fig. 2.19: The momentum error in the left panel is calculated in the GEAR4$_2$ corrector. The momenta in the right panel are $p4$ and $p6$. The coordinate error is labeled with open circles. Timesteps are $dt = (\ \frac{2\pi}{(50)}, \frac{2\pi}{2(50)}, ..., \frac{2\pi}{2^5(50)}\ )$.

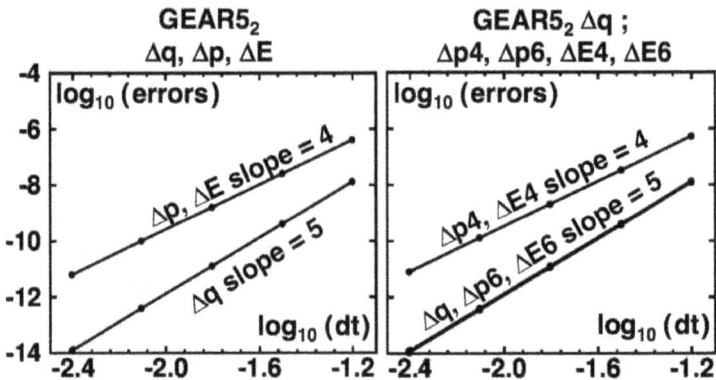

Fig. 2.20: The momentum in the left panel is calculated in the GEAR5$_2$ corrector. The momenta in the right panel are $p4$ and $p6$. The timesteps are $dt = (\ \frac{2\pi}{2(50)}, \frac{2\pi}{2^2(50)}, ..., \frac{2\pi}{2^4(50)}\ )$.

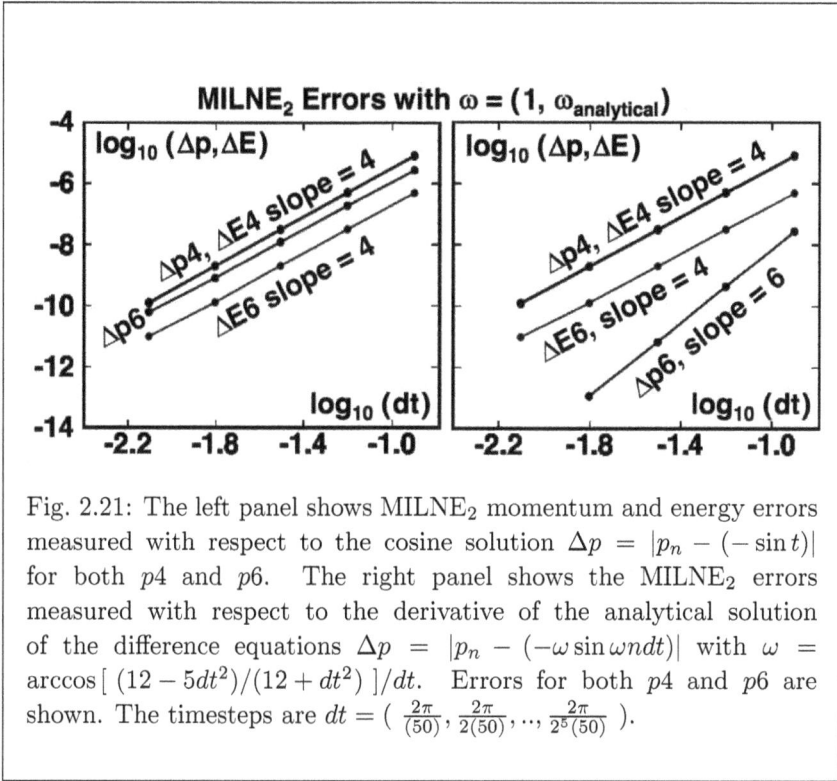

Fig. 2.21: The left panel shows MILNE$_2$ momentum and energy errors measured with respect to the cosine solution $\Delta p = |p_n - (-\sin t)|$ for both $p4$ and $p6$. The right panel shows the MILNE$_2$ errors measured with respect to the derivative of the analytical solution of the difference equations $\Delta p = |p_n - (-\omega \sin \omega n dt)|$ with $\omega = \arccos\left[\,(12 - 5dt^2)/(12 + dt^2)\,\right]/dt$. Errors for both $p4$ and $p6$ are shown. The timesteps are $dt = (\frac{2\pi}{(50)}, \frac{2\pi}{2(50)}, \cdots, \frac{2\pi}{2^5(50)})$.

Figure 2.21 displays the momenta and corresponding energy errors for the MILNE$_2$ algorithm using the $p4$ and $p6$ momentum approximations. In the left panel the momentum and energy errors are measured with respect to the cosine solution $\dot{q} = -\sin(t)$. The right panel displays the errors in terms of the derivative of the analytical coordinate. The results differ because the frequencies are different. The frequencies are $\omega = 1$ for the cosine solution and $\omega dt = \arccos[\,(12 - 5dt^2)/(12 + dt^2)\,]$ for the analytical solution derived in the previous section on page 80. Although the $p4$ momentum is fourth-order it is an order of magnitude less accurate than the coordinate. The $p6$ momentum is 3 orders of magnitude more accurate than the coordinate. We conclude from this that the energy estimate is more nearly correct using the $p4$ momentum.

2.10   A Fourth-Order Bit-Reversible Algorithm

Here I revisit the issues of reversibility and bit-reversible algorithms discussed in Section 2.3.2 on page 50. Milne developed a fourth-order predictor
algorithm that is explicit, centered, time-reversible, and stable. These fa-
vorable attributes suggest testing the algorithm in its bit-reversible form.[30]
Bit-reversible algorithms are particularly useful for analyses of forward and
reversed Lyapunov instability. With typical reference and satellite trajectories separations of order 0.000001 second-order algorithms are inade-
quate. Fourth-order coordinate accuracy paired with a six-point definition
of momentum provides a qualitatively-better phase-space picture than can
Levesque and Verlet's bit-reversible method.[11] Levesque-Verlet integration
is only second-order accurate.

   See Milne's book, pages 140 and 99, respectively, for his fourth-order
coordinate algorithm and his sixth-order momentum definition :

$$q_{n+2} - q_{n+1} - q_{n-1} + q_{n-2} = [\, 5\ddot{q}_{n+1} + 2\ddot{q}_n + 5\ddot{q}_{n-1} \,](dt^2/4) \; ; \; \delta q \simeq \tfrac{17}{240}dt^6 \; .$$

The six-point momentum definition is :

$$p6_n = \frac{3}{2}\frac{(q_{n+1}-q_{n-1})}{2dt} - \frac{3}{5}\frac{(q_{n+2}-q_{n-2})}{4dt} + \frac{1}{10}\frac{(q_{n+3}-q_{n-3})}{6dt} \; .$$

To rewrite the coordinate algorithm in integer form, first compute the floating point contributions of the three accelerations to the motion :

$$a_{n+1} = \tfrac{5}{4}\ddot{q}_{n+1}dt^2 \; ; \; a_n = \tfrac{2}{4}\ddot{q}_n dt^2 \; ; \; a_{n-1} = \tfrac{5}{4}\ddot{q}_{n-1}dt^2 \; .$$

Next convert these products to integers :

$$IA_{n+1} = \mathrm{INT}(a_{n+1}) \; ; \; IA_n = \mathrm{INT}(a_n) \; ; \; IA_{n-1} = \mathrm{INT}(a_{n-1}) \; .$$

Using these integers gives the final form of the bit-reversible algorithm :

$$IQ_{n+2} = IQ_{n+1} + IQ_{n-1} - IQ_{n-2} + (IA_{n+1} + IA_n + IA_{n-1}) \; .$$

Notice that each of the accelerations is truncated separately. Bit-reversible
methods must be based on centered algorithms with identical truncations
and identical coordinate sequences both forward and backward in time.

   Linear stability analysis can be applied to this problem using a solution
of the form $q \propto e^{iwt}$. The result is

$$\cos(2\omega dt) - \cos(\omega dt) + [\, 5\cos(\omega dt) + 1 \,](dt^2/4) = 0 \; .$$

   This simplifies to

$$2C^2 + [\, (5dt^2/4) - 1) \,]C + (dt^2/4) - 1 = 0 \; ; \; \text{where } C = \cos(\omega dt) \; .$$

Here there is a phase-shift quartic in dt. The corresponding frequency error
can be calculated from this quadratic equation and calculating $\delta\omega = 1 - \omega$.

   The following program segment implements the bit-reversible Milne
idea. A compiler option selects 8-byte integers giving 15-digit precision.

```
      IMPLICIT DOUBLE PRECISION (A-H,O-Z)
      INTEGER IQMM,IQM,IQO,IQP,IQPP
      INTEGER IAP,IAO,IAM,I1,I2,I3,I4
      ITMAX = 100
      TWOPI = 2.D0*3.141592653589793D0
      DT = TWOPI/ITMAX
      QMM  =  DCOS(-2.D0*DT)
      QM   =  DCOS(-1.D0*DT)
      Q0   =  DCOS( 0.D0*DT)
      QP   =  DCOS(+1.D0*DT)
CONVERT COORDINATES TO 15-DIGIT INTEGERS
      IQMM   = QMM*(10.D0**15)
      IQM    = QM *(10.D0**15)
      IQO    = Q0 *(10.D0**15)
      IQP    = QP *(10.D0**15)
C TIMESTEP LOOP
      DO IT = 1,ITMAX
      TIME = IT*DT
COMPUTE INGREDIENTS OF THREE ACCELERATIONS
      IAP  = 0.25D00*DT*DT*(5.D0*IQP)
      IAO  = 0.25D00*DT*DT*(2.D0*IQO)
      IAM  = 0.25D00*DT*DT*(5.D0*IQM)
COORDINATE UPDATES FOR FIVE SUCCESSIVE TIMES
      IQPP = IQP + IQM - IQMM - (IAP + IAO + IAM)
      IQMM = IQM
      IQM = IQO
      IQO = IQP
      IQP = IQPP
      END DO


COORDINATE REVERSAL
      I1 = IQMM
      I2 = IQM
      I3 = IQO
      I4 = IQP
      IQMM = I4
      IQM  = I3
      IQO  = I2
      IQP  = I1
```

```
    DO IT = 1,ITMAX
CONTINUE REVERSAL WITH SAME STATEMENTS AS IN THE FORWARD LOOP
    END DO
```

At the end of a timestep the five coordinates are converted to floating point to calculate the momentum and energy. Integer coordinate values are converted to floating point by multiplying by 10^{-15}. There must be seven saved points rather than five for the $p6$ momentum calculation. The program can be modified to include two more points QMMM and QPPP.

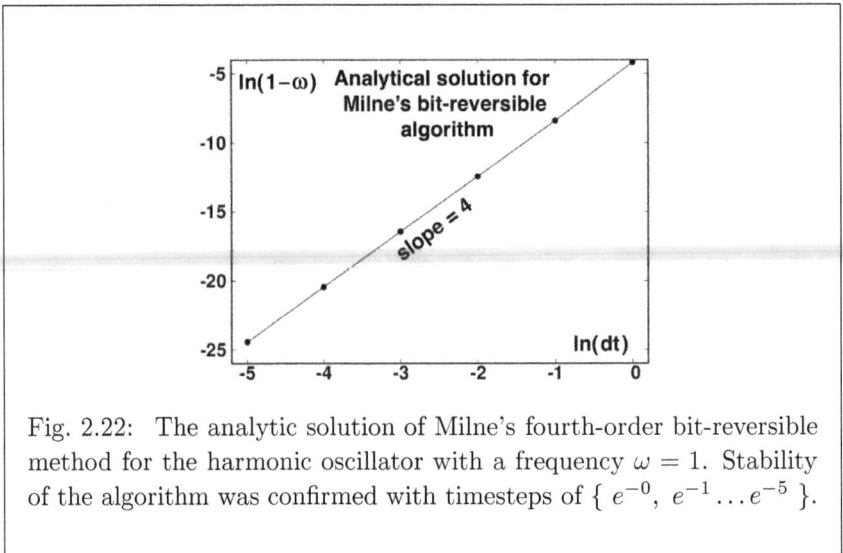

Fig. 2.22:   The analytic solution of Milne's fourth-order bit-reversible method for the harmonic oscillator with a frequency $\omega = 1$. Stability of the algorithm was confirmed with timesteps of $\{ e^{-0}, e^{-1} \ldots e^{-5} \}$.

The results of six simulations of this bit-reversible algorithm for the harmonic oscillator are shown in **Figure 2.22**. No doubt this algorithm will be proved useful to studies related to Second Law irreversibility as in the inelastic collision problem shown in **Figures 2.23-2.24**.

Bit-reversible algorithms have been extremely valuable in illustrating the irreversibility of nonlinear dynamical systems described by reversible equations of motion. **Figures 2.23 and 2.24** illustrate the forward and reversed motions of an inelastic collision of two cold 400-particle crystal-lites.[31] Snapshots of the forward motion $[\,0 < t < 100\,]$ are shown in **Figure 2.23**. Snapshots of the reversed motion are shown in **Figure 2.24**, starting at $t = 100$ with a *negative* timestep and ending at $t = 0$.

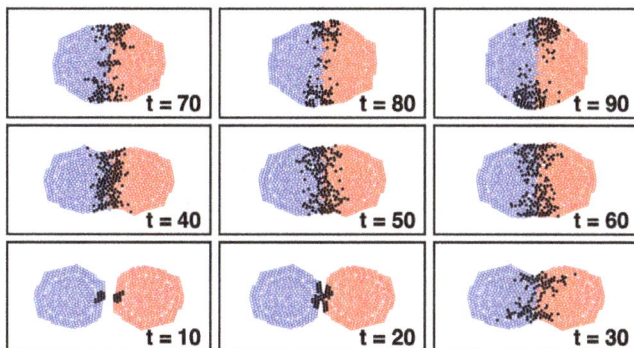

Fig. 2.23: The inelastic collision of two 400-particle cold crystallites. The forward motion begins at time $t = 0$ and ends at time $t = 100$.

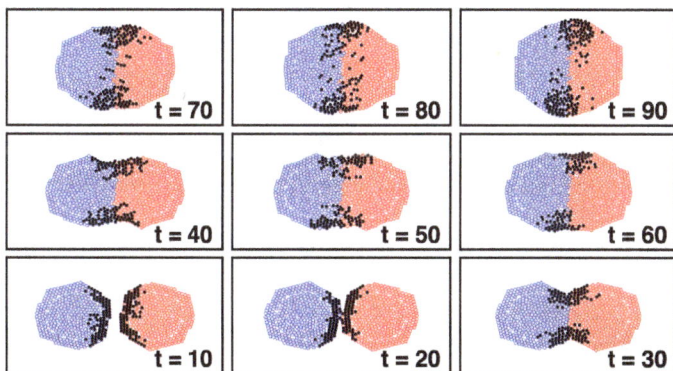

Fig. 2.24: The reverse collision begins at time $t = 100$ and ends at time $t = 0$. These configurations match those of Figure 2.23 perfectly.

The bit-reversible coordinates in the figures are identical to machine accuracy in both the forward and backward trajectories. The blackened particles in the two sets of snapshots are those whose motion makes an above-average contribution to the squared phase-space separation

$$\delta^2 \equiv \sum^{800} [\; \delta_x^2 + \delta_y^2 + \delta_{p_x}^2 + \delta_{p_y}^2 \;] \; .$$

The motion occurs in an 800-particle phase-space with 3200 dimensions. The blackened particles indicate the least-stable directions in phase space. These correspond very nicely to those regions which are undergoing irreversible "plastic" strain. Detailed studies of such problems show the connection between Lyapunov instability and the direction of time satisfying the Second Law of Thermodynamics. The most interesting analysis of this reversed, but irreversible, deformation, should apply near time 50. In the forward direction we expect to find the largest entropy production along the plane (actually a line in two dimensions) of contact between the two solid balls. In the reversed direction we expect to find the entropy production localized along the lateral surfaces of the balls as they are being pulled apart. We expect that such a qualitative agreement with macroscopic stress-strain analysis will provide a promising link between microscopic reversibility and macroscopic irreversibility. We will return to this topic in **Lecture 8**.

Exercise : Write a program to calculate the fourth-order bit-reversible trajectory of the standard harmonic oscillator program for a few periods. Calculate the reverse trajectory and compare the results to verify your program is correct. Calculate the six-point momentum and the corresponding energy in floating point arithmetic. Generate a log-log plot of the coordinate, momentum, and energy errors as a function of the timestep. Use the six timesteps $dt = (2\pi/25), ...(2\pi/800)$.

2.11   Relative Algorithm Accuracy and Efficiency

This section provides a summary of the relative errors and efficiency for the integration algorithms discussed in this lecture. Our standard problem is one period of the harmonic oscillator with initial condition $\{q_0, p_0\} = (1, 0)$.

Algorithm efficiency is always a consideration when selecting a method, especially for large problems. The most expensive part of a molecular dynamics calculation is the force evaluation. When comparing the efficiency of two methods of the same order the timestep of the more efficient method (that with the least number of force evaluations) can be made larger by the ratio of the number of force evaluations. This assumes that the method is sufficiently accurate and also stable for the larger timestep.

Accuracy and efficiency are not the only criteria distinguishing these methods. The choice of an integrator also depends on the type of problem being solved. Runge-Kutta and Gear methods are needed for chaotic problems and nonequilibrium problems with velocity-dependent forces. Monotonically increasing or decreasing energy drift is present with these integrators. The symplectic integrators are Hamiltonian-based and are useful for equilibrium calculations, Newtonian mechanics and for problems in which energy drift is undesirable.

Consider first a comparison of integration methods for first-order differential equations. All of the second-order methods are effectively symplectic. Errors for the second-order methods (including the related Størmer-Verlet method) are shown in **Figures 2.7 and 2.8**, pages 56-57. The trajectory and energy errors are identical for Velocity-Verlet, Symplectic-Leapfrog and Størmer-Verlet algorithms. The coordinates for the Symplectic-Leapfrog and Velocity-Verlet methods are the same as Størmer-Verlet. The Rahman method is less advantageous because the errors are larger and additional force evaluations are likely needed for the corrector step. The energy error in the Rahman method is zero within machine precision.

The p2 and p4 centered momenta have proven useful as auxiliary momenta for the coordinates calculated for second-order differential equations. By using p4 with the Størmer-Verlet method the energy error can be reduced by a factor of three relative to the corresponding Symplectic-Leapfrog and Velocity-Verlet errors.

The errors in the second-order Monte-Carlo method are about five orders of magnitude smaller than those coming from the other second-order methods. The method could justifiably be compared with the fourth-order methods.

We evaluated a selection of fourth-, fifth- and sixth-order methods for first-order differential equations. The algorithms included Gear predictor-corrector, symplectic, and Runge-Kutta methods. The coordinate, momentum and energy errors are shown in **Figures 2.25 and 2.26**.

The trajectory calculations for the RK4 integrator are almost two orders of magnitude more accurate than $GEAR4_1$. RK4 is also simpler to use than the Gear algorithms. The $GEAR4_1$ convergence for the standard oscillator required four corrector iterations for the largest timestep and one for the smallest timestep in the Figures.

For the standard oscillator the $GEAR5_1$ q and p are more accurate than RK4 for a timestep smaller than $(2\pi/200)$. The energy using $GEAR5_1$ is considerably less accurate (by an order of magnitude). The Gear algorithms for first-order differential equations are easy to program compared to their counterparts for second-order (Newtonian) differential equations.

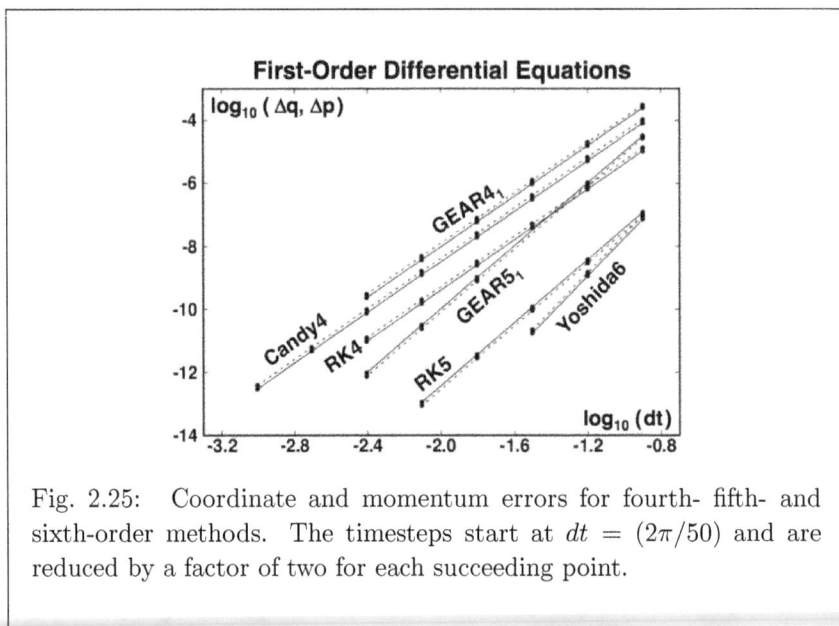

Fig. 2.25: Coordinate and momentum errors for fourth- fifth- and sixth-order methods. The timesteps start at $dt = (2\pi/50)$ and are reduced by a factor of two for each succeeding point.

For nonlinear chaotic problems it is more challenging to analyze errors as there is no analytical solution to compare with the results. However, it is useful to know that the phase-space chaotic sections for the Nosé-Hoover oscillator are a very sensitive tool to use for determining an accurate timestep. The GEAR4$_1$ integrator is able to reproduce the Nosé-Hoover sections that we have tested carefully with RK4 and RK5 integrators. More often the accuracy for nonlinear problems is determined by using a higher-order method to serve as the "known" solution as we demonstrated in the case of the nonlinear cell model. Quadruple precision is necessary for analyzing the errors in the higher-order methods.

The MILNE$_1$ predictor-corrector algorithm appeared to be a very accurate integrator for first-order differential equations. But remember that predictor-corrector methods have implicit terms that can give rise to instabilities. The Milne fourth-order predictor-corrector is computationally stable for the standard oscillator but unstable for the Nosé Hoover oscillator. It should not be used. It should be tested for stability when applied to nonlinear problems.

The symplectic integrators in **Figure 2.25** and **Figure 2.26** are the Candy-Rozmus and Yoshida integrators. Other fourth-order symplectic algorithms which are worth a look can be found in Gray, Noid, and

Sumpter's[16] paper. Their Table II provides an informative guide for selecting a symplectic method for many-body problems. The number of force evaluations for the Candy-Rozmus method is three compared to seven for the Yoshida method. Several of Yoshida's symplectic methods are well described in the two references cited here. I recommend consulting those when considering which methods to evaluate for your own problem. I also recommend testing the second-order symplectic methods.

All of the symplectic methods are easy to program. The higher-order methods usually have constants with many digits which are a bit of a hassle to enter correctly into a program.

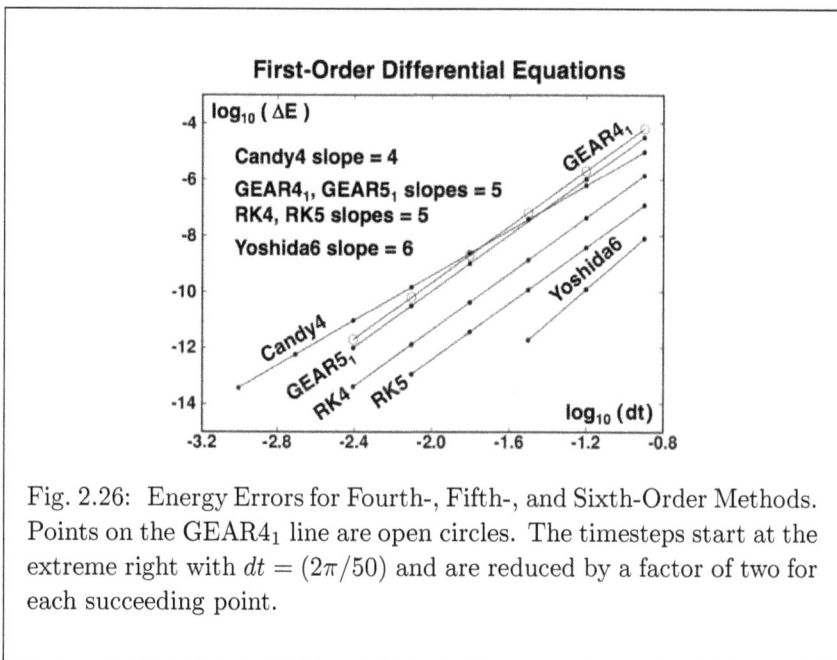

Fig. 2.26: Energy Errors for Fourth-, Fifth-, and Sixth-Order Methods. Points on the GEAR4$_1$ line are open circles. The timesteps start at the extreme right with $dt = (2\pi/50)$ and are reduced by a factor of two for each succeeding point.

Exercise : Consider a particle confined between two reflectors located at $x = \pm 1$, by potentials δx^4. Begin with $(x, v) = (0, 1)$ and compute ten collisions using $dt = 0.1$, 0.01, 0.001. Compare the energy conservation for the GEAR$_1$, Runge-Kutta, and Candy-Rozmus integrators.

Next let us review the algorithms we have studied for solving second-order differential equations. We have found that the $p2$, $p4$, and $p6$ centered momentum approximations improve the accuracy of these methods as shown in **Figures 2.27 and 2.28**. The fourth- and fifth-order methods include MILNE$_2$, GEAR4$_2$, and GEAR5$_2$. MILNE$_2$ and GEAR4$_2$ agree within a few percent with a $p4$ momentum and they are both computationally stable algorithms. The momentum is less accurate for the two algorithms than is the coordinate. Momenta are needed in order to approximate the energy, pressure, and temperature as well as the equilibrium Green-Kubo correlation functions used to compute transport coefficients.

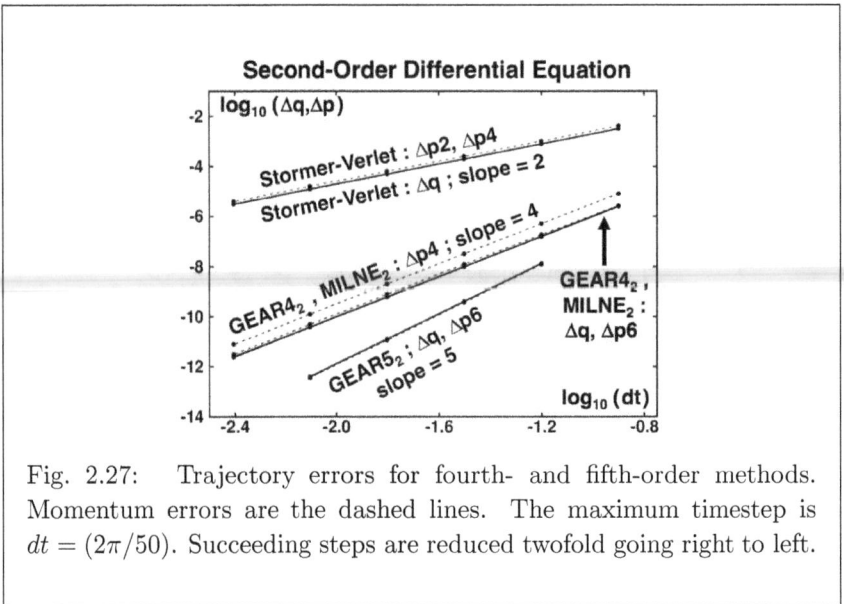

Fig. 2.27:   Trajectory errors for fourth- and fifth-order methods. Momentum errors are the dashed lines. The maximum timestep is $dt = (2\pi/50)$. Succeeding steps are reduced twofold going right to left.

The centered momenta $p4$ and $p6$ considerably improve the accuracy of the energy estimates compared to the other momentum approximations we investigated. For the second-order methods the fourth-order momentum improves the energy estimate by a factor of three. The energy estimates for the fourth-order algorithms are improved by more than an order of magnitude with a sixth-order centered momentum. The GEAR$_5$ energy is fourth-order with a fourth-order momentum and fifth-order with the sixth-order momentum. The more accurate energy estimates are useful both as a diagnostic and also for improved accuracy for nonlinear systems.

Second-Order Differential Equation

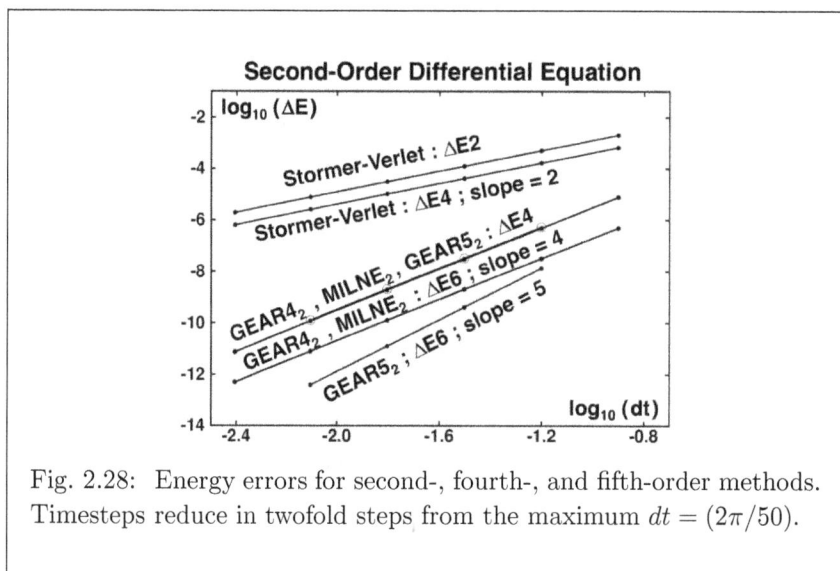

Fig. 2.28: Energy errors for second-, fourth-, and fifth-order methods. Timesteps reduce in twofold steps from the maximum $dt = (2\pi/50)$.

2.12 Summary of Lecture 2

Explicit numerical methods for molecular dynamics reviewed here vary from the venerable second-order Størmer-Verlet algorithm to an eighth-order symplectic algorithm. The higher-order algorithms often need quadruple precision for evaluating errors. Double and quadruple precision programs are both written with the following declarative in all of the subprograms.
`IMPLICIT DOUBLE PRECISION (A-H,O-Z)`
The gfortran compile line for double precision is :
`gfortran -O -o xcode code.f`
The gfortran compile line for quadruple precision is :
`gfortran -O -o xcode -freal-8-real-16 code.f`
Double precision is about fifty times faster than quadruple precision.

We adopted the harmonic oscillator to analyze these algorithms' phase, amplitude, and energy errors. Three Approximations related to Størmer-Verlet are Velocity-Verlet, Symplectic-Leapfrog, and Rahman's. The first two generate the same coordinates as Størmer-Verlet. None of these methods addresses the velocity-dependent forces occurring in nonequilibrium problems.

Lyapunov instability is the physical phase-space instability characterizing chaotic nonlinear problems. The ϕ^4 model illustrates the Lyapunov-unstable exponential growth of errors compared to the local short-time powerlaw growth characterizing numerical integrators. The maximum Lyapunov exponent describes the growth rate of the separation of two nearby phase-space trajectories. The growth rate at each timestep $\dot{\delta}_t$ is measured and the two trajectories' separation is then rescaled to the original length. The local value of the maximum Lyapunov exponent is $\lambda_1(t) \equiv \ln(|\delta_{t+dt}|/|\delta_t|)/dt$.

The equations of motion for dynamical systems are time-reversible. Nevertheless, nonlinear chaotic systems display irreversible behavior caused by Lyapunov instability. The stability differences distinguishing forward and reversed trajectories of these systems can be studied using bit-reversible numerical algorithms. Time reversibility can be achieved numerically by using large-integer versions of time-centered difference approximations. Levesque and Verlet's second-order bit-reversible algorithm is based on an integer form of the Størmer-Verlet method. Similarly a fourth-order bit-reversible method can be based on Milne's fourth-order predictor algorithm for second-order differential equations. The inelastic collision of two balls clearly demonstrates time-symmetry breaking by contrasting the more-unstable particles in the two time directions. These calculations provide understanding of the fundamental physics underlying the Second Law of Thermodynamics.

Fourth- and fifth-order Runge-Kutta methods offer improved trajectory accuracy along with relatively small but systematic and monotonic energy drift, a nuisance in long-time simulations. The most important applications for these integrators are for modelling nonequilibrium systems, nonlinear flows, and chaos.

Symplectic methods are particularly well suited to long-time calculations of planetary motion as well as large-molecule dynamics in chemical or biological systems. These methods satisfy Liouville's flow equation in phase space, conserving the phase volume as well as satisfying lower-dimensional area-preserving Poincaré invariants. These integrators are not applicable for non-Hamiltonian systems that are thermostatted.

Stiff differential equations require special numerical methods for treating the large fluctuations on the righthand sides of the differential equations. Physical problems with two or more timescales and problems in chemical kinetics can be stiff. Stiff problems can be solved numerically using implicit predictor-corrector techniques or the simpler adaptive methods.

We completed our discussion of integration algorithms by comparing the efficiency and accuracy of integration methods for first- and second-order differential equations. These included second-, fourth-, fifth- and sixth-order methods.

Recently Li and Liao[32] have explored large-scale integration techniques with thousands of digits and thousands of Taylor's series terms. Their "Clean Numerical Simulation" method provides accurate solutions over much longer times (by factors of millions) at the cost of much greater computational support. That these calculations are feasible is interesting. No doubt more than a few will prosper in the future keeping step with advances in computer technology.

We have compared integration methods by measuring trajectory accuracy, energy accuracy, and reversibility. We developed and applied a useful smooth nonlinear cell-model test problem well-suited to comparing the performance of integrators.

We surveyed a small collection of the integration techniques in this lecture, specifically selecting those that are most often used in molecular dynamics calculations. There are many other algorithms used in other fields that are missing. I encourage students to use these same ideas and tools to evaluate useful algorithms not included in this lecture.

In later lectures we will study nonequilibrium steady-state systems. They are typically chaotic. For those simulations we mostly adopt the Runge-Kutta methods. For simulations requiring accurate long-time reversed trajectories integer-based versions of Milne's algorithms are useful.

References

1. Wm. E. Milne, *Numerical Calculus* (Princeton University Press, Princeton, New Jersey, 1949). Reprinted by Princeton University Press in 2015.
2. D. Frenkel and B. Smit, *Understanding Molecular Simulation from Algorithms to Applications*, Second Edition (Academic Press, New York, 2002).
3. Wm. G. Hoover, *Computational Statistical Mechanics* (Elsevier, Amsterdam, 1991), available free at http://www.williamhoover.info/book.pdf.
4. Wm. G. Hoover, *Smooth Particle Applied Mechanics, The State of the Art* (World Scientfic, Singapore, 2006).
5. Wm. G. Hoover and C. G. Hoover, *Time Reversibility, Computer Simulation Algorithms, Chaos*, Second Edition (World Scientific, Singapore, 2012).
6. Wm. G. Hoover and C. G. Hoover, *Simulation and Control of Chaotic Nonequilibrium Systems* (World Scientific, Singapore, 2015).
7. J. C. Sprott, *Chaos and Time-Series Analysis* (Oxford University Press, Oxford, 2003). Other excellent books are currently listed on Sprott's web site.

8. T. Tél and M. Gruiz, *Chaotic Dynamics, An Introduction Based on Classical Mechanics* (Cambridge University Press, Cambridge, 2006).

9. B. D. Todd and P. J. Daivis, *Nonequilibrium Molecular Dynamics* (Cambridge University Press, Cambridge, 2017).

10. A. Rahman, "Correlations in the Motion of Atoms in Liquid Argon", Physical Review A **136**, 405-411 (1964).

11. D. Levesque and L. Verlet, "Molecular Dynamics and Time Reversibility", Journal of Statistical Physics **72**, 519-537 (1993).

12. Wm. G. Hoover, O. Kum, and N. E. Owens (now Fulda), "Accurate Symplectic Integrators *via* Random Sampling", The Journal of Chemical Physics **103**, 1530-1532 (1995).

13. Wm. G. Hoover, J. C. Sprott, and C. G. Hoover, "Adaptive Runge-Kutta Integration for Stiff Systems: Comparing the Nosé and Nosé-Hoover Oscillator Dynamics", American Journal of Physics **84**, 786-794 (2016).

14. J. D. Meiss, "Symplectic Maps, Variational Principles, and Transport", Reviews of Modern Physics **64**, 795-848 (1992).

15. H. Yoshida, "Recent Progress in the Theory and Application of Symplectic Integrators", Celestial Mechanics and Dynamical Astronomy **56**, 27-43 (1993).

16. S. K. Gray, D. W Noid, and B. G. Sumpter, "Symplectic Integrators for Large Scale Molecular Dynamics Simulations : A Comparison of Several Expicit Methods", The Journal of Chemical Physics **101**, 4062-4072 (1994).

17. J. Candy and W. Rozmus, "A Symplectic Integration Algorithm for Separable Hamiltonian Functions", The Journal of Computational Physics **92**, 230-256 (1991).

18. H. Yoshida, "Construction of Higher Order Symplectic Integrators", Physics Letters A **150**, 262-268 (1990).

19. K. Aoki and D. Kusnezov, "Nonequilibrium Statistical Mechanics of Classical Lattice ϕ^4 Field Theory", Annals of Physics **295**, 50-80 (2002).

20. Wm. G. Hoover and C. G. Hoover, "Instantaneous Pairing of Lyapunov Exponents in Chaotic Hamiltonian Dynamics and the 2017 Ian Snook Prize", Computational Methods in Science and Technology **23**, 73-79 (2017).

21. G. Benettin, L. Galgani, A. Giorgilli, and J.-M. Strelcyn, "Lyapunov Characteristic Exponents for Smooth Dynamical Systems and for Hamiltonian Systems; A Method for Computing All of Them", Meccanica **15**, 9-30 (1980).

22. I. Shimada and T. Nagashima, "A Numerical Approach to Ergodic Problem of Dissipative Dynamical Systems", Progress in Theoretical Physics **61**, 1605-1616 (1979).

23. Wm. G. Hoover and H. A. Posch, "Direct Measurement of Lyapunov Exponents", Physics Letters **113A**, 82-84 (1985).

24. I. Goldhirsch, P.-L. Sulem, and S. Orszag, "Stability and Lyapunov Stability of Dynamical Systems: A Differential Approach and a Numerical Method", Physica D **27**, 311-337 (1987).

25. Wm. G. Hoover and C. G. Hoover, "Comparison of Very Smooth Cell-Model Trajectories Using Five Symplectic and Two Runge-Kutta Integrators", Computational Methods in Science and Technology **21**, 109-116 (2015).

26. Ch. Schlier and A. Seiter, "High-Order Symplectic Integration: An Assessment", Computer Physics Communications **130**, 176-189 (2000).

27. C. Wm. Gear, *Numerical Initial Value Problems in Ordinary Differential Equations* (Prentice-Hall, Incorporated, Englewood Cliffs, New Jersey, 1971).

28. G. Venneri and Wm. G. Hoover, "Simple Exact Test for Well-Known Molecular Dynamics Algorithms", Journal of Computational Physics **73**, 468-475 (1987).

29. D. Beeman, "Some Multistep Methods for Use in Molecular Dynamics Calculations", Journal of Computational Physics **30**, 130-139 (1976).

30. Wm. G. Hoover and C. G. Hoover, "Bit-Reversible Version of Milne's Fourth-Order Time-Reversible Integrator for Molecular Dynamics", Computational Methods in Science and Technology **23**, 299-303 (2017).

31. Wm. G. Hoover and C. G. Hoover, "Time Symmetry Breaking in Hamiltonian Mechanics", Computational Methods in Science and Technology **19**, 77-87 (2013).

32. X. Li and S. Liao, "Comparison Between Symplectic Integrators and Clean Numerical Simulation for Chaotic Hamiltonian Systems"= arχiv:1609.09344.

Chapter 3

Molecular Dynamics with Thermostats

/ Review of Molecular Dynamics / Kinetic Temperature / Mechanics of the Ideal-Gas Thermometer / Is Configurational Temperature Useful ? / Boltzmann equation and Krook-Boltzmann equation / Boltzmann's H Theorem and Temperature / Dog-Flea Model / Molecular Dynamics versus Gibbs' Statistical Mechanics / Nosé-Hoover Mechanics / Wang-Yang Knots / Isokinetic Dynamics *via* Gauss' and Hoover-Leete Mechanics / Summary of Lecture 3 /

3.1 Brief Review of Molecular Dynamics

"In the beginning" molecular dynamics was nearly all Newtonian, localized at a handful of governmental scientific laboratories in the United States. Enrico Fermi's Group at Los Alamos,[1] Berni Alder and Tom Wainwright at Livermore,[2] George Vineyard's group at Brookhaven,[3] and Aneesur Rahman at the Argonne Laboratory[4] all devoted themselves to exploring and expanding the capabilities of "fast" computers to solve "manybody problems". This work was pioneering, assessing the capabilities of these new tools for the design, execution, and analysis of "computer experiments". Numerical experiments fleshed out the theoretical structures which had been built by Maxwell, Boltzmann, and Gibbs two or three generations earlier. Computer models were put to the task of interpreting laboratory experiments with real-world analogs. I was fortunate to finish my Ph D in 1961, at just the right time to begin a career in computational physics.

Some 20 years later my supervisor/colleague at the Lawrence Radiation Laboratory was Mark Wilkins, a brilliant self-taught expert in computational continuum mechanics. Mark emphasized the synergy among the three routes to knowledge used by physicists and engineers :

[1] Theories: such as the gas law $PV = NkT$, or continuity, $\dot{\rho} = -\rho \nabla \cdot v$;
[2] Simulation: computational fluid mechanics, or molecular dynamics ;
[3] Experiment: laboratory measurements of the thermal and mechanical behavior of materials, equation-of-state measurements, often using chemical and/or nuclear explosives to reach and characterize novel material properties at high temperatures and pressures.

Mark emphasized that in order to be "interesting" a scientific project needed to involve at least two of these three ingredients : theory, simulation, and experiment. With the advent of computers, at the height of the Second World War, high-speed simulations partnered with theory and experiment in improving the many models used to describe material behavior in the world around us.

From today's perspective the goals of that pioneering work, more than half a century ago, were primitive. Carol described two aspects of the early molecular dynamics work at equilibrium : [1] measuring the pressure P and temperature T for Newtonian systems with fixed energy E and volume V ; [2] measuring the pair distribution function $g(r)$ in connection with modelling that equation of state work with atomistic forces. Nonequilibrium studies included measuring the diffusion coefficient D as well as estimating the shear and bulk viscosities (η, η_v) and thermal conductivity κ from Green and Kubo's autocorrelation function formulæ.

In 1984 a significant addition to the computational toolkit was discovered by Shuichi Nosé.[5,6] He extended atomistic dynamics to conform to Gibbs' canonical and isobaric ensembles. Nosé found a way to carry out dynamical simulations using temperature and/or pressure as independent variables rather than Newton's energy and/or volume. This Lecture will make that pioneering work accessible. We will show that the ideal-gas thermometer provides a mechanical basis for temperature and thermometry within atomistic mechanics. The kinetic-theory definition of temperature in terms of kinetic energy goes back to Maxwell and Boltzmann. Boltzmann's equation for the evolution of the dynamical state of a gas grounds the concepts of temperature and entropy in the classical atomistic model of the nearly-ideal gas.

Gibbs' statistical mechanics and molecular dynamics are brought into correspondence through the dilute gas model and form the basis for our studies here. Because we are particularly interested in the nonequilibrium studies we'll see how Boltzmann's equation goes beyond equilibrium properties to provide estimates for nonequilibrium flows governed by viscosity and conductivity. We will illustrate Nosé's thermal mechanics in the simplest possible case, a one-dimensional harmonic-oscillator model. We begin by reviewing the temperature concept, starting out with Maxwell and Boltzmann's kinetic theory of dilute gases, so dilute as to be "ideal".

3.2   Kinetic Temperature, $kT_{xx} = \langle (P_{xx}V/N) = (p_x^2/m) \rangle$

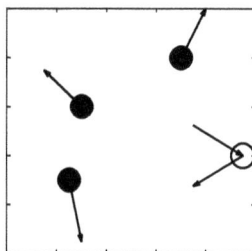

Fig. 3.1:   Any particle moving at (v_x, v_y) contributes (mv_x^2/V) and (mv_y^2/V) to the wall forces whose average values measure P_{xx} and P_{yy} . For simplicity ignore intermolecular interactions and focus on P_{xx}. During the roundtrip time, $(2L_x/v_x)$ the white particle shown at the wall in the figure transfers momentum $2mv_x$ to that wall. Such a collision makes the contribution (mv_x^2/L_xL_y) to the force per unit length P_{xx} *in* the x direction *on* the x wall shown in the **Figure**. Pressure can also be defined as the momentum flux so that $P_{xx}V$ is the summed up flow of $m\dot{x}$ momentum in the x direction per unit volume (mv_x^2/V) for all of the particles in V. In this figure four such terms contribute to P_{xx}.

In the absence of particle-particle collisions the ideal-gas law, $PV = NkT$, follows from the sketch of wall collisions shown in **Figure 3.1**. Pressure is the force per unit area on the wall $\sum m\dot{x}^2/V$ and the ideal gas law is the connection between pressure and temperature for the "ideal gas". A sufficiently dilute sample of *any* gas can act as an ideal-gas thermometer. Adding temperature to the set of mechanical state variables (N, E, V, P) makes it possible to develop *thermo*dynamics.

The mechanical concept of pressure as force per unit area guarantees that two fluids in mechanical equilibrium with a third have the same pressure. Mechanical equilibrium parallels thermal equilibrium, as expressed by the "Zeroth Law of *Thermo*dynamics" : if a body [the thermometer] is in *thermal* equilibrium with two fluids those two have the same *temperature*. The ideal-gas law relates the pressure P of such an idealized gas to its "kinetic temperature" T, $P = (NkT/V)$. (N/V) is the number density, and k is Boltzmann's constant, honoring Ludwig Boltzmann's research in the fundamental kinetic theory of low-density gases.[7] To start out we will sketch Boltzmann's explanation of the effect of interparticle collisions at low density. His kinetic theory connects dynamics to statistical mechanics.

Josiah Willard Gibbs (1839-1903) developed statistical mechanics at about the same time as did Ludwig Boltzmann (1844-1906). Gibbs focused on probability and entropy rather than the details of collisions. Thermodynamic entropy had been defined earlier by Rudolf Clausius (1822-1888) as a state function obtained by integrating the reversible heat transfer, divided by temperature, in any reversible process converting one thermodynamic state to another: $\int dS \equiv (\int dQ_{rev}/T)$. By 1884 both Boltzmann and Gibbs had discovered that the new state function, the thermodynamic entropy S, was also, within an arbitrary additive constant, and an arbitrary multiplicative one [Boltzmann's constant] the logarithm of the "phase-space" volume describing the macroscopic system state in question. For a state of known composition N, volume V, and energy E the entropy is

$$S(N, E, V) = k \ln[\int \cdots \int \prod dqdp \ ] \ .$$

The phase volume in question consists of all the states with the specified volume V and energy E of a specified number of atoms, or molecules in the polyatomic case N. This phase-volume definition of entropy corresponds to the thermodynamic concept of entropy in four ways : the phase-volume entropy

[1] is an extensive state function ;
[2] of an isolated system will increase if constraints are removed ;
[3] is otherwise constant (a consequence of Liouville's Theorem) ;
[4] responds to changes in the volume and the energy as described by the combined First and Second Laws of Thermodynamics :

$$dE = dQ - dW \ ; \ dE \leq TdS - PdV \ ; \ (dQ_{rev}/T) \equiv dS \ .$$

The energy change dE reflects both incoming heat dQ and work done by the system on its surroundings dW. The equality governs idealized "reversible" equilibrium processes. The inequality describes the inexorable increase of entropy associated with nonequilibrium ("irreversible") processes.

It is significant that a differential phase volume, a tiny hypercube or hypersphere $\otimes = \prod(dqdp)$, is a conserved quantity in flows following Hamilton's equations of motion,

$$\{ \dot{q} = +(\partial \mathcal{H}/\partial p) \; ; \; \dot{p} = -(\partial \mathcal{H}/\partial q) \} \longrightarrow$$

$$(\dot{\otimes}/\otimes) = \sum [(\partial \dot{q}/\partial q) + (\partial \dot{p}/\partial p)] = 0 .$$

You can show the last equality easily from Hamilton's equations of motion by noting that $(\partial^2 \mathcal{H}/\partial q \partial p)$ and $(\partial^2 \mathcal{H}/\partial p \partial q)$ are equal. It is the entropy *inequality*, away from equilibrium, that is not so easy to understand. We will come back to discussing changes in the comoving phase volume in the course of describing Nosé's thermomechanics in Section 3.9. Let us first become more familiar with the concepts of entropy and temperature.

3.3 The Classical Mechanics of the Ideal-Gas Thermometer

The early Newtonian simulations were carried out at constant energy and constant volume. Simulations at constant temperature, with varying energy, or at constant pressure, with varying volume, became desirable and stimulating goals ever since molecular dynamics became routine, in the 1960s. We will concentrate on isothermal algorithms, the simpler and more usual of the two. There is a comprehensive literature on achieving both goals. To begin, what *is* temperature?

A systematic exposition of temperature and equilibrium thermodynamics begins with the Zeroth Law, the basis of thermometry: "If a *thermometer* is in thermal equilibrium with two other bodies, those two are likewise in thermal equilibrium with each other." *Thermo*dynamics requires choosing a thermometer. Our mechanical model thermometer is a "bag of gas", a localized collection of many light particles that measures temperature through elastic collisions with other bodies. See **Figure 3.2**. The tiny particles in the bag are so light and so numerous that their distribution is always the equilibrium Maxwell-Boltzmann one. We will see that this distribution follows from Boltzmann's equation.

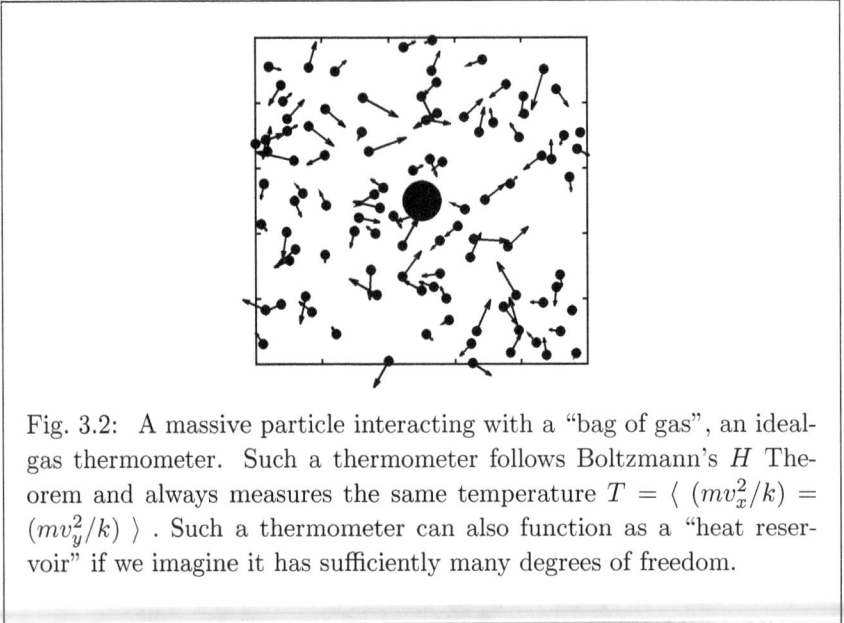

Fig. 3.2: A massive particle interacting with a "bag of gas", an ideal-gas thermometer. Such a thermometer follows Boltzmann's H Theorem and always measures the same temperature $T = \langle (mv_x^2/k) = (mv_y^2/k) \rangle$. Such a thermometer can also function as a "heat reservoir" if we imagine it has sufficiently many degrees of freedom.

Gibbs' idea that any system will reach its "most likely" maximum-entropy state at equilibrium relates the Maxwell-Boltzmann distribution of velocities to thermodynamics. Two thermodynamic systems which can interact thermally but are otherwise isolated will come to "thermal equilibrium" through heat transfer. Heat transfer stops once Gibbs' maximum entropy state is reached, with $(\partial S/\partial E)_1 = (\partial S/\partial E)_2$. If we suppose that the first of the two bodies is a D-dimensional "ideal gas", with a known Gibbs' entropy :

$$S_1 = S_{ideal} = Nk \ln[(V/N)(E/N)^{D/2}] + \text{constant} ,$$

we find the common value

$$(\partial S_1/\partial E_1)_{V_1} = (\partial S_2/\partial E_2)_{V_2} = (NkD/2E_{ideal}) = (1/T) ,$$

for any system with which the ideal-gas thermometer is in thermal equilibrium. From the macroscopic standpoint the ideal gas can be weakly coupled to any small portion of the second body, body number 2, so that the temperature measured by the thermometer is a local point property.

A similar maximum-entropy argument applies to an ideal-gas-plus-fluid system where the gas and fluid are separated by a conducting movable piston rather than a motionless wire. See **Figure 1.3** on page 11. Then the equilibrium condition for the piston position gives :

$$S = S_1(V_1) + S_2(V_2) \text{ and } V = V_1 + V_2 \rightarrow$$

$$(\partial S_1/\partial V_1)_{E_1} = (\partial S_2/\partial V_2)_{E_2} = (P/T) ,$$

with $dE = TdS - PdV \rightarrow T = (\partial E/\partial S)_V$ and $(P/T) = (\partial S/\partial V)_E$, establishing both of the necessary links of mechanics to thermodynamics.

Let us look at the details of the brief elastic interaction of an ideal-gas thermometer particle of mass m with a much more massive fluid particle with mass $M \gg m$. For simplicity we illustrate this idea in one dimension[8] where conservation of momentum and energy relate the particle velocities after collision, $(\dot{x}', \dot{X}')$ to those before, $(\dot{x}, \dot{X})$:

$$m\dot{x}^\alpha + M\dot{X}^\alpha = m\dot{x}'^\alpha + M\dot{X}'^\alpha \text{ for } \alpha = 1, 2 \longleftrightarrow$$

$$\dot{X}' = \frac{M-m}{M+m}\dot{X} + 2\frac{m}{M+m}\dot{x} \; ; \; \dot{x}' = \frac{m-M}{M+m}\dot{x} + 2\frac{M}{M+m}\dot{X} .$$

Although a bit tedious it is straightforward to compute the average value of the heavy particle's velocity change, $\langle \dot{X}' - \dot{X} \rangle$.

The analytic approach, "kinetic theory", assigns collision probabilities to each ideal-gas velocity $\dot{x}$ proportional to the relative speed, $|\dot{x} - \dot{X}|$. The *averaged* effect over the whole Maxwell-Boltzmann distribution becomes an integral over $d\dot{x}$. Expanding the integral as a power series in the square root of the mass ratio $\sqrt{(m/M)}$ shows that the first nonvanishing term is quadratic, proportional to (m/M) and to the heavy-particle velocity $\dot{X}$:

$$\langle \ddot{X} \rangle = \frac{\int |\dot{X} - \dot{x}|(\dot{X}' - \dot{X})e^{-m\dot{x}^2/2kT}d\dot{x}}{\int |\dot{X} - \dot{x}|e^{-m\dot{x}^2/2kT}d\dot{x}} \longrightarrow \langle \ddot{X} \rangle \propto -4(m/M)\dot{X} .$$

Thus the ideal-gas thermometer, on the average, provides a *frictional* force countering the velocity of the heavy particle, eventually bringing its most likely (averaged) velocity to zero.

A similar calculation holds for the rate of energy loss (or gain) of the heavy particle :

$$\frac{\int (M/2)|\dot{X} - \dot{x}|(\dot{X}'^2 - \dot{X}^2)e^{-m\dot{x}^2/2kT}d\dot{x}}{\int |\dot{X} - \dot{x}|e^{-m\dot{x}^2/2kT}d\dot{x}} \longrightarrow \langle \dot{E} \rangle \propto 4(m/M)(kT - M\dot{X}^2) .$$

Evidently an ideal-gas thermometer, in addition to its frictional slowing of a massive particle also introduces fluctuations with the long-time-averaged effect of eliminating heat transfer between the particle and the thermometer if and only if the two have the same kinetic temperature, $\langle (m\dot{x}^2/k) \rangle = \langle (M\dot{X}^2/k) \rangle$.

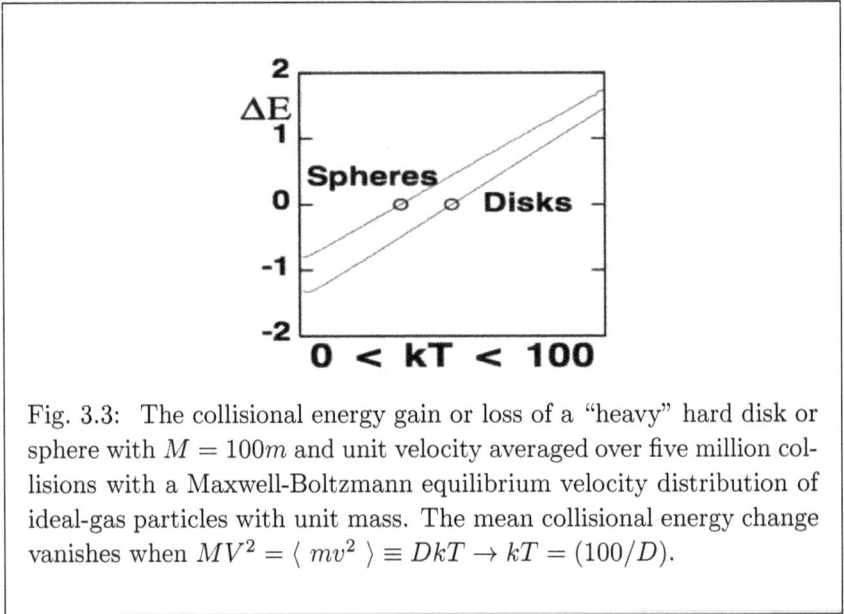

Fig. 3.3: The collisional energy gain or loss of a "heavy" hard disk or sphere with $M = 100m$ and unit velocity averaged over five million collisions with a Maxwell-Boltzmann equilibrium velocity distribution of ideal-gas particles with unit mass. The mean collisional energy change vanishes when $MV^2 = \langle\, mv^2 \,\rangle \equiv DkT \rightarrow kT = (100/D)$.

Similar kinetic theory calculations apply in two and three dimensions. To confirm this Carol and I carried out dynamical simulations for hard disks and hard spheres.[9] **Figure 3.3** shows that the average heavy-particle energy change ΔE varies nearly linearly with temperature, and vanishes when the bath and heavy-particle kinetic temperatures are equal.

Further, an ideal-gas thermometer composed of parallel hard cubes constrained against rotation, makes possible thought experiments measuring a symmetric temperature *tensor* with $T_{ij} = T_{ji}$. Although any physical property that varies with temperature could in principle be a "thermometer" the kinetic temperature is an ideal choice in more ways than one ! Let us briefly consider just one of the many alternatives, a "configurational" temperature depending upon particle coordinates rather than momenta.

3.4   Is Configurational Temperature a Useful Alternative?

It is clear that temperature needs to be measured in a "comoving" and corotating frame. Otherwise, just adding a constant linear or angular velocity to all the particles in a system would change its temperature. For this reason there have been several papers exploring "configurational temperatures", based on coordinates rather than velocities. The simplest of these temperatures is based on the ratio of two canonical-ensemble averages. To start we consider an integration by parts, differentiating $e^{-\mathcal{H}/kT}$, as Landau and Lifshitz did in their *Statistical Physics* textbook :

$$\int dq(1/kT)(\nabla_q^2\mathcal{H})e^{-\mathcal{H}/kT} = \int dq(1/kT)^2(\nabla_q\mathcal{H})^2 e^{-\mathcal{H}/kT}$$

$$\longrightarrow kT_c = \langle\,(\nabla\mathcal{H})^2\,\rangle/\langle\,\nabla^2\mathcal{H}\,\rangle\ .$$

In the integration by parts [$\int u\,dv = uv - \int v\,du$] v is $e^{-\mathcal{H}/kT}$ and u is $\nabla_q\mathcal{H}$. The only assumption, other than smoothness, is that the Hamiltonian diverges (so that uv vanishes) at the integration limits. In working out the averages, $\langle\,\dots\,\rangle$ the normalizing integral is $\int dq e^{-\mathcal{H}/kT}$.

We favor the kinetic temperature over this configurational alternative because the kinetic temperature is consistent with a mechanical measurement. The kinetic temperature can be "measured" both at and away from equilibrium. An additional reason to bypass configurational temperature is that the denominator in the Landau-Lifshitz expression can change signs, or even vanish, (as in the case of a gravitational field). Further, blind application of the configurational concept in a rotating system would give a false temperature gradient generated by centrifugal forces. To appreciate this difficulty have a look at the particular example of the steadily-rotating five-atom molecule in **Figure 3.4**.

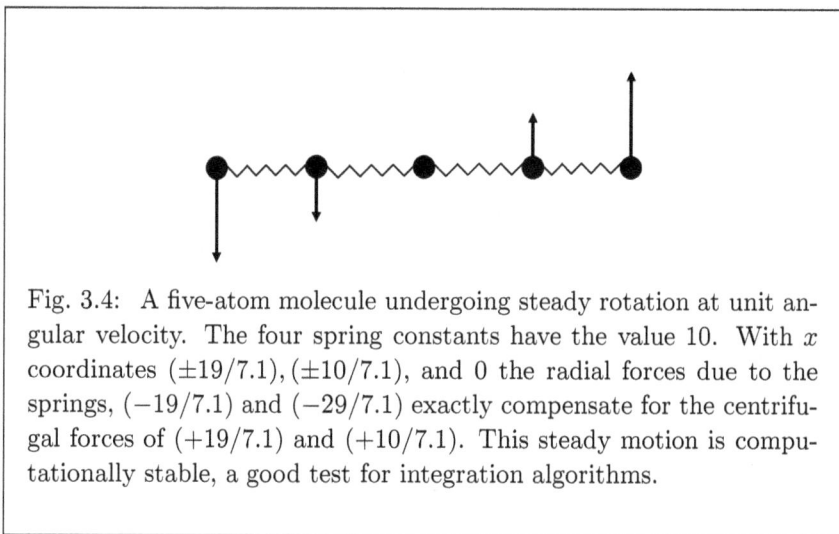

Fig. 3.4: A five-atom molecule undergoing steady rotation at unit angular velocity. The four spring constants have the value 10. With x coordinates $(\pm 19/7.1), (\pm 10/7.1)$, and 0 the radial forces due to the springs, $(-19/7.1)$ and $(-29/7.1)$ exactly compensate for the centrifugal forces of $(+19/7.1)$ and $(+10/7.1)$. This steady motion is computationally stable, a good test for integration algorithms.

The concept of temperature becomes a bit of a puzzle in the case of polyatomic molecules. **Figure 3.4** shows five particles bound by four Hooke's-

Law springs of unit restlength with force constant 10, in steady rotation at unit frequency. The Hamiltonian is :

$$\mathcal{H} = \sum (p^2/2) + 5[\ (r_{12}-1)^2 + (r_{23}-1)^2 + (r_{34}-1)^2 + (r_{45}-1)^2\ ]\ .$$

The spring and centrifugal forces on particles 4 and 5 have the following forms in steady rotation :

$$\ddot{r}_4 \equiv 0 = 10[\ r_5 - 2r_4\ ] + r_4 = 0\ ;$$

$$\ddot{r}_5 \equiv 0 = 10[\ r_4 + 1 - r_5\ ] + r_5 = 0\ .$$

Solving for the coordinates of particles 4 and 5 gives $r_4 = (10/7.1)$ and $r_5 = (19/7.1)$. Likewise $r_1 = -(19/7.1)$ and $r_2 = -(10/7.1)$. Is there a sensible definition of the temperature of this "molecule"? I think not. Evidently there is no unique temperature for arbitrarily rotating, vibrating, and translating molecules. This complicates studies of nucleation, where atomistic fluids in a thermodynamically unstable initial state begin to combine into molecules or drops of liquid. There is no difficulty in carrying out the simulations, while there are puzzles in interpreting the results.

3.5   The Boltzmann and Krook-Boltzmann Equations

Prior to numerical solutions of manybody problems two special cases had monopolized the attention of theoreticians, the nearly-ideal gas, described by the Boltzmann equation and the nearly-harmonic crystal, described by the weak interaction of phonons (normal modes). From the standpoint of ergodicity these models are quite different. Two-body collisions are easy to calculate as did Maxwell and Boltzmann. Collisional models of dilute gases reproduce readily understandable phenomena in addition to pressure and temperature: viscosity, conductivity, shockwave velocities, and the rate of approach to equilibrium.

The (quadratic) Boltzmann equation and the simpler (linear) model of it, the Krook-Boltzmann equation, describe the evolution of the probability distribution for a nonequilibrium gas. $f(r,p)drdp$ is the probability that a gas molecule lies in the region dr about r with a momentum p within dp. Two-body collisions govern the evolution of the gas, which is described by the changing density in the six-dimensional phase space of a single atom :

$$(\partial f/\partial t) = -\nabla_r \cdot (fv) - \nabla_v \cdot (fF) + (\partial f/\partial t)_{\text{collisions}}\ .$$

The time derivative *following* the motion depends upon collisions of pairs of particles :

$$\dot{f} = (\partial f/\partial t) + v \cdot \nabla_r f + (F/m) \cdot \nabla_v f = (\partial f/\partial t)_{\text{collisions}} \simeq (f_{eq} - f)/\tau\ .$$

For illustrative purposes we are adopting the Krook-Boltzmann approximation, $(f_{eq} - f)/\tau$ for the collision term. This simplified version of Boltzmann's equation requires a collision time τ, which is easy to estimate. A pair of particles in a periodic box of area A or volume V undergoes collisions at a rate

$$\Gamma_{2D \text{ or } 3D} = (1/A)|v_1 - v_2|2\sigma \text{ or } (1/V)|v_1 - v_2|\pi\sigma^2 .$$

Here σ is the range of the forces.

A numerical approach to solving such a gas-phase problem, called the Direct Simulation Monte Carlo Method by Græme Bird,[10] who developed and applied the method, can be implemented by carrying out a two-step process. Each particle is first propagated in time according to the *collisionless* motion equations :

$$\{ \dot{r} = v ; \dot{v} = (F/m) \} .$$

The external force F includes both boundary and gravitational forces. Dividing the system into computational zones allows the effect of collisions to be simulated zone by zone. This completes the timestep. Collisions are weighted according to the relative velocity of the particles selected. The Boltzmann equation would be modelled by choosing particle pairs from each zone. The Krook-Boltzmann equation would collide particles from each zone with particles chosen from the Maxwell-Boltzmann distribution matching the density, mean velocity, and kinetic energy of the zone in question. An even simpler model would choose *random* pairs of post-collision velocities in each zone, conserving both momentum and energy. In two dimensions :

$$v = \sqrt{(\dot{x}_1 - \dot{x}_2)^2 + (\dot{y}_1 - \dot{y}_2)^2} ; \theta = 2\pi\mathcal{R} ;$$

$$\dot{x}_\pm \longrightarrow (1/2)(\dot{x}_1 + \dot{x}_2) \pm (v/2)\sin(\theta) ;$$

$$\dot{y}_\pm \longrightarrow (1/2)(\dot{y}_1 + \dot{y}_2) \pm (v/2)\cos(\theta) ,$$

where $\mathcal{R}$ is a pseudorandom number between 0 and 1 .

Exercise: Show, either analytically or numerically, that the new values of the velocity components $(\dot{x}_\pm, \dot{y}_\pm)$ conserve momentum and energy.

For small gradients an analytic approach is feasible. Chapman and Cowling's text does this for the Boltzmann equation.[11] We will apply the simpler Krook-Boltzmann model to the calculation of the shear viscosity coefficient in a two-dimensional dilute gas with a velocity gradient $v_y = \dot{\epsilon}x$ as well as the heat conductivity of such a gas with an isobaric temperature gradient $(\partial T/\partial x)_P$.

3.5.1 *Krook-Boltzmann Equation* $\longrightarrow$ *Shear Viscosity*

We assume that the velocity gradient $\dot{\epsilon} = (du_y/dx)$ is small relative to the single-particle collision rate. This makes it possible to solve the Krook-Boltzmann equation analytically. We start with the local-equilibrium distribution :

$$f_{eq} \propto e^{-m(v_y-\dot{\epsilon}x)^2/2kT} \rightarrow v_x(\partial f/\partial x) = \dot{\epsilon} f_{eq}mv_xv_y/kT = (f_{eq} - f_{KB})/\tau \ .$$

Choosing terms through first order in the strain rate $\dot{\epsilon}$ provides the Krook-Boltzmann approximation to the nonequilibrium distribution function :

$$f_{KB} = f_{eq}[\, 1 - (\dot{\epsilon}\tau)(mv_xv_y/kT) \,] \longrightarrow$$

$$P_{xy} = (N/V)\langle\, (mv_xv_y) \,\rangle_{KB} = -\eta\dot{\epsilon} = -(NkT/V)\tau\dot{\epsilon} \rightarrow \eta \simeq \sqrt{mkT}/\sigma \ .$$

The scattering cross section in two dimensions is just a length, here σ. In the three-dimensional case the cross section is quadratic in the range of the forces and $\eta \propto (\sqrt{mkT}/\sigma^2)$. The finding that the viscosity coefficient is independent of the density was a major success of kinetic theory. Because the distance between collisions varies inversely with density and the difference in momentum carried is linear in density the two effects cancel.

3.5.2 *Krook-Boltzmann Equation* $\longrightarrow$ *Heat Conduction*

A similar treatment for the thermal conductivity requires a temperature that varies linearly with x. This *temperature gradient* implies a corresponding *density gradient* so as to avoid a *pressure gradient* : $d(\rho T)/dx = 0$. A pressure gradient would launch a weak shockwave travelling at the speed of sound, a much faster process than the diffusive transport of momentum or energy, which moves as the square root of the time.

In the case of conductivity the nonequilibrium distribution function, as calculated from the Krook-Boltzmann equation, is proportional to the temperature gradient and to $v_x(mv^2 - 4kT)$ in two dimensions and

$v_x(mv^2 - 5kT)$ in three. The mass current associated with these perturbations vanishes as they maintain a pressure independent of x :

$$\langle\, v_x\,\rangle_{\text{neq}} \propto \langle\, v_x^2(mv_x^2 + mv_y^2 - 4kT)\,\rangle_{\text{eq}} = 0 \text{ [2D] ;}$$

$$\langle\, v_x\,\rangle_{\text{neq}} \propto \langle\, v_x^2(mv_x^2 + mv_y^2 + mv_z^2 - 5kT)\,\rangle_{\text{eq}} = 0 \text{ [3D] .}$$

Like the viscosity, the heat conductivity is density independent in accord with experiment (and for exactly the same reason). Again the transport coefficent is independent of the density, due to the cancellation of two effects: the distance traveled with the density of travelers.

Although Maxwell-Boltzmann kinetic theory was successful in these qualitative predictions the failure of diatomic oxygen and nitrogen molecules to exhibit a heat capacity characteristic of six degrees of freedom (three translational, two rotational, and one vibrational) so that $(C_v/k) = 3(k/2) + 2(k/2) + k = (7k/2)$ was not understood until the development of quantum mechanics in the early 20th century.

3.6 Boltzmann's *H* Theorem and Temperature

Another major success of the Boltzmann equation was its agreement with the Second Law of Thermodynamics, predicting the approach of entropy for an isolated system to its maximum (equilibrium) value. Although Boltzmann "proved" the monotonicity of the entropy change formally any numerical faithful simulation of the physics underlying the Boltzmann equation indicates the presence of fluctuations rather than monotone behavior. The *H*-Theorem proof with the Krook-Boltzmann equation is less complex in that it doesn't require combining forward and backward collisions. But the ideas are similar and both formulations of kinetic-theory collisions lead to the Gaussian Maxwell-Boltzmann distribution. By definition the local equilibrium f_{eq} or nonequilibrium f_{KB} distribution function share common values of density, mean velocity, and temperature :

$$\langle\, v^\alpha\,\rangle_{eq} = \langle\, v^\alpha\,\rangle_{KB} \text{ for } \alpha = 0,\ 1,\ 2 \ .$$

This allows us to combine the Krook-Boltzmann equation's formulation of the entropy with the corresponding equilibrium average :

$$(\dot{S}/Nk)_{KB} = (1/\tau) \prod(\textstyle\int dq \int dp)[f_{KB} - f_{eq}] \ln[\ f_{KB}\ ] =$$

$$(1/\tau) \prod(\textstyle\int dq \int dp)[f_{KB} - f_{eq}] \ln[\ f_{KB}/f_{eq}\ ] \geq 0 \ .$$

Notice that the integrand is *always positive* unless $f_{KB} = f_{eq}$ for all values of the coordinates and momenta $\{\, q, p\,\}$.

Boltzmann carried out a similar, but more elaborate, proof of entropy increase using his quadratic equation for $\dot{f}$. He first expressed the time derivative of $(S/Nk) = \langle\, \ln f\, \rangle$ in terms of "gain" and "loss" collisions :

$$(\dot{S}/Nk) = \int\int\int\int dr dr_1 dv dv_1 |v - v_1| \ln(f)[\, f f_1 - f' f_1'\,]\ .$$

Here f and f_1 are the densities of particles at r and r_1 having velocities v and v_1 . Because for every collision of the type on the left

$$[\, (r, v, r_1, v_1) \to (r', v', r_1', v_1')\,]\ \xrightarrow{r-i}\ [\, (r', v', r_1', v_1') \to (r, v, r_1, v_1)\,]$$

there is a "reversed-inverse" collision as illustrated in **Figure 3.5** the loss and gain terms for the collision and its reversed-inverse can be combined and symmetrized, for each of the four (r, v) pairs involved in the before and after description of each collision type :

$$(\dot{S}/Nk) = \int\int\int\int dr dr_1 dv dv_1 |v - v_1|(1/4) \ln(f f_1/f' f_1')[\, f f_1 - f' f_1'\,] \geq 0\ !$$

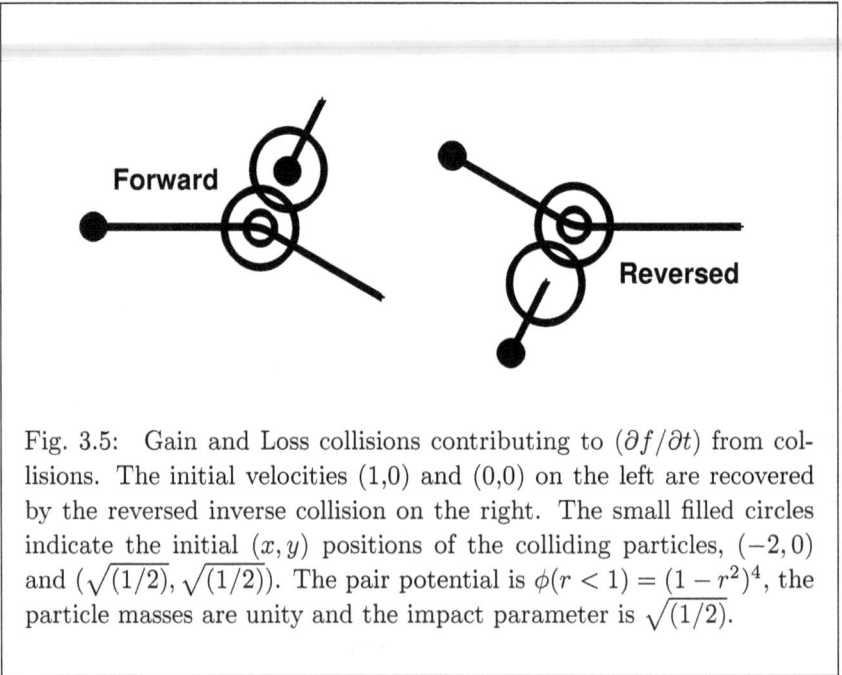

Fig. 3.5:   Gain and Loss collisions contributing to $(\partial f/\partial t)$ from collisions. The initial velocities $(1,0)$ and $(0,0)$ on the left are recovered by the reversed inverse collision on the right. The small filled circles indicate the initial (x, y) positions of the colliding particles, $(-2, 0)$ and $(\sqrt{(1/2)}, \sqrt{(1/2)})$. The pair potential is $\phi(r < 1) = (1 - r^2)^4$, the particle masses are unity and the impact parameter is $\sqrt{(1/2)}$.

The result, $[\, (\dot{S}/Nk) \geq 0\,]$ is fully as remarkable as was Gibbs' and Boltzmann's identification of phase volume with entropy. It "proves" the increase of entropy unless the distribution function is the equilibrium one and it proves that the logarithm of the distribution function is conserved in that case, allowing the coefficients of the conserved mass, momentum, and energy to be chosen to match the density, stream velocity, and temperature giving the Maxwell-Boltzmann Gaussian velocity distribution.

Though Boltzmann's Theorem was a tremendous achievement his proof soon drew criticism and caused Boltzmann grief. His friend Josef Loschmidt pointed out that any trajectories causing entropy to increase could be reversed by changing the signs of all the momenta $\{ p \rightarrow -p \}$. His friend Ernst Zermélo pointed out that any phase point on a Hamiltonian orbit in a closed system would eventually recur due to the conservation of phase volume. Both objections suggest that the H Theorem is false. It is important to keep in mind that the H "Theorem", like any other theorem, only applies to numerical models of nature and not to nature itself, which is certainly not obeying any mathematical theorems beyond simple arithmetic. In fact the H theorem *is* false provided that we choose to support Loschmidt and Zermélo by using the Levesque-Verlet integer version of dynamics :

$$\{ q_{t+dt} - 2q_t + q_{t-dt} = [(dt)^2 F(q_t)/m]_{integer} \} .$$

Provided that the integer coordinates on the lefthand side and the force term on the righthand side are computed in the same way the forward and backward trajectories generated by this algorithm are strictly time reversible to the very last bit. As the integer phase-space is finite both Loschmidt's and Zermélo's objections to the H theorem are valid.

From the standpoint of fourth-order Runge-Kutta integration it is clear that the Theorem is correct : roundoff error and Lyapunov instability together prevent reversal as well as a finite number of phase points (due to the dissipation in the integrator). Discussions of these contradictions have persisted to the present day despite the presence of fluctuations and despite their lack of relevance for dynamics with the exception of Levesque and Verlet's integer version of Störmer's centered-difference algorithm and Milne's fourth-order predictor algorithm (the top equation on page 81).[12]

Early in the 20th century the Ehrenfests developed a picturesque model, simpler than Boltzmann's, and analogous to the Levesque-Verlet algorithm by its use of an integer phase space. This "dog-flea" [or "Hund-Flöhe"] model was designed to show (through explicit inclusion of fluctuations) just why the H theorem was false. It is worth our while to consider it.

3.7   The Ehrenfests' Dog-Flea Model for Fluctuations

The ingredients of the "dog-flea" model are two dogs, N numbered fleas with N an even integer, and a random number generator. These are enough to understand the successes and failures of the H Theorem while also learning a bit about the generator. Starting with all the fleas on one dog the random number generator chooses the next flea to jump, ending up with $N-1$ on the flea-rich dog and a single flea on the other. The second jump results in a reversal of the first in only one of the possible N jumps, so that the decrease toward $(N/2)$ is *statistical* rather than inevitable. The simplest case, $N=2$ has two equally-likely outcomes. Either the same flea jumps both times (returning to the original condition) or each flea jumps once. Either possibility ends up with two fleas on one dog, none on the other. Enlarge the model to include four fleas. After a single jump the flea population is [3,1]. After two jumps we have [2,2] with probability 3/4 or [4,0] with probability 1/4. After three jumps we have [3,1] with probability one, after four jumps either [4,0] with probability 1/4 or [2,2] with probability 3/4. This is the time when it becomes logical to switch from mental effort to computer simulation.

    It is straightforward to use a random number generator to trace the evolution of the populations jump by jump. Our favorite generator provides 2^{22} random numbers and then repeats. The pseudorandom numbers include all nonegative fractions of the form $(n/2^{22})$ where n varies from 0 to $4\,194\,303 = 2^{22} - 1$:

```
FUNCTION RUND(INTX,INTY)
IMPLICIT DOUBLE PRECISION (A-H,O-Z)
I = 1029*INTX + 1731
J = I + 1029*INTY + 507*INTX - 1731
INTX = MOD(I,2048)
J = J + (I - INTX)/2048
INTY = MOD(J,2048)
RUND = (INTX + 2048*INTY)/4194304.0
RETURN
END
```

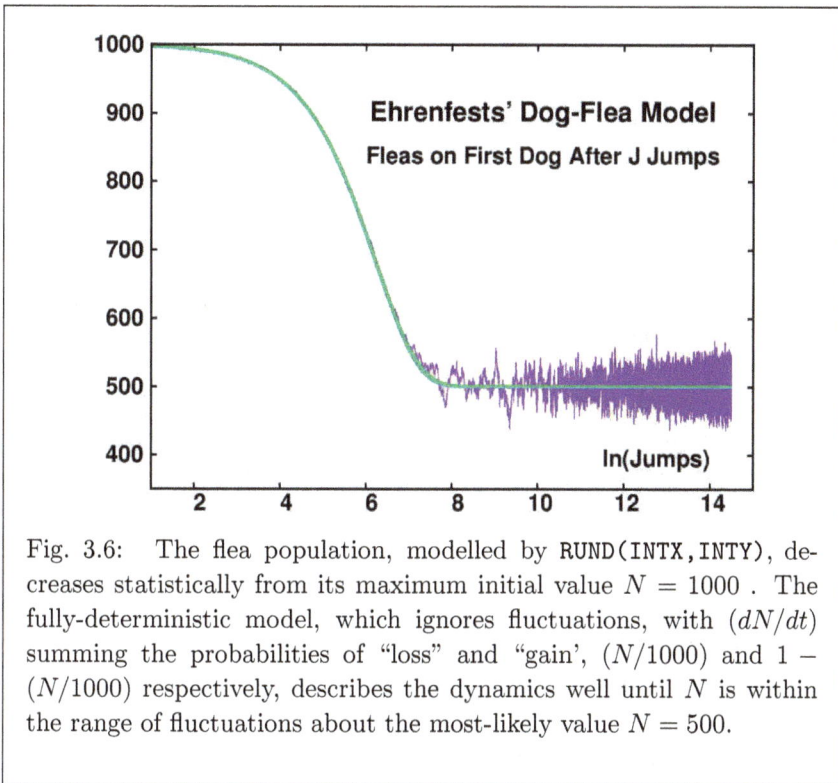

Fig. 3.6: The flea population, modelled by RUND(INTX,INTY), de-creases statistically from its maximum initial value $N = 1000$. The fully-deterministic model, which ignores fluctuations, with (dN/dt) summing the probabilities of "loss" and "gain', $(N/1000)$ and $1 - (N/1000)$ respectively, describes the dynamics well until N is within the range of fluctuations about the most-likely value $N = 500$.

Each of the two seeds for the generator can be set from 0 to $2047 = 2^{11} - 1$. Notice that a simulation with 2^{22} fleas would be a total failure. As the generator would choose each flea once, at the end of 2^{22} jumps all the fleas would be on the other dog. More elaborate failures could be constructed by using $2^{21}, 2^{20}, \ldots$ fleas. Hiding the recurrence of the random number generator by choosing 1000 fleas with 2500 jumps is a modest goal. To achieve it with a simple example we choose the number of the next flea to jump to be $1000\mathcal{R} + 1$, where $\mathcal{R}$ is the next pseudorandom number from RUND(INTX,INTY) with initial seeds (intx,inty) $= (0,0)$.

In **Figure 3.6** the number of fleas on the initially flea-laden dog is plotted as a function of the number of jumps. The Figure illustrates the typical $\sqrt{N}$ fluctuations resulting once the populations have become equal. In the present case it is on jump 2230 that the fleas become for the first time equally divided between the two dogs. There is an extensive literature on the distribution of time intervals separating such moments. The gist of the problem (like that of the H Theorem) can be modelled by ignoring fluctuations, treating the flea populations as continuous functions of time :

$$(dN/dt) = 1 - 0.002N \longrightarrow N(t) = 500[\ 1 + e^{(-t/500)}\ ]\ .$$

This model never quite reaches equilibrium but gives a good account of the *approach* to equilibrium, again just like the H Theorem.

The dog-flea model illustrates quite well why Boltzmann's H Theorem merited objections. Although on the average it is always more likely to equalize the populations at $(N/2)$ per dog, the excess probability is vanishingly small (for large N) near the most likely state. Fluctuations of the order of $\sqrt{N}$ seem likely. The Figure shows that excursions of order 50 are commonplace with 1000 fleas. Boltzmann was not bothered by Zermélo's recurrence paradox. With 60 distinguishable fleas and one jump per second it would take a bit over 36 billion years [more than twice any estimates of the age of the Universe] to go through all possible distributions, with no duplications, of the fleas on the dogs. The Moral is that ergodicity is seldom achieved even in simple models so that the statistical approach demanded by a clear-eyed look at Lyapunov instability with many many states, is the only path open to us. We will look at the details of Lyapunov instability in **Lectures 9 and 10**.

3.8   Molecular Dynamics *versus* Gibbs' Statistical Mechanics

Evidently the kinetic temperature, as modelled by the ideal-gas thermometer, is a reliable concept. We can consider its use in molecular dynamics so as to make contact with Gibbs' statistical mechanics. To what extent is Newton's molecular dynamics consistent with Gibbs' statistical mechanics? This natural question was very much on the minds of the men developing molecular dynamics. A harmonic chain is a clearcut counterexample. Such a system, with its N independent "normal modes" has no possibility to attain Gibbs' microcanonical distribution with equal energies in all of its N normal modes. Enrico Fermi expected that adding small quadratic or cubic forces to the linear forces of a harmonic chain would promote equipartition. In his exploratory work at Los Alamos Fermi (1901-1954), along with Pasta and Ulam, found instead no equipartition. They carried out several simulations and summarized their work in a Los Alamos Report "Studies of Nonlinear Problems" written shortly after Fermi's death.[1]

Typically the Newtonian chain dynamics was started from rest, with its initial displacements corresponding to a sine wave. In the ensuing motion only a few of the (typically N=64) normal modes were excited. The chain dynamics would gradually transfer energy among a very few modes before returning closely to its initial condition. Later work showed that even longer runs returned even more closely to the initial condition. These simulations exhibit a common characteristic of isoenergetic Hamitonian systems, a division of the phase space into infinitely many noncommunicating parts. Even today there is no æsthetic modification of Hamiltonian mechanics designed to achieve Gibbs' smooth microcanonical distribution. Fermi, Pasta, and Ulam's dynamical simulations were not quasiergodic. They noted this difficulty but did not try to surmount it. Attaining isoenergetic ergodicity remains a tantalizing research goal.

In 1953 Metropolis, the Rosenbluths, and the Tellers published an account of eight of their Monte Carlo simulations of the equation of state of 224 hard disks at expansions from close-packing varying from about 4% to about fourfold.[13] They used a periodic square container. Their values of $(PV/NkT) - 1$ had estimated statistical errors of about 3%. To quote their next-to-last sentence: "There is no indication of a [solid-fluid] phase transition." More careful work would have revealed that the two phases do coexist.

During the next decade the hard-disk Monte Carlo work was continued at Los Alamos by Bill Wood. The main product of that work was his overly-long report, "Monte Carlo Calculations of the Equation of State of Systems of 12 and 48 Hard Circles".[14] The final section of Wood's summary, pages 306-307 is entitled "Is it worthwhile ?" Evidently boring, the Monte Carlo work was automatically quasiergodic by design as the procedure of moving particles "randomly" could eventually reach any state. Wood's view of a phase transition was "possibly indicative" [page 3], as well as "tentative" [page 36], "not conclusive" "possible artifact" [page 298]. He exuded "skepticism" [page 306], even after seeing Alder and Wainwright's van der Waals' loop for disks.

Molecular dynamics was different : on a collisional timescale the sequence of states made sense and followed Newton, though Newton may not have thought of periodic boundary conditions. The summed-up velocity was a constant of the motion, with zero the only reasonable choice. This correlation of velocities increases the collision rate by terms of order $(1/N)$. There is a second finite-system effect which compensates for this. A disk moving to the right encounters a leftmoving environment with a number density $(N - 1)/V$ rather than (N/V) so that the two effects cancel and the number-dependence of molecular dynamics is reduced. The effect of the velocity constraint $\sum v \equiv 0$ is to increase the nonideal part of the pressure by an additional contribution of order $(1/N)$.

Berni Alder and I carried out a detailed comparison of the molecular dynamics pressures with $N = 12$ disks[15] to Wood's corresponding Monte Carlo results.[14] Within the statistical fluctuations of the data we found quantitative agreement with the pressure correction derived from simple kinetic theory. There is more. The dynamical pressure for disks and spheres can be determined in either one of two ways, from the collision rate or from the Virial Theorem. The first simply counts collisions while the second averages the collisional momentum transfer. I worked out the $(1/N)$ effects for the difference between the two approaches for hard disks to compare to Berni and Tom's hard-disk data. The kinetic-theory corrections made all three routes to the pressure intelligibly consistent. We established that Gibbs' ensembles and Newton's dynamics agreed quantitatively once the lack of center-of-mass motion was properly taken into account. A more thorough repeat of this work for an even smaller system, say $N = 4$, with today's much faster machines, would be both easy to do and useful.

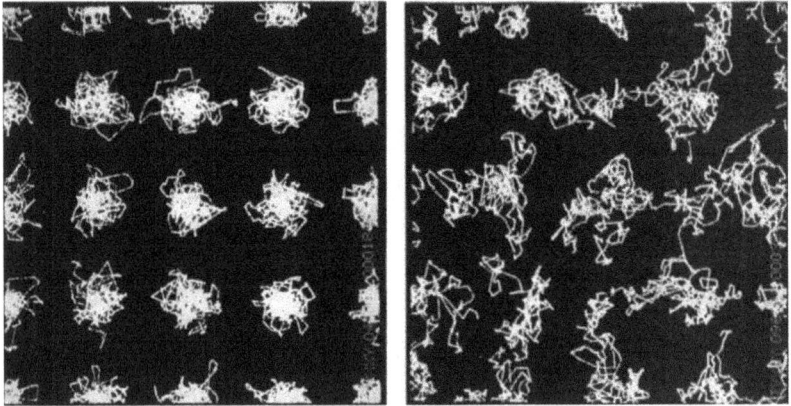

Fig. 3.7: Typical 32-particle traces of the solid and fluid phases for periodic hard spheres from Berni Alder and Tom Wainwright's Scientific American simulations in the 1950s.[16] Their simulations provided realistic views of fluid structure and led to the development of successful perturbation theories of liquids based on hard-sphere models.

By 1956 Alder and Wainwright were able to report on their successful implementation of Newton's Laws for disks and spheres on the state-of-the-art computers at Livermore. See **Figure 3.7** for typical time exposures of solid and fluid trajectories. Their dynamical simulations made it possible to evaluate transport coefficients from the decay of equilibrium fluctuations and to check on the accuracy of Boltzmann's H theorem in quantifying the "approach to equilibrium". Apart from fluctuations the Theorem passed this test.

The quantitative agreement between the dynamical and statistical determinations of pressure was the first to show that Gibbs' distribution can be achieved with Hamiltonian dynamics. The number dependence of small systems' properties agreed with predictions based on Gibbs' ergodic ensembles. Pair distribution functions from repulsive force theories and simulations formed the basis for perturbation theories similar to van der Waals' picture of liquids. This synergy of computer experiments with statistical mechanics was a great success. By the 1970s the pre-computer mysteries of equilibrium "simple liquids" had been put to rest. Two-, or even three-phase, systems could be simulated quickly and accurately. The properties of systems with somewhat different forces could then be understood with perturbation theory, treating the difference in forces as the perturbation. With the confirmation that the Monte Carlo and molecular dynamics results were consistent at constant energy it was natural to seek out a dynamics which could reproduce Gibbs' canonical-ensemble results, an isothermal molecular dynamics. A Japanese post-doctoral fellow working in Canada took a giant step toward solving that problem.

3.9 Nosé, Nosé-Hoover, and Dettmann-Morriss Mechanics

In 1984 Shuichi Nosé invented a temperature-dependent Hamiltonian with *scaled* momenta $\{ (p/s) \}$.[5,6] His dimensionless scale factor s could vary with time so as to provide a canonical distribution for both the scaled kinetic energy and the configurational energy. To simplify notation we consider the simplest possible interesting application, a harmonic oscillator with unit mass and force constant. We also set Boltzmann's constant and the temperature equal to unity in what follows. The reader should have no problem generalizing these one-particle ideas to a manybody system of N mass points. There is a little more work to be done in discussing polyatomic situations with vibrational and rotational degrees of freedom. The many references to Nosé's ideas have explored these generalizations.

See Davidchack's arχiv 1412.7067 paper mentioned in **Lecture 1.**[17]

Consider Nosé's Hamiltonian with just $\# = 2$ degrees of freedom, the oscillator coordinate-momentum pair (q, p) and the additional "time-scaling" pair (s, ζ) :

$$2\mathcal{H}_N = (p/s)^2 + q^2 + \# \ln(s^2) + \zeta^2 .$$

Because the canonical distribution includes states with all energies, $0 < \mathcal{H} < \infty$, or even $-\infty < \mathcal{H} < +\infty$, Nosé had the idea to include a scale factor s making it possible for the scaled momentum (p/s) to cover the whole range $-\infty < p < +\infty$. The equations of motion from his Hamiltonian are :

$$\{ \dot{q} = (p/s^2) ; \ \dot{p} = -q ; \ \dot{s} = \zeta ; \ \dot{\zeta} = [(p^2/s^3) - (2/s)] \} ,$$

Let us rewrite them in terms of his scaled momentum (p/s) for which we will still use the notation p :

$$\{ \dot{q} = (p/s) ; \ \dot{p} = -(q/s) - (\zeta p/s) ; \ \dot{s} = \zeta ; \ \dot{\zeta} = (p^2/s) - (2/s) \} .$$

In order that the oscillator move at the scaled momentum p rather than (p/s) it is tempting to "scale the time", multiplying all of these rates by s. Nosé did so, getting the new set of time-scaled equations :

$$\{ \dot{q} = p ; \ \dot{p} = -q - \zeta p ; \ \dot{s} = s\zeta ; \ \dot{\zeta} = p^2 - 2 \} .$$

By using judicious initial conditions s can be confined to the range $0 < s < 1$. Notice that there is now no need to solve the motion equation for s. The $\{q, p, \zeta\}$ furnish a complete description of the motion. Given the value of the Hamiltonian s can be computed directly whenever the momentum vanishes :

$$2\mathcal{H}_N \overset{p=0}{=} q^2 + \zeta^2 + 2\ln s^2 \overset{p=0}{\longrightarrow} s^2 = e^{\mathcal{H}} e^{-q^2/2} e^{-\zeta^2/2} .$$

It is interesting to notice that Nosé's time scaling leaves the four-dimensional (q, p, s, ζ) trajectory *unchanged* but simply changes the *rate* of motion of the system point along the trajectory. By keeping track of s one can convert back and forth between the unscaled and scaled trajectories but in fact this is neither necessary nor desirable. We will see that the scaled trajectories nearly satisfy the canonical distribution, though with some interesting caveats.

Two simple examples of time scaling should help clarify the concept. The simplest is just changing the definition of rates "per unit time". If the rate at which daily newspapers pile up is $1/\text{day} = 7/\text{week} = 365/\text{year}$, we can express time scaling as follows :

$$\dot{N}_d = 1 ; \ \dot{N}_w = 7 ; \ \dot{N}_y = 365 .$$

To return to mechanics we can halve the speed and reduce the acceleration of a harmonic oscillator fourfold with the Hamiltonian

$$\mathcal{H} = (q^2/2) + (p^2/2) - 1 = 0 \overset{scaling}{\longrightarrow} \mathcal{H} = (1/2)[\ (q^2/2) + (p^2/2) - 1 = 0\ ] \ ,$$

by doubling the timescale or halving the frequency. The original solutions,

$$q = \sin(t) \ ; \ p = \cos(t) \rightarrow$$

each proceed at half the rate when we divide the Hamiltonian by 2 :

$$\{\ \dot{q} = +(p/2) \ ; \ \dot{p} = -(q/2) \rightarrow q = \sin(t/2) \ ; \ p = \cos(t/2)\ \} \ .$$

Nosé's idea is conceptually more difficult in that the time scaling varies along the orbit in a way dictated by the orbit. Fortunately this feedback-directed time-scaling is not essential to the notion of thermostatting. In fact we can avoid it altogether by proceeding from Liouville's Theorem rather than from scaled Hamiltonian mechanics. Let us do so.

3.9.1 *Nosé-Hoover Mechanics from Liouville's Theorem*

Remember, thinking back to Gibbs, that the flow through (q, p) phase space satisfies a continuity equation, with the physical meaning being conservation of probability, and thereby entropy. The rate at which a phase-space flow changes the local probability at a fixed (qp) location is given by a continuity equation :

$$(\partial f/\partial t) = -(\partial f \dot{q}/\partial q) - (\partial f \dot{p}/\partial p) \ .$$

Where $\dot{q}$ and $\dot{p}$ satisfy Hamilton's motion equations :

$$(\partial \dot{q}/\partial q) + (\partial \dot{p}/\partial p) = (\partial^2 \mathcal{H}/\partial p \partial q) - (\partial^2 \mathcal{H}/\partial q \partial p) = 0 \ .$$

The result of the last three lines gives the *comoving* change in density *along the trajectory* : $\dot{f} = (\partial f/\partial t) + \dot{q}(\partial f/\partial q) + \dot{p}(\partial f/\partial p) \equiv 0$. These ideas motivated my derivation of the Nosé-Hoover analog of Nosé's motion equations.[18] Introducing this deterministic nonHamiltonian thermostatting provides trajectories consistent with Gibbs' canonical distribution.

This simpler path to constant-temperature canonical-ensemble motion equations follows from an augmented set of motion equations including the friction coefficient ζ :

$$\dot{q} = p \ ; \ \dot{p} = F - \zeta p \ ; \ \dot{\zeta} = [\ (p^2/T) - 1)\ ] \ .$$

Applying the phase-space continuity equation shows that a Gaussian distribution for the friction coefficient is consistent with the canonical distribution function :

$$f \propto e^{-q^2/2T} e^{-p^2/2T} e^{-\zeta^2/2} \; ; \; (\partial f/\partial t) =$$

$$-f[\,(\partial \dot{q}/\partial q) + (\partial \dot{p}/\partial p) + (\partial \dot{\zeta}/\partial \zeta)\,] - \dot{q}(\partial f/\partial q) - \dot{p}(\partial f/\partial p) - \dot{\zeta}(\partial f/\partial \zeta) =$$

$$-f[\,0 - \zeta + 0\,] - f[\,-(pq/T) + (-q - \zeta p)(-p/T) + (\,(p^2/T) - 1\,)(-\zeta)\,] \equiv 0 \,.$$

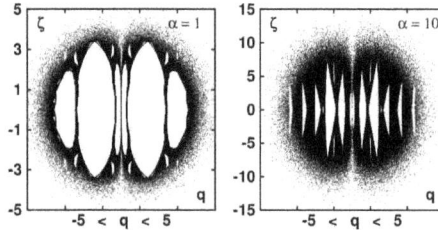

Fig. 3.8: $\zeta(q)$ cross sections for two billion fourth-order Runge-Kutta timesteps using $dt = 0.001$. The Nosé-Hoover friction coefficient evolves according to $\dot{\zeta} = \alpha(p^2 - 1)$. The number of points in the two sections are 503,158 for $\alpha - 1$ and 597,183 for the stiffer choice $\alpha = 10$. Both simulations have initial conditions $(q,p,\zeta) = (0,5,0)$.

The Nosé-Hoover oscillator with an initial condition of $(q,p,\zeta) = (2.4,0,0)$ is 50 percent greater in efficiency than the Nosé oscillator, judged by the number of penetrations of the $p = 0$ plane. **Figure 3.8** shows the cross sections from a billion adaptive Runge-Kutta timesteps where the timestep is doubled whenever the error is less than 10^{-14} and halved whenever the error is greater than 10^{-12} . The "error" is $\sqrt{\delta_q^2 + \delta_p^2 + \delta_\zeta^2}$.

The Nosé and Nosé-Hoover oscillator Lyapunov exponents in the chaotic sea can be estimated by evolving δ_q, or δ_p, or δ_ζ . In the initial conditions $(2.4,0,0)$ a tiny perturbation, 10^{-15} for double precision and 10^{-30} for quadruple, was added to q or p or ζ. The displacements are plotted as functions of time in **Figure 3.9**, using RK4, and **Figure 3.10**, using RK5. Because the fluctuations in the local exponents are large the slopes of the curves in the figures are not likely to reflect long-time averages. The data from the two integrators suggest a Lyapunov exponent around 0.02 while the *long-time* average is 0.0139, using the methods I describe in **Lecture 10**.

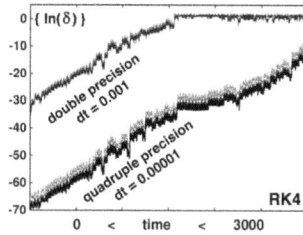

Fig. 3.9: Three curves [logarithm of displacement versus time for $(\delta_q, \delta_p, \delta_\zeta)$] are plotted for double and quadruple precision using RK4.

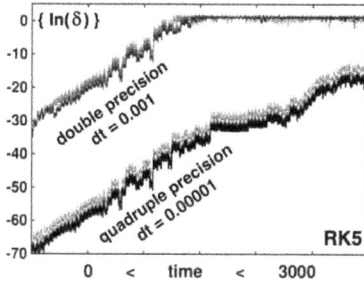

Fig. 3.10: Three curves [logarithm of displacement versus time for $(\delta_q, \delta_p, \delta_\zeta)$] are plotted for double and quadruple precision using RK5.

Exercise : To demonstrate the value of adaptive integrators compare fourth-order Runge-Kutta integrators with [1] a fixed timestep and [2] an adaptive timestep, as just described in the text, for the stiff Nosé (*not* Nosé-Hoover) oscillator motion equations :

$$\dot{q} = (p/s^2) \; ; \; \dot{p} = -q \; ; \; \dot{s} = \zeta \; ; \; \dot{\zeta} = 100[\, (p^2/s^3) - (2/s)\,] \; .$$

Use the chaotic initial condition $(q, p, s, \zeta) = (0, 5, 1, 0)$.

3.9.2 *Nosé-Hoover Dynamics from Dettmann-Morriss*

In Section 1.7.2 on page 22 of **Lecture 1** we reviewed Dettmann and Morriss' simplification of Nosé's time-scaling operation. They multiplied his
Hamiltonian by s and set the result equal to zero, obtaining Nosé-Hoover
dynamics by this two-step process. It is worthwhile to consider a slight
generalization of our Section 1.7 account, including here the temperature
T as well as a thermostat relaxation time τ in the Hamiltonian :

$$2\mathcal{H}_D = sq^2 + (p^2/s) + Ts\ln(s^2) + (s\xi^2/T\tau^2) \equiv 0 \longrightarrow$$

$$\{\ \dot{q} = (p/s)\ ;\ \dot{p} = -sq\ ;\ \dot{s} = s(\xi/T\tau^2)\ ;\ \dot{\xi} = (p/s)^2 - T\ \}\ .$$

Now, introducing the change of variable $(p/s) \longrightarrow p$ gives

$$\{\ \dot{q} = p\ ;\ \dot{p} = -q - p(\xi/T\tau^2)\ ;\ \dot{\xi} = p^2 - T\ \}\ .$$

Then, replacing $(\xi/T\tau^2) \longrightarrow \zeta$ gives

$$\{\ \dot{q} = p\ ;\ \dot{p} = -q - \zeta p\ ;\ \dot{\zeta} = (p^2 - T)/(T\tau^2) = [\ (p^2/T) - 1\ ]/\tau^2\ \}\ .$$

The distribution function consistent with this generalized Hamiltonian is

$$f(q,p,\zeta,\mathcal{H},T) = (2\pi)^{-3/2}(\tau/T)e^{-q^2/2T}e^{-p^2/2T}e^{-\zeta^2\tau^2/2}\ .$$

Notice that the friction coefficient has the units of frequency and that the
product $\zeta\tau$ is dimensionless. With the possibility of comparing double- and
quadruple-precision simulations, with and without variable timesteps, we
have the means to calculate accurate solutions of these oscillator models.
A pedagogical account is given in Reference 19.

3.10 Wang-Yang Knotted Nosé-Hoover Oscillator Solutions

In 2015 Lei Wang and Xiao-Song Yang considered somewhat stiffer versions
of the Nosé-Hoover friction coefficient[20] :

$$\{\ \dot{q} = p\ ;\ \dot{p} = -q - \zeta p\ ;\ \dot{\zeta} = (p^2 - 1)/\tau^2 = \alpha(p^2 - 1)\ \}\ ;\ \alpha = (1/\tau)^2 = 10\ .$$

In the simplest version of the Nosé-Hoover oscillator with $\alpha = 1$ about
six percent of the Gaussian measure is chaotic.[21] With Wang and Yang's
larger value, $\alpha = 10$, about 54 percent of the probability density lies in the
chaotic sea. Their paper shows a wide variety of solutions including six
independent tori, obtained using different initial conditions. By plotting
the $(6 \times 5)/2 = 15$ pairs of these topologically distinct tori Wang and Yang
were able to show that they are all "knotted". In addition to these two-tori
knots they also found examples of a trefoil knot (the simplest knot that can

be supported with a continuous rope). Wang and Yang's initial conditions provide the torus shown in **Figure 3.11** side-by-side with a more symmetric version of the trefoil.

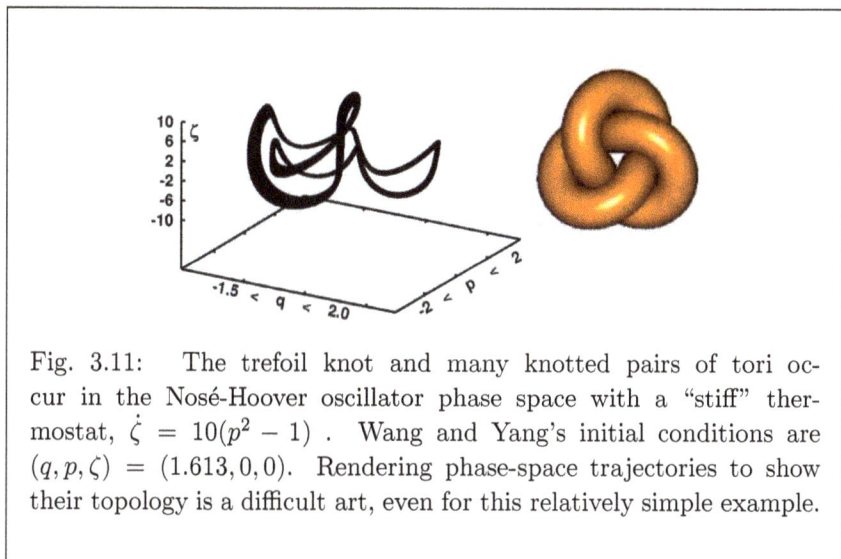

Fig. 3.11: The trefoil knot and many knotted pairs of tori occur in the Nosé-Hoover oscillator phase space with a "stiff" thermostat, $\dot{\zeta} = 10(p^2 - 1)$. Wang and Yang's initial conditions are $(q, p, \zeta) = (1.613, 0, 0)$. Rendering phase-space trajectories to show their topology is a difficult art, even for this relatively simple example.

3.11 Isokinetic Dynamics, Gauss' and Hoover-Leete

Another approach to "constant-temperature" dynamics is to make the kinetic energy a constant of the motion. This idea misses out on canonical temperature fluctuations and cannot be applied to the one-dimensional oscillator but has proved useful in cell-model simulations and in theoretical investigations. A two-dimensional problem can serve us as a demonstration test bed. We place a point mass at the origin of a 2×2 square with $-1 < x, y < +1$ with velocity (0.6,0.8). See again **Figure 2.16** on page 73. There are scatterers at the vertices of the square which repel the moving particle when it comes within unit distance with a potential $\phi = (1 - r^2)^4$.

Two ways to constrain the kinetic temperature of the moving particle come to mind. Gauss' Principle of Least Constraint directs that a constraint force should be as small as possible in a mean-squared sense: $\sum F_c^2$ should be minimized. A variational calculation shows that the force is linear in the velocities :

$$\dot{p}_x = F_x - \zeta p_x \ ; \ \dot{p}_y = F_y - \zeta p_y \ ; \ \zeta = \sum F \cdot p / \sum p^2 \ [\text{ Gauss }] \ .$$

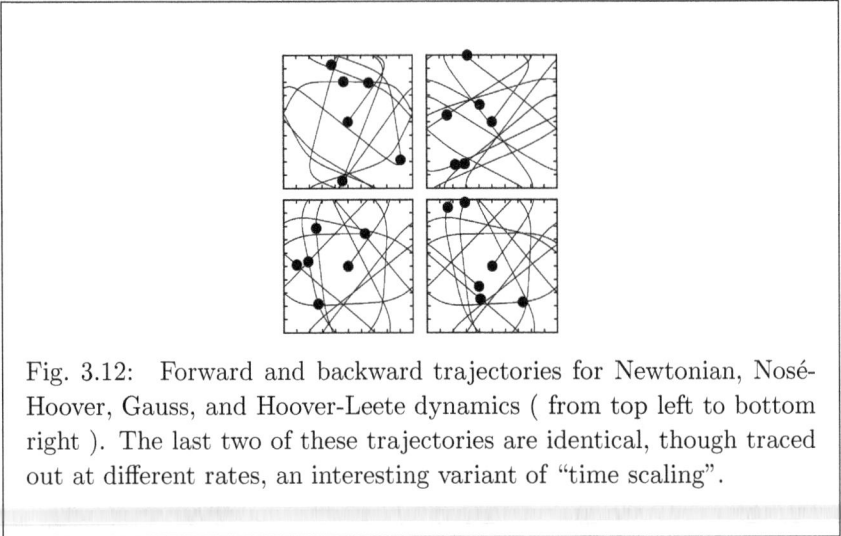

Fig. 3.12: Forward and backward trajectories for Newtonian, Nosé-Hoover, Gauss, and Hoover-Leete dynamics (from top left to bottom right). The last two of these trajectories are identical, though traced out at different rates, an interesting variant of "time scaling".

An alternative is to use a Lagrange multiplier rather than a constraint force. That approach was worked out by Tom Leete and me as an extension of his 1979 Master's thesis work at the University of West Virginia.[22] This approach provides *different* equations of motion as well as a novel Hamiltonian :

$$\mathcal{H}_{HL} = 2\sqrt{K_0 K_p} + \Phi - K_0 \text{ where}$$

$$K_0 = \sum(\dot{q}^2/2) \ ; \ K_p = \sum(p^2/2) \ [\text{ Hoover} - \text{Leete }] \ .$$

$$\{ \ \dot{q} = p\sqrt{K_0/K_p} \ ; \ \dot{p} = F \ \} \ .$$

Gauss' mechanics constrains $\sum(p^2/m)$ while Hoover-Leete mechanics constrains $\sum(m\dot{q}^2)$.

Figure 3.12 compares the trajectories of a moving mass (initially at the center of the square) to a time of 25. The filled circles indicate times that are multiples of 5. At the final time of 25 the sign of dt in the Runge-Kutta integrator was changed and the trajectories returned to the origin with negligible error, demonstrating the reversibility of all four forms of dynamics : classical Newtonian, Nosé-Hoover, Gauss, and Hoover-Leete.

Notice that the last two of these produce exactly the same trajectory but at different rates ! Though this is not generally true, the velocity scaling of the momentum for a single particle is equivalent to using a Lagrange multiplier. In a more complicated manybody problem the scalings of the various particles' momenta are different.

Exercise : Formulate and solve a problem showing that the Gauss and Hoover-Leete motion equations keep the kinetic energy constant.

3.12   Summary of Lecture 3

Molecular dynamics and Monte Carlo provide equilibrium solutions of the manybody problem, with molecular dynamics able to provide nonequilibrium simulations quantifying the approach to equilibrium. Thermodynamics introduces a new variable into dynamics, the ideal-gas Temperature. Boltzmann's treatment of kinetic theory shows that such an ideal-gas thermometer maintains a Maxwell-Boltzmann distribution making it possible to define nonequilibrium temperatures. Boltzmann's work also provides a route to the transport coefficients for dilute gases. The monotonic increase of his entropy, for isolated systems, ignores the effect of fluctuations. Nosé's introduction of a time-reversible friction coefficient controlling energy feedback provides a dynamics consistent with Gibbs' canonical ensemble and with chaotic fluctuations. The harmonic oscillator shows that this approach is not necessarily ergodic and opens up many interesting problems in nonlinear dynamics.

References

1. E. Fermi, J. Pasta, and S. Ulam, "Studies of Nonlinear Problems", Los Alamos Scientific Laboratory Report LA-1940 (1955).
2. B. J. Alder and T. E. Wainwright, "Molecular Dynamics by Electronic Computers", Proceedings of the International Union of Pure and Applied Physics Symposium on the *Statistical Mechanical Theory of Transport Properties* (Brussels, 1956) (Interscience Publishers, New York, 1958).
3. J. B. Gibson, A. N. Goland, M. Milgram, and G. H. Vineyard, "Dynamics of Radiation Damage", Physical Review **120**, 1229-1253 (1960).
4. A. Rahman, "Correlations in the Motion of Atoms in Liquid Argon", Physical Review A **136**, 405-411 (1964).
5. S. Nosé, "A Molecular Dynamics Method for Simulations in the Canonical Ensemble", Molecular Physics **52**, 255-268 (1984).

6. S. Nosé, "A Unified Formulation of the Constant Temperature Molecular Dynamics Methods", The Journal of Chemical Physics **81**, 511-519 (1984).
7. L. Boltzmann, *Lectures on Gas Theory* (1896, in German), English translation by Stephen G. Brush (University of California Press, Berkeley, 1964).
8. Wm. G. Hoover, B. L. Holian, and H. A. Posch, "Comment I on 'Possible Experiment to Check the Reality of a Nonequilibrium Temperature' ", Physical Review E **48**, 3196-3198 (1993).
9. Wm. G. Hoover and C. G. Hoover, "Nonequilibrium Temperature and Thermometry in Heat-Conducting ϕ^4 Models", Physical Review E **77**, 041104 (2008).
10. G. A. Bird, *The DSMC Method* (CreateSpace Amazon.com, 2013).
11. S. Chapman and T. G. Cowling, *The Mathematical Theory of Nonuniform Gases* (MacMillan, New York, 1940).
12. Wm. G. Hoover and C. G. Hoover, "Bit-Reversible Version of Milne's Fourth-Order Time-Reversible Integrator for Molecular Dynamics", Computational Methods in Science and Technology **23**, 299-303 (2017).
13. N. Metropolis, A. W. Rosenbluth, M. N. Rosenbluth, A. H. Teller, and E. Teller, "Equation of State Calculations by Fast Computing Machines", The Journal of Chemical Physics **21**, 1087-1092 (1953).
14. W. W. Wood, "Monte Carlo Calculation of the Equation of State of Systems of 12 and 48 Hard Circles", Los Alamos Scientific Laboratory Report LA-2827 (1963).
15. Wm. G. Hoover and B. J. Alder, "Studies in Molecular Dynamics IV. The Pressure, Collision Rate, and Their Number Dependence for Hard Disks", The Journal of Chemical Physics **16**, 686-691 (1967).
16. B. J. Alder and T. E. Wainwright, "Molecular Motions", Scientific American **202(4)**, 113-126 (1959).
17. R. L. Davidchack, "Discretization Errors in Molecular Dynamics Simulations with Deterministic and Stochastic Thermostats", Journal of Computational Physics **229**, 9323-9346 (2010) = arχiv 1412.7067.
18. Wm. G. Hoover, "Canonical Dynamics: Equilibrium Phase-Space Distributions", Physical Review A **31**, 1695-1697 (1985).
19. Wm. G. Hoover, J. C. Sprott, and C. G. Hoover, "Adaptive Runge-Kutta Integration for Stiff Systems: Comparing Nosé and Nosé-Hoover Dynamics for the Harmonic Oscillator", American Journal of Physics **84**, 786-794 (2016).
20. L. Wang and X-S. Yang, "The Invariant Knot Type and the Interlinked Invariant Tori in the Nosé-Hoover Oscillator", The European Physical Journal B **88**, 78 (2015).
21. P. K. Patra, Wm. G. Hoover, and C. G. Hoover, "The Equivalence of Dissipation from Gibbs' Entropy Production with Phase-Volume Loss in Ergodic Heat-Conducting Oscillators", International Journal of Bifurcation and Chaos **26**, 1650089 (2016).
22. Wm. G. Hoover, "Atomistic Nonequilibrium Computer Simulations", Physica **118A**, 111-122 (1983).

Chapter 4

Simple Systems with Thermal Constraints

/ The Lucy Fluid, Prototypical Simple System / Free Expansion / Shock-wave Temperature / Simulation with Newtonian Mechanics / Simulation with Gauss' Mechanics / Simulation with Hoover-Leete Mechanics / Simulation with Nosé-Hoover Mechanics / Multi-Moment Thermostatted Oscillators / The ϕ^4 Model for Heat Conduction / Harmonic Chain Dynamics / Bit-Reversible Levesque-Verlet Dynamics / Summary of Lecture 4 /

4.1   The Lucy Fluid, a Prototypical Simple System

Now that we have the numerical skills necessary to solve differential equations of motion as well as those needed to include thermal constraints in the equations, we can choose a particle model to which these ideas can be applied. For numerical and conceptual simplicity our fluid model has very smooth pairwise forces and lives in a periodic two-dimensional box. The fluid is the "Lucy fluid", named after one of the originators of Smooth Particle Applied Mechanics,[1] a Professor at Cambridge University. His interest was stellar dynamics, like molecular dynamics but on a larger scale. He introduced a "weight function" $w(r < h)$, smooth at the origin and cutoff $(r = h)$ and with a spatial integral of unity. We will use Lucy's weight function as our pair-potential function :

$$\phi(r < h) = w(r < h) = (5/\pi h^2)[\, 1 - 6(r/h)^2 + 8(r/h)^3 - 3(r/h)^4 \,] \,.$$

If we multiply by $2\pi r$ and integrate from 0 to h we can confirm that the weight function is normalized so that it can be used for averaging as well as for moving particles about. The force associated with the potential is

$$-\nabla\phi = F(r < h) = (60/\pi h^3)[\, (r/h) - 2(r/h)^2 + (r/h)^3 \,] \,.$$

Both the force and its derivative vanish at $r = h$, eliminating the usual trajectory errors that come from discontinuities in the equations of motion.

We will take up the simulation of fluid motion using Lucy's pair potential with four types of mechanics : Newtonian, Gauss, Hoover-Leete, and Nosé-Hoover. **Figure 4.1** illustrates the potential, its derivative, and its one-dimensional spatial integral. With current computers code development takes a few hours and hundred-particle simulations just a few minutes. There is no problem doing all of this in three spatial dimensions but we shall nearly always stick to two in order to save time and to simplify the graphical display of our results. To begin we consider some applications of the Lucy fluid which can serve as benchmarks in code development.

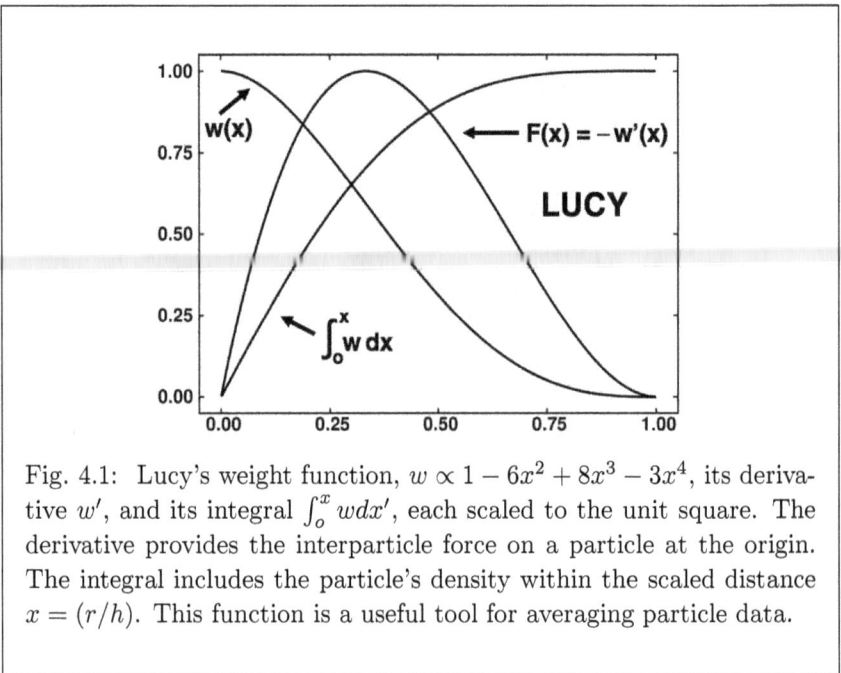

Fig. 4.1: Lucy's weight function, $w \propto 1 - 6x^2 + 8x^3 - 3x^4$, its derivative w', and its integral $\int_o^x w\,dx'$, each scaled to the unit square. The derivative provides the interparticle force on a particle at the origin. The integral includes the particle's density within the scaled distance $x = (r/h)$. This function is a useful tool for averaging particle data.

4.1.1 *Free Expansion with Periodic Boundary Conditions*

In 1999 Harald Posch and I decided to study the free expansion problem confined to a periodic box.[2] To a good approximation the atomistic dynamics of a Lucy Fluid resembles the fluid dynamics of an ideal gas so we used that model. We considered systems ranging in size from 16×16 with 256 particles initially occupying the central 8×8 square up to 256×256 with 65,536 particles expanding from a 128×128 square to fill up a 256×256

square. We were interested in seeing if we could capture the difference between the entropy according to Liouville's Theorem (where there is no change at all) and the "coarse-grained" entropy from the thermodynamic or Boltzmann-Equation point of view. In the thermodynamic case there is an increase of $Nk \ln(4)$ corresponding to the fourfold volume expansion at constant energy.

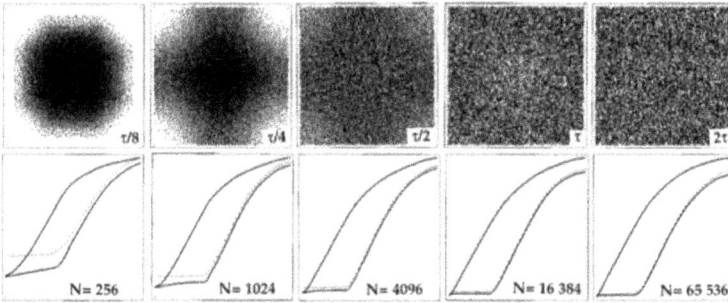

Fig. 4.2: At the top we have snapshots from the fourfold free expansion of 16,384 Lucy particles. τ is the sound traversal time of the 128×128 periodic box. The entropy change for half a sound traversal time is shown below for 256 to 65,536 particles. The upper curves show the erroneous result of including the mean velocity in the thermal entropy. The dashed curves show the result of a grid-based entropy calculation using Lucy's weight function. See Reference 2 for details.

To calculate entropy we used the ideal-gas result, $(S/Nk) = \ln(T/\rho)$. The density ρ_r at any point r can be calculated by summing the contribution of nearby particles' weight functions. The mean velocity v_r at any point is calculated by a weighted average of the $\{ v_i \}$ and the temperature is calculated by a weighted average of the velocity fluctuations:

$$\rho_r \equiv \sum_i^N w(r - r_i) \ ; \ v_r \equiv \sum_i^N w(r - r_i)v_i / \sum_i^N w(r - r_i) \ ;$$

$$2T_r \equiv [\sum_i^N w(r - r_i)v_i^2 / \sum_i^N w(r - r_i)] - [\sum_i^N w(r - r_i)v_i / \sum_i^N w(r - r_i)]^2 \ .$$

Here we are following our typical practice of setting the particle mass and Boltzmann's constant equal to unity to simplify the notation.

The need for subtracting the local velocity in calculating the fluctuations is quite clear in these simulations. Without subtraction the thermodynamic or Boltzmann entropy begins to rise immediately. *With* subtraction the entropy increase occurs later, when the expanding fluid (which moves at close to the sound velocity) collides with itself at the periodic box boundaries.

Figure 4.2 shows the evolution of the expanding fluid and its entropy. The increase takes little time, just two sound traversal times, much faster than the diffusive processes associated with viscosity and heat transfer.

4.1.2 *Periodic Heat Flow and Shear Flow in a Lucy Fluid*

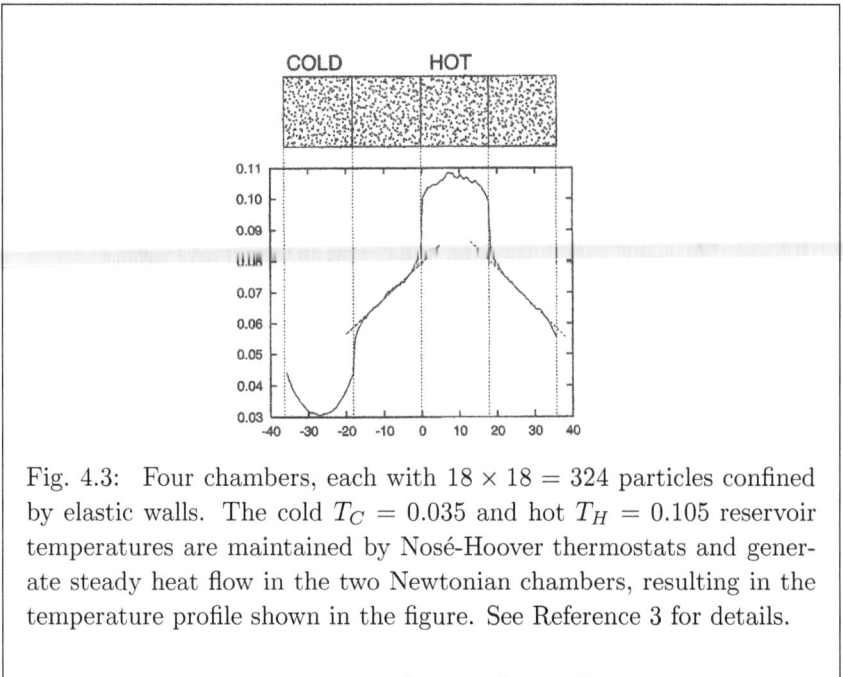

Fig. 4.3: Four chambers, each with $18 \times 18 = 324$ particles confined by elastic walls. The cold $T_C = 0.035$ and hot $T_H = 0.105$ reservoir temperatures are maintained by Nosé-Hoover thermostats and generate steady heat flow in the two Newtonian chambers, resulting in the temperature profile shown in the figure. See Reference 3 for details.

In 1996 Harald Posch and I set out to measure the Lucy Fluid heat conductivity[3] as a step in understanding Rayleigh-Bénard convective flow in the Lucy Fluid. **Figures 4.3 and 4.4** show the geometry of the two heat-flow problems. Applying cold and hot thermostats to the particles in two of the four square regions soon results in a nice linear temperature profile. The resulting temperatures are determined from the mean-squared velocities. Fourier's Law for heat transfer, $Q_x = -\kappa(dT/dx)$, where Q is the heat flux vector (flow of energy per unit "area" and time) and κ is the "thermal" (or "heat") conductivity provides κ. A similar setup,[4] with particles thermostatted relative to an upward velocity in one box and a downward one in another provides values of the shear viscosity coefficient η. For a Newtonian fluid $P_{xy} = -\eta(dv_y/dx)$ defines the shear viscosity, likewise required for the modelling of Rayleigh-Bénard flow.

The Rayleigh-Bénard flow in **Figure 4.4** was a part of Oyeon Kum's Ph D work at UCDavis. It illustrates simultaneous heat flow and shear flow. Three rows of "cold" particles are tethered at the top of the box with three rows of "hot" particles tethered at the bottom. For sufficiently low values of the viscosity and conductivity convection results. The example in the Figure corresponds to a regular stationary flow.

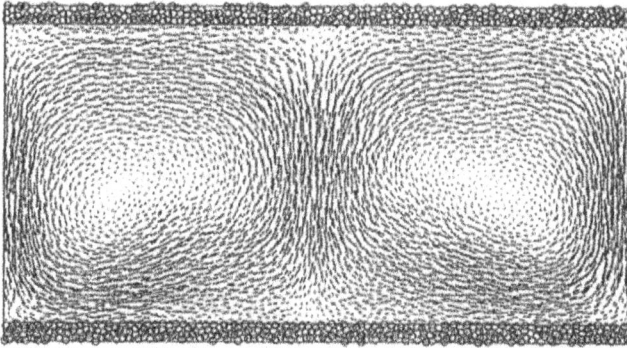

Fig. 4.4: A Rayleigh-Bénard convection simulation using 5000 Lucy particles with periodic vertical boundaries. Three rows of particles at the bottom are maintained at three times the temperature of the three rows at the top. This particle simulation is in good agreement with corresponding smooth-particle simulations. The simulation shown here corresponds to a Rayleigh Number of 40,000. The Rayleigh Number is proportional to the area of the system. Convection sets in at Rayleigh numbers greater than about 1700. See Reference 4 for details.

4.1.3 *Temperature Maximum in a Stationary Shockwave*

In a 1997 simulation Oyeon, Carol, and I carried out a shockwave simulation for the Lucy Fluid.[5] In **Figure 4.5** we see cold fluid entering at the left with a speed 1.35. That flow is imposed by moving a square lattice of cold particles at constant speed to the right until they are beyond a distance equal to the range of the forces from the left boundary where particles enter. As particles near the right boundary, coming within the range of the forces from the boundary, they are assigned a longitudinal velocity of 0.90 and are discarded once they reach the edge of the system. Conservation of mass flux (density multiplied by velocity) dictates that the density at the right boundary must be 50% greater than that at the entrance on the left.

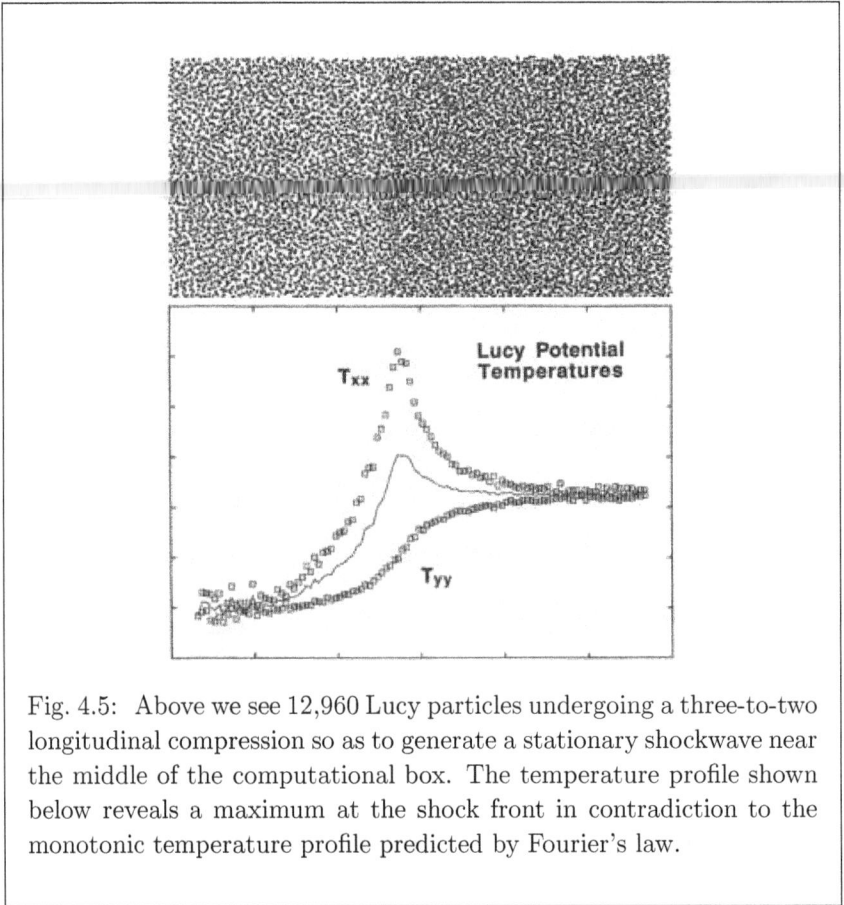

Fig. 4.5: Above we see 12,960 Lucy particles undergoing a three-to-two longitudinal compression so as to generate a stationary shockwave near the middle of the computational box. The temperature profile shown below reveals a maximum at the shock front in contradiction to the monotonic temperature profile predicted by Fourier's law.

This is in fact observed. A shockwave, a steady transformation of velocity, density, pressure, energy, and temperature, soon results and can be seen in the center of the system. The profiles of the density and pressure and energy show a relatively gradual rise from the cold entrance values to the hot exiting values. The temperature here is specially interesting as it is a tensor rather than a scalar. Because the flow of heat is definitely from right to left in the Figure it is clear that nonlinear heat transport is an interesting research area. These example problems, worked out with a simple model force law demonstrate our abilities to solve nonequilibrium problems with molecular dynamics. Let us now turn to the details of the thermostatting, the boundary conditions, and the interpretation of nonequilibrium results.

4.2   Lucy Fluid Simulations for Four Varieties of Mechanics

In **Lecture 3** we introduced three new varieties of mechanics which control temperature, Gauss' and Hoover-Leete, where the kinetic temperature is constant, and Nosé-Hoover, which follows Gibbs' canonical distribution of the energy. To develop our techniques let us compare these three with Newtonian mechanics, all applied to the same thermodynamic system, 100 Lucy particles in a periodic box. We adopt fourth-order Runge-Kutta integration with a timestep $dt = 0.01$ in all four cases. In each case we will apply a particular mechanics within a 10×10 box using 100 Lucy particles with $h = 3$. For reproducibility we start with a square lattice (which is typically unstable to shear) by the following construction of the initial conditions :

```
INDEX = 0
DO 30 I = 1,10
DO 30 J = 1,10
INDEX = INDEX + 1
X(INDEX) = I - 5.5d00
Y(INDEX) = J - 5.5d00
30 CONTINUE
```

In this arrangement each particle interacts with 24 neighbors, four each at mean squared distances of 1, 2, 4, and 8 and eight of them at $r^2 = 5$. Because Lucy's potential is a weight function we can calculate the density at each particle's location by adding its "self" contribution $w(0) = (5/\pi h^2) = (5/9\pi) \simeq 0.1768388$ to the summed-up contributions of the 24 neighbors closer than $r = h = 3$.

Notice particularly that the periodic boundaries require care. In the event that $|x_{ij}|$ or $|y_{ij}|$ exceeds 5, half the box length, there is a nearer periodic image of the pair obtained by reaching across the periodic boundary. For each IJ pair of particles the distance RIJ is determined as follows :

```
IF(XIJ.LT.-5) XIJ = XIJ + 10
IF(XIJ.GT.+5) XIJ = XIJ - 10
IF(YIJ.LT.-5) YIJ = YIJ + 10
IF(YIJ.GT.+5) YIJ = YIJ - 10
RIJ = DSQRT(XIJ*XIJ + YIJ*YIJ)
```

and is used to calculate the contribution of that pair to the total potential energy. With this choice of boundary conditions the result of periodicity is that the initial densities are all equal to 1.0029300 so that the total potential energy is about 50.14650. Because, apart from the self contributions, the usual potential energy of the lattice is just the sum of pair potentials, equal to half the densities of the particles, we *define* the initial potential energy for the Lucy fluid as half the density sum,

$$\rho_i \equiv w(0) + \sum w(r_{ij} < 3) \; ;$$

$$\Phi_0 \equiv N(w(0)/2) + \sum_{i<j} \phi(r_{ij} < h) = (1/2) \sum \rho_i \simeq 50.14650 \; .$$

We will carry out simulations with $(E/N) \equiv 1$ so that the initial kinetic energy will be EKINO = 100 - EPOTO .

We start with velocities from our random number generator RUND(INTX,INTY)[6,7] initially using the two seeds INTX,INTY both equal to zero. This random number generator, in addition to its simplicity, is time-reversible, as shown by Federico Ricci-Tersenghi.[7] As velocities are generated we sum their components so as to make the mean velocity zero :

```
SUMPX = 0
SUMPY = 0
INDEX = 0
DO 30 I = 1,10
DO 30 J = 1,10
INDEX = INDEX + 1
PX(INDEX) = RUND(INTX,INTY) - 0.5d00
PY(INDEX) = RUND(INTX,INTY) - 0.5d00
```

```
   SUMPX = SUMPX + PX(INDEX)
   SUMPY = SUMPY + PY(INDEX)
30 CONTINUE
```

Often it is desirable to start with a particular kinetic temperature. This is necessary in our case as two of the mechanics we will employ (Gauss' and Hoover-Leete[8]) are designed to keep the kinetic energy constant. In order that all four varieties of mechanics are compared under the same conditions we will apply a scaling of the squared velocities after first forcing the velocity sums $\sum p_x$ and $\sum p_y$ to vanish :

```
   SUMXX = 0
   SUMYY = 0
   INDEX = 0
DO 40 I = 1,10
DO 40 J = 1,10
   INDEX = INDEX + 1
   PX(INDEX) = PX(INDEX) - SUMPX/100
   PY(INDEX) = PY(INDEX) - SUMPY/100
   SUMXX = SUMXX + PX(INDEX)**2
40 SUMYY = SUMYY + PY(INDEX)**2
```

To bring the two squared velocity sums of p_x^2 and p_y^2 to their desired initial values each p_x is multiplied by $\sqrt{(K_0/\sum p_x^2)}$ and each p_y by $\sqrt{(K_0/\sum p_y^2)}$.

4.2.1 *Newtonian Mechanics of the Lucy Fluid*

Newtonian mechanics is straightforward. The variables $\{ x, y, p_x, p_y \}$ are stored in a vector of length $4N$. The Runge-Kutta integrator requires new values of each of these $4N$ variables four times during each timestep. First the force arrays are initiated: FX(i) = 0 ; FY(I) = 0 . Next the nearest-image values of XIJ, YIJ, and RIJ are calculated, and whenever the resulting RIJ is less than h, 3 in our case, the forces on particles I and J are updated :

```
FX(I) = FX(I) - (XIJ/RIJ)*WP(RIJ)
FY(I) = FY(I) - (YIJ/RIJ)*WP(RIJ)
FX(J) = FX(J) + (XIJ/RIJ)*WP(RIJ)
FY(J) = FY(J) + (YIJ/RIJ)*WP(RIJ)
```

It is essential to use the correct sign of `XIJ = X(I) - X(J)` and `YIJ` while
`RIJ` is always positive. `WP(RIJ)` is the derivative of the weight function (or
potential function) at a distance `RIJ` and is never positive for the purely-
repulsive Lucy Fluid. A timestep of 0.01 is perfectly adequate for the
simulations. You should find that with this choice the initial energy 100,
is conserved to 9-figure accuracy in a double-precision run with 100 000
timesteps. The mean value of the kinetic energy over the last half of the run
is 45.58 while the potential energy has average value 54.42. The mean value
of the kinetic energy is a significant result in that we should use it in comparing our Newtonian results with those from the Gauss and Hoover-Leete
simulations, as those make the kinetic energy a constant of the motion,
with only the potential energy fluctuating.

4.2.2 *Gauss' Mechanics of the Lucy Fluid*

In the initial conditions for the Gauss' Mechanics' 100-particle fluid problem we use the same square lattice with potential energy $\Phi_0 \simeq 50.14650$.
In scaling the momenta the initial kinetic energy should be 45.58 rather
than 49.85350. The Gauss equations of motion are $\{\dot{p} = F - \zeta p\}$. After
summing all of the pair contributions to the forces, the constraint force for
kinetic energy, $-\zeta p$, needs to be added on. The corresponding additional
FORTRAN required to convert Newton's mechanics to Gauss' is

```
FZP = 0
TWOK = 0
DO I = 1,100
FZP = FZP + PX(I)*FX(I) + PY(I)*FY(I)
TWOK = TWOK + PX(I)*PX(I) + PY(I)*PY(I)
ENDDO
DO I = 1,100
FX(I) = FX(I) - (FZP/TWOK)*PX(I)
FY(I) = FY(I) - (FZP/TWOK)*PY(I)
ENDDO
```

In this application of Gauss' mechanics you should find that the total energy, over the last half of the run, has an average value of 100.07, again
using a double-precision timestep of $dt = 0.01$.

4.2.3 *Hoover-Leete Mechanics of the Lucy Fluid*

Hoover-Leete Mechanics[8] has unusual equations of motion. It includes two
different versions of the kinetic energy, one, $K(\dot{q})$, which depends upon the

velocities $\{\ \dot{q}\ \}$, is a constant of the motion; the other kinetic energy $K(p)$ depends upon the momenta, and is different. $K(p)$ fluctuates and reacts directly to the forces. The velocities are continuously scaled according to the current value of the momentum-based kinetic energy $K(p)$:

$$\{\ \dot{q} = p\sqrt{(K(\dot{q})/K(p))}\ ;\ \dot{p} = F(q)\ \}\ [\text{ Hoover} - \text{Leete }]\ .$$

The kinetic energy EKQ is fixed by the initial conditions while the EKP can be evaluated in the force subroutine. The resulting equations of motion modify the propagation of the coordinates rather than the momenta, as the forces are calculated from the pair potential without any modification of the Newtonian forces :

```
DO 30 I = 1,100
   XDOT(I)   = PX(I)*DSQRT(EKQ/EKP)
   YDOT(I)   = PY(I)*DSQRT(EKQ/EKP)
   PXDOT(I) = FX(I)
30 PYDOT(I) = FY(I)
```

Here you should find an average value of the total energy over the last half of the run equal to 99.93, just as close as the Gauss value but with an algorithmic dependence of the opposite sign relative to Newtonian mechanics. The fact that these discrepancies are quite small relative to $(1/N)$ suggests that the constraint of constant temperature, whether imposed by Gauss' constraint force or by Leete's Lagrange multiplier plays little havoc with the dynamics and is perfectly suitable as a thermostatting technique if it is desired to avoid fluctuations.

4.2.4 *Nosé-Hoover Mechanics of the Lucy Fluid*

Nosé spoke of his mechanics as living in an "extended" phase space with the new pair of variables (s, ζ) requiring two additional dimensions. The simpler Nosé-Hoover mechanics requires only one additional variable, ζ, which can be added on at the end of the vector array holding the coordinates and momenta. The programming closely resembles that for Gauss' dynamics except that the friction coefficient ζ, rather than being an explicit function of the coordinates and momenta, satisfies its own differential equation,

$$\dot{\zeta} = [\ \sum(p_x^2 + p_y^2)\ ] - 91.16 \text{ or } \dot{\zeta} = \left[\ \sum[\ (p_x^2 + p_y^2)\ ]/(91.16)\ \right] - 1\ .$$

Although these two alternative forms cause the response time of the thermostat to vary by an order of magnitude, $\sqrt{91.16}$, *either* of them provides a stable trajectory, necessarily leading to an exact replication of the desired long-time-averaged kinetic energy.

4.2.5 *Four Simulations of the Lucy Fluid*

The simplicity of the numerical thermostats makes it easy to include them in a dynamics program. Typically no more than ten additional lines of FORTRAN are required. Another conclusion from this work is more puzzling and applies to all four forms of mechanics. Lucy's weight function makes it possible to evaluate local values of the density, velocity, and kinetic energy (using $\sum w_{ir} v_i^\alpha / \rho_r$ for $\alpha = 0$, 1, and 2) on a regular grid, facilitating the production of contour plots and movies. Doing so one finds that the grid values of the mean-squared velocity are somewhat lower, 10% or so, than the unsmoothed sums, $\sum v_i^2$. Evidently this is an excellent area for new research in developing a continuum description more nearly faithful to the atomistic one, for use in defining the local temperature.

With a good understanding of manybody mechanics under isothermal conditions let us return to small systems for a look at more specialized thermostats in which moments other than the first three are constrained, all in an effort to bridge the gap between dynamics and Gibbs' statistical mechanics.

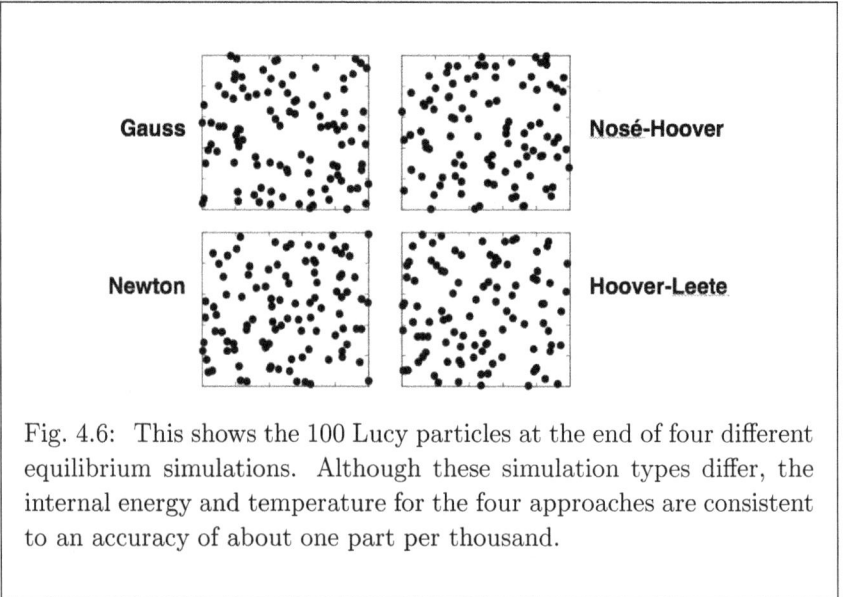

Fig. 4.6: This shows the 100 Lucy particles at the end of four different equilibrium simulations. Although these simulation types differ, the internal energy and temperature for the four approaches are consistent to an accuracy of about one part per thousand.

4.3 Multi-Moment Thermostatted Oscillators

The harmonic oscillator is a terrific model for the investigation of thermostats. Its proper distributions in Gibbs' microcanonical and canonical ensembles are simple analytic functions. Because the canonical distribution can be characterized by any one of its even moments it is natural to consider thermostatting moments other than the second. **Figure 4.7** shows what happens if, rather than the second, the fourth and sixth moments are constrained by integral control.

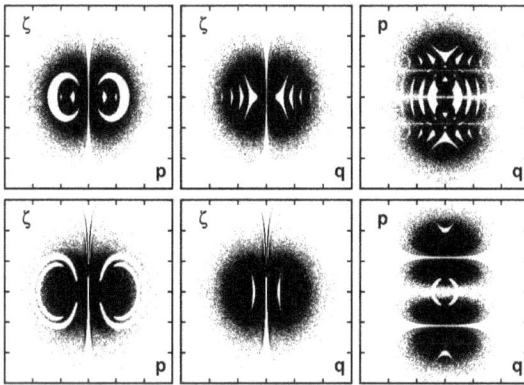

Fig. 4.7: It is possible to thermostat moments other than the second. Here the top row shows sections in which $\langle\, p^4 - 3p^2 \,\rangle$ is thermostatted with initial conditions $(q, p, \zeta) = (0, 5, 0)$ and the bottom row $\langle\, p^6 - 5p^4 \,\rangle$ with $(q, p, \xi) = (0, 2, 0)$. Just as in the Nosé-Hoover oscillator case where $\zeta = p^2 - 1$ all of these problems have a mixed phase space with both toroidal and chaotic solutions. In each of the six sections shown here the variables plotted range from -6 to +6.

In order that the differential equations are consistent with Gibbs' distribution with $\langle\, p^2 \,\rangle = 1$ these constraints require the following forms :

$$\dot{q} = p \; ; \; \dot{p} = -q - \zeta p^3 \; ; \; \dot{\zeta} = p^4 - 3p^2 \; [\text{ Fourth Moment }] \; ;$$

$$\dot{q} = p \; ; \; \dot{p} = -q - \zeta p^5 \; ; \; \dot{\zeta} = p^6 - 5p^4 \; [\text{ Sixth Moment }]$$

Figure 4.7 shows the "Poincaré Sections" that result, plotting the first 300,000 points which intersect the $q = 0$, $p = 0$, and $\zeta = 0$ planes. With a Runge-Kutta timestep of 0.0001 simply plotting a (p, ζ) point whenever the product of $q_{t+dt}q_t$ is negative provides a sufficiently sharp image. These examples show that fixing a single moment is not sufficient to recover Gibbs' distribution. It is natural to try controlling more than just a single moment.

Configurational moments (which can alternatively be thought of as *force* moments) are likewise available for thermostatting. Karl Travis and Carlos Braga[9] pointed out that thermostatting the oscillator's mean-squared-force provides differential equations isomorphic to the Nosé-Hoover thermostat. All one has to do is to make the replacement $(+q, +p) \longrightarrow (-p, +q)$:

$$\{ +\dot{q} = p \; ; \; +\dot{p} = -q - \zeta p \; ; \; \dot{\zeta} = p^2 - 1 \} \longrightarrow$$

$$\{ -\dot{p} = q \; ; \; +\dot{q} = +p - \zeta q \; ; \; \dot{\zeta} = q^2 - 1 \}.$$

Exercise: Demonstrate that thermostatting $\langle p^4 \rangle$ or $\langle p^4 - 3p^2 \rangle$ fails to give a Gaussian distribution for the momenta. Recall that for the Gaussian $e^{-p^2/2}$ the first three even moments are $\langle p^{2,4,6} \rangle = 1, 3, 15$.

With failure to obtain ergodicity by constraining a single oscillator moment it is natural, perhaps only in retrospect, to extend the thermostatting idea to the simultaneous control of two or more moments. Several of these ideas and their fruits are worth noting. We start with the simultaneous control of coordinates and momenta.

4.3.1 *Bhattacharya-Ezra-Patra-Sergi's Two-Moment Idea*

The failure to obtain the canonical distribution for the oscillator with single-moment control leads naturally to the idea of simultaneous control of two or more moments. Sergi and Ezra[10] preceded Patra and Bhattacharya[11] in suggesting simultaneous integral control of fluctuations of the force and of the momentum:

$$\dot{q} = p - \xi q \; ; \; \dot{p} = -q - \zeta p \; ; \; \dot{\xi} = q^2 - 1 \; ; \; \dot{\zeta} = p^2 - 1 \; [\text{ BEPS }].$$

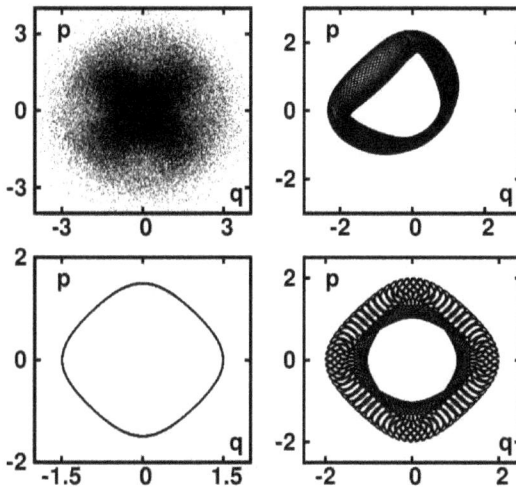

Fig. 4.8: These (q,p) projections (not sections) were generated with the Bhattacharya-Ezra-Patra-Sergi equations of motion. They include 100,000 points from fixed-timestep runs of 100,000,000 timesteps with $dt = 0.001$. In the top left chaotic section the initial values of (q, p, ζ, ξ) were all 0.0001. The upper right torus had initial values $(1, 0, 1, 0)$ while the near-periodic orbit at the lower left began at $(1.5, 0, 0, 0)$. The lower right torus' initial values are $(2, 0, 0, 0)$. The differential equations $\{ \dot{q} = +p - \xi^n q \; ; \; \dot{p} = -q\zeta^n p \}$ have much larger chaotic seas with $n = 3$ than with $n = 1$. The lack of circular symmetry here makes it clear that the linear thermostats do not reproduce the canonical distribution for the oscillator.

Figure 4.8 with its evident lack of circular symmetry, shows the failure of the simplest two-moment plan to reproduce the canonical distribution. Numerical experiments show that the quadratic thermostat controls do achieve the correct values for $\langle\, q^2, p^2\, \rangle$, unity for both. Of course they must. Integral feedback's success is guaranteed provided that the solution algorithm does not diverge, exhibiting "computational instability". On the other hand the fourth moments, $\langle\, q^4, q^2 p^2, p^4\, \rangle$, do not reproduce the correct canonical values 3,1,3 . The (q,p) plane projection of the chaotic solution shows that the distribution does not have circular symmetry but instead favors the directions where $q = \pm p$. We conclude that this stimulating idea in its most straightforward form is unsuccessful.

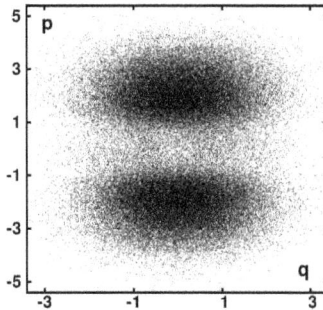

Fig. 4.9: One million points from a variable-timestep run of one billion points showing the result of simultaneous *cubic* control of $\langle (q^2, p^2) \rangle$. The lack of circular symmetry in this figure is misleading as this distribution *does have* circular symmetry. An explanation should occur to the diligent reader if he is familiar with the idea of adaptive integration.

Another strategy, described by Bauer, Bulgac, and Kusnezov in their pair of valuable references from the 1990s,[12] is to include *cubic* thermostat forces, for example :

$$\dot{q} = p - \xi^3 q \; ; \; \dot{p} = -q - \zeta^3 p \; ; \; \dot{\zeta} = p^2 - 1 \; ; \; \dot{\xi} = q^2 - 1 \; [\text{ BBK }] .$$

It is easy to show that these four equations are consistent with a four-dimensional distribution, Gaussian in (q, p, ζ^2, ξ^2). The (q, p) plot in **Figure 4.9** along with calculations of the moments shows that the enhanced nonlinearity of cubic thermostat forces can be useful. It is amusing that the sampling in the figure (every thousandth point) is responsible for an apparent lack of symmetry. A reader should be able to modify the sampling so as to regain a symmetric projection from exactly the same data.

4.4   More Ergodic Solutions of the Oscillator Problem

Brad Holian and I continued the double-moment idea by separately thermostatting the kinetic temperature and its fluctuations for the oscillator, using integral feedback for both moments.[13]

$$\dot{q} = p \; ; \; \dot{p} = -q - \zeta p - \xi p^3 \; ; \; \dot{\zeta} = p^2 - 1 \; ; \; \dot{\xi} = p^4 - 3p^2 \; [\text{ HH }] .$$

This approach appears to be ergodic : phase-space cross sections showed no holes; the various moments, through the sixth order, had no statistically significant deviation from the Gaussian values. Further, the largest Lyapunov exponent, the most useful measure of chaos, was statistically independent of the choice of initial conditions.

Figure 4.10 shows the combined effect of thermostatting both the kinetic energy and its fluctuation. This idea along with an alternative "chain thermostat" developed by Martyna, Klein, and Tuckerman[14] were both successful in providing an ergodic coverage of the harmonic oscillator phase space. The latter group had the idea of thermostatting not only the momentum but also the thermostat variable itself :

$$\dot{q} = p \; ; \; \dot{p} = -q - \zeta p \; ; \; \dot{\zeta} = p^2 - 1 - \xi\zeta \; ; \; \dot{\xi} = \zeta^2 - 1 \; [\text{ MKT }] \; .$$

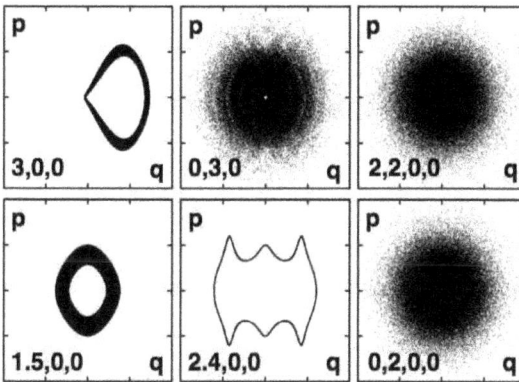

Fig. 4.10: The kinetic energy (first column) or its square (second column), or both (third column) are constrained through integral feedback forcing $\langle p^2 - 1 \rangle$ or $\langle p^4 - 3p^2 \rangle$ or both to vanish. The initial conditions for (q, p, ζ) or (q, p, ξ) or (q, p, ζ, ξ) are given for all six simulations. The ranges for the variables shown are from -4 to +4.

Exercise : Show that the MKT thermostat equations given above are consistent with the canonical distribution multiplied by a Gaussian distribution of both ζ and ξ . This approach was confirmed ergodic. It was the subject of the Snook Prize problem of 2014, arχiv.1408.0256.

One summer afternoon, while watering trees in Ruby Valley, I had the idea of trying "weak" control as opposed to the certainty of integral control, using a single thermostat for the control of two different moments. A few hours later I found a simple example that worked for the oscillator[15] :

$$\dot{q} = p \; ; \; \dot{p} = -q - 0.05\zeta p - 0.32\zeta p^3 \; ;$$

$$\dot{\zeta} = 0.05(p^2 - 1) - 0.32(p^4 - 3p^2)$$

[0532 Model].

This became the first single-thermostat model to provide ergodicity for the harmonic oscillator. It left open the question of finding an analogous solution for the stiffer Quartic and Mexican Hat potentials :

$$\phi_Q = (q^4/4) \; ; \; \phi_{MH} = -(q^2/2) + (q^4/4) \; .$$

I repeated the process of looking for coefficients of $(p^2 - 1)$, $(p^4 - 3p^2)$, and even $(p^6 - 5p^4)$ with a Monte Carlo search, but was unable to find a successful combination. This failure to find an ergodic friction coefficient for these potentials led Carol and me to set this problem as a target for the 2016 Snook Prize.

Diego Tapias, Allesandro Bravetti, and David Sanders found the prize-winning solution, using a hyperbolic-tangent form for the thermostatting force.[16] An example which appears to be ergodic in the Mexican-Hat case has the form :

$$\dot{q} = p \; ; \; \dot{p} = q - q^3 - 6.9p \tanh(6.9\zeta) \; ; \; \dot{\zeta} = p^2 - 1 \; .$$

Because this problem is relatively stiff it is a good challenge for the various adaptive integration strategies I introduced in **Lecture 1** .

Adaptive integration solved another leftover problem from 1996. At Harald Posch's Family home in the mountains of Murau, Austria Brad Holian, Harald, and I made an effort to thermostat the first three even moments of the harmonic oscillator :

$$\dot{q} = p \; ; \; \dot{p} = -q - \zeta p - \xi p^3 - \omega p^5 \; ; \; \dot{\zeta} = p^2 - 1 \; ; \; \dot{\xi} = p^4 - 3p^2 \; ; \; \dot{\omega} = p^6 - 5p^4 \; .$$

We were unsuccessful. The solutions always diverged! Some twenty years later Carol and I used adaptive integration and easily got a solution. The timestep was allowed to vary within a narrow "error band" where the error was the mean squared error between a single Runge-Kutta timestep dt and the result of two successive steps of $(dt/2)$:

```
ERROR = 0
DO I = 1,4
ERROR = ERROR + (XX(I)-YY(I)**2
ENDDO
IF(ERROR.LT.10.0D00**(-24) DT = DT*2
IF(ERROR.GT.10.0D00**(-20) DT = DT/2
```

The mean value of the timestep, averaged over a run of one billion steps was about 0.001, so that the triply-thermostatted oscillator was relatively easy to solve. In 1996 we had had no experience with adaptive integration and drew the wrong conclusion that the equations were intractable. One can also compare solutions using RK4 and RK5 timesteps in order to decide whether or not a new timestep is desirable.

Exercise : The first three even moments of the oscillator coordinate were 1.001, 3.010, and 15.145 while the first three even moments of the momentum were 1.00000, 3.00000, 15.00001. Why are the latter more nearly accurate?

4.5 The ϕ^4 Model for Heat Conduction

Let us break away from equilibrium simulations and consider a nonequilibrium problem. To study heat flow the velocity moments of selected degrees of freedom can be constrained to *different* temperatures. If successful, this idea can generate and maintain a nonequilibrium steady state.[17] The ϕ^4 Model is ideal for such problems. The "ϕ^4" model gets its name from the quartic tethering potential which keeps solid-phase particles near their lattice sites. There is in addition a nearest-neighbor Hooke's-Law potential with a restlength of unity. The Hamiltonian is the sum of three terms :

$$\mathcal{H} = \sum (p^2/2) + \sum (\delta_i^4/4) + \sum (|r_{ij}| - 1)^2/2 ,$$

where δ_i is the displacement of the ith particle from its lattice site.

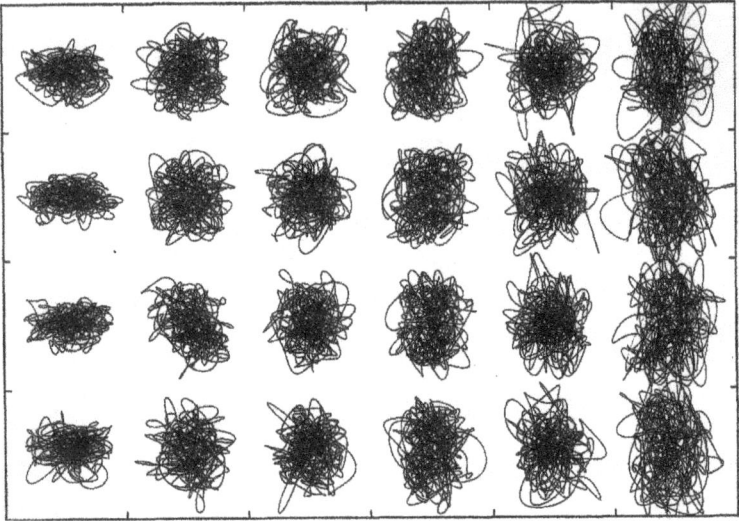

Fig. 4.11: Shows a 24-particle ϕ^4 crystal with a column of cold particles to the left and a column of hot particles to the right. Such prototypical models, with a variety of lengths and widths were used to compare the relative usefulness of seven different thermostats in Reference 17, Physica **187D**, 253-267 (2004). A "better" thermostat will exhibit less number-dependence in its conductivity than do its fellows.

We can formulate a nonequilibrium heat flow problem by thermostatting some particles at a cold temperature 0.005, and others at a hotter one, 0.015. **Figure 4.11** shows particle trajectories for a two-dimensional system with bounding columns of cold and hot particles at its sides. The systems are periodic in the vertical direction. Such problems were used to test the dependence of the heat current on the thermostat type used at the boundaries. By using different system lengths we found that the number-dependence of the conductivity was relatively small using the Nosé-Hoover thermostat and concluded that on the bases of simplicity, convergence, and utility that the Nosé-Hoover thermostat was best of the seven tested.

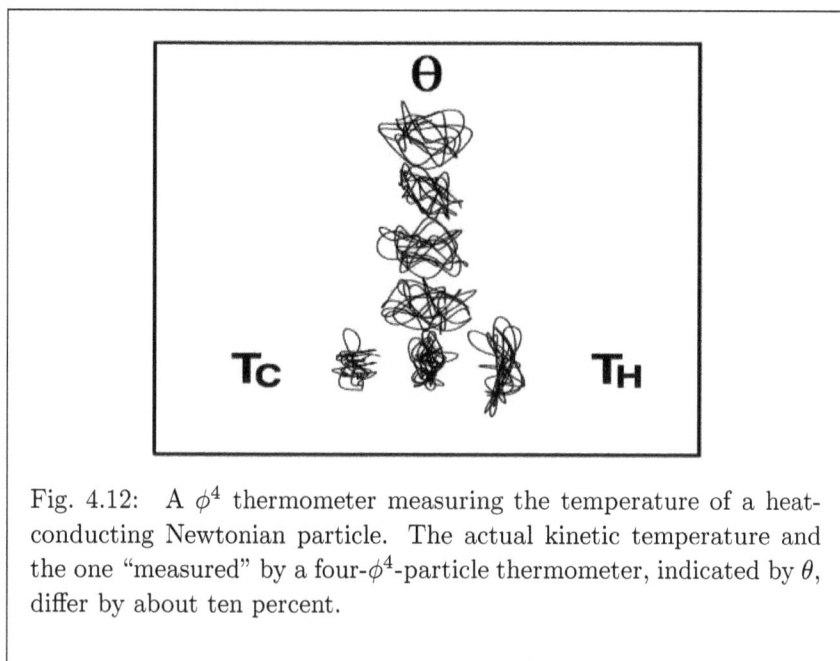

Fig. 4.12: A ϕ^4 thermometer measuring the temperature of a heat-conducting Newtonian particle. The actual kinetic temperature and the one "measured" by a four-ϕ^4-particle thermometer, indicated by θ, differ by about ten percent.

The ϕ^4 model works well even in one dimension, where most models fail, and can also be used as a solid-state thermometer.[18] **Figure 4.12** shows a 4-particle thermometer attached to a heat transfer problem with one thermostatted ϕ^4 particle at $T_C = 0.01$ and another at $T_H = 0.03$. The "thermometer temperature" T_θ varies only a little with length. It is 0.024. We used from 4 to 21 particles in the thermometer. The kinetic temperature of the Newtonian conducting particle is actually 0.021, indicating that nonequilibrium thermometry is a dynamic rather than a thermodynamic problem. The discrepancy could likely be reduced by changing the force constant or the power law linking the thermometer to the Newtonian particle. This is no doubt a good research topic.

We will come back to the ϕ^4 model in **Lecture 10** in illustrating the microscopic analog of the Second Law of Thermodynamics. **Figure 4.13** shows that the model is mainly chaotic over a wide range of energies per particle ranging from 10^{-2} to 10^5. It can provide a flexible and informative view of heat flow in one-, two-, and three-dimensional systems. You might wonder about the need for the quartic tethering potential. Why not just use harmonic forces?

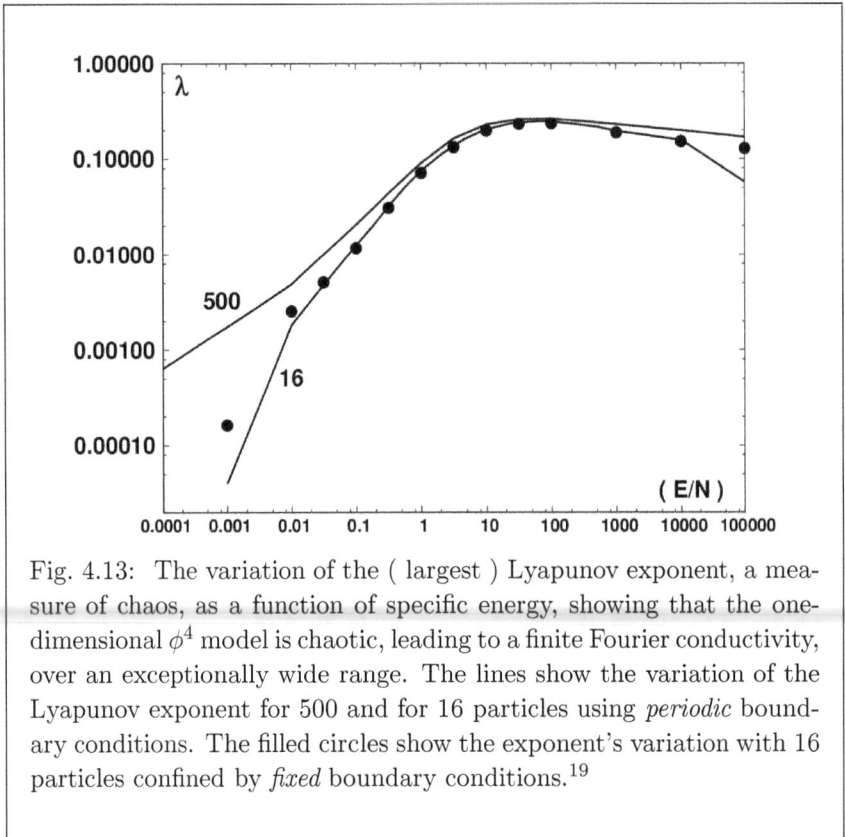

Fig. 4.13: The variation of the (largest) Lyapunov exponent, a measure of chaos, as a function of specific energy, showing that the one-dimensional ϕ^4 model is chaotic, leading to a finite Fourier conductivity, over an exceptionally wide range. The lines show the variation of the Lyapunov exponent for 500 and for 16 particles using *periodic* boundary conditions. The filled circles show the exponent's variation with 16 particles confined by *fixed* boundary conditions.[19]

4.6   Harmonic Chains with Initial Velocities of +1 and -1

Now we consider two purely-harmonic chains with 200 or 2000 rightmoving particles approaching 200 or 2000 leftmoving ones. The forces are Hooke's Law with $\ddot{x}_i = x_{i+1} - 2x_i + x_{i-1}$. If you know about shockwaves you might well expect to see two of them here, one running to the left and one to the right at speeds somewhat greater than the sound speed. The **Figure** shows the displacements and velocities at times of 100 and 1000, when we would expect to see about half of the particles affected by the left-meets-right collision. Half of the particles are relatively close to their lattice sites but there is no real shockwave. Instead the disorderly region covers most of the "hot" compressed material in the center. After a single sound-traversal time (400 or 4000) the entire chain is in a rarefied state.

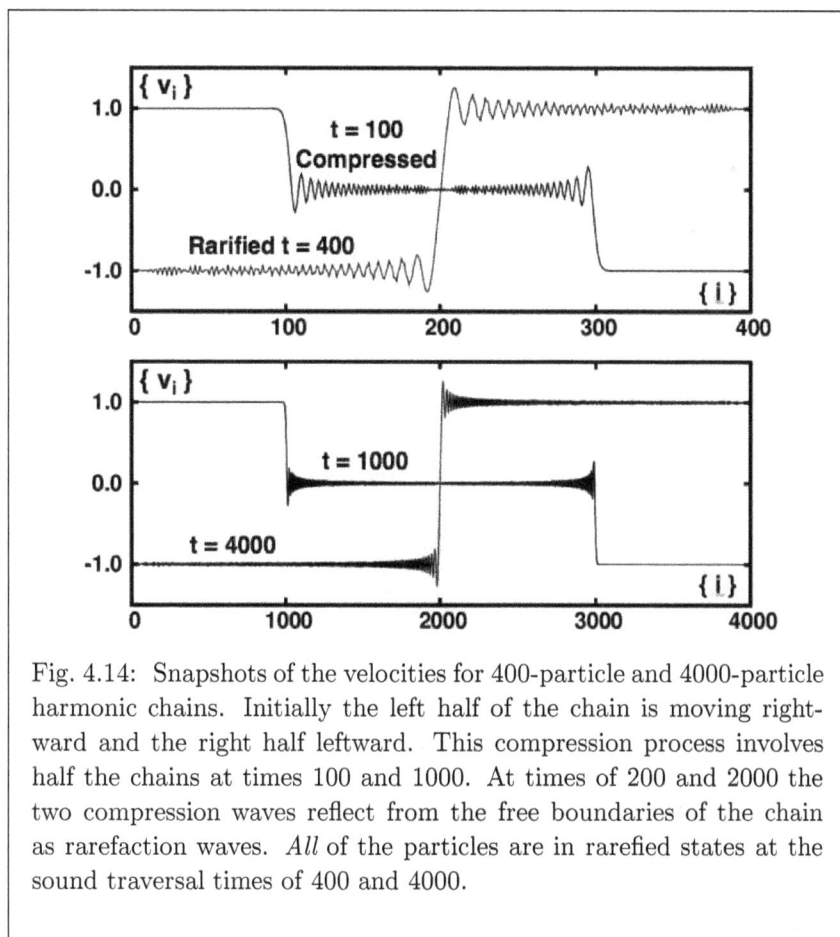

Fig. 4.14: Snapshots of the velocities for 400-particle and 4000-particle harmonic chains. Initially the left half of the chain is moving rightward and the right half leftward. This compression process involves half the chains at times 100 and 1000. At times of 200 and 2000 the two compression waves reflect from the free boundaries of the chain as rarefaction waves. *All* of the particles are in rarefied states at the sound traversal times of 400 and 4000.

What happens if the initial conditions are reversed, with the two halves separating rather than dividing ? The displacement and velocity plots become their mirror images. This reversibility is a defect of the harmonic model. For the chain expansion and compression are perfectly symmetric. The chain has no mechanism for scattering waves into heat. The two calculations, with 400 at the top and with 4000 shown at the bottom of **Figure 4.14**, look essentially the same. It turns out, as you may well have expected, that there is an analytic expression for the motion of the particles in a harmonic chain. An analytic description of N-particle dynamics can be described by Nth-order Bessel functions. It is amusing that the simplest way to determine those Bessel functions is to simulate the N-particle chain!

Exercise : Substitute a wave with an amplitude $Ae^{i\omega t}e^{ikx}$ into the motion equation for the harmonic chain and show that long waves with small $k = 2\pi/\lambda$, where *this* λ is wavelength, travel at a speed of unity. For this exercise assume that the rest distance between the particles is unity. In a two-dimensional triangular lattice (six nearest-neighbors and hexagonal symmetry) the zero-pressure sound speed is considerably greater if the *same* Hooke's-Law forces, with a restlength of unity, are used. Why is that?

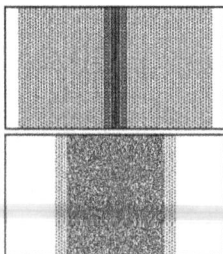

Fig. 4.15: Two stages in the collision of two cold zero-pressure zero-temperature mirror-image blocks of particles at unit density with a repulsive pair potential $\phi(r < 1) = (10/\pi)(1-r)^3$. The initial velocities are ± 0.965. For more details see Reference 20.

In the more-realistic two-dimensional simulation of **Figure 4.15** two blocks of 1600 particles each are propelled toward one another with sufficient speed to reach twofold compression. The bottom picture shows that the overall length is approaching half the original. Notice that the shock-front is very thin, just a few particle diameters in width. The dissipation in the wave comes from the conversion of velocity in the x direction to "heat" with thermal motion in both directions. We will return to shock-waves in **Lectures 8 and 11** and discover how to measure and predict their structures.

I wish to mention one additional very significant topic having to do with the simulation of irreversible processes and related to Loschmidt's Reversibility Paradox, the one which points out that reversible equations of motion should be capable of going backward in time just as well as forward.

4.7   Levesque and Verlet's Bit-Reversible Algorithm

In **Lecture 2** Carol explained that by using *integer arithmetic*, with a fixed maximum length, it is possible to reverse any Hamiltonian trajectory infinitely far into the past or the future :

$$\{ \; q_{t+dt} - 2q_t + q_{t-dt} = [Fdt^2]_{integer} \; \}$$

This idea was introduced by Levesque and Verlet and applied to a simple smooth-particle continuum problem by Oyeon Kum and me.[21, 22] The time-reversible symmetry of the motion equations is obvious here. Consistent rounding of the righthandside to integers is all that is required to guarantee that one can reproduce the integer coordinates exactly, starting arbitrarily far in the past or in the future. This algorithm is perfectly capable of reversing the two-block shockwave of **Figure 4.15** all the way back to the initial condition of two separated cold blocks.

   In the laboratory what actually happens when a compressed block is released into vacuum is very different to what happens in a shockwave. We saw it in the free-expansion problem. The material expands nearly isentropically (because there is no time for heat transfer and the gradients become so small that viscosity is inactive) with the expansion spread out over what is called a rarefaction fan. The denser part of the wave runs away from the more rarefied part. When matter is compressed, as in the collision of the blocks shown in **Figure 4.15** the typical increase of sound velocity with density implies that the denser part of the pressure wave moves faster, catching up with the slower less dense foot of the wave. In a strong shockwave the width of the wave is only a few mean free paths, thousands of Ångstroms in air and a few Ångstroms in water. For fans of the Systèm International d'Unités an Ångstrom is a tenth of a nanometer. This unit is a fine one, about the size of an atom and honoring the Swedish spectroscopist Anders Ångstrom (1814-1874).

4.8   Summary of Lecture 4

We introduced the subject of thermostatted mechanics by summarizing applications of the idea to simulations including shear and heat flows and their combination in Rayleigh-Bénard problems. Four types of mechanics were compared, three of them thermostatted, and shown to be in overall agreement, to three-figure accuracy, for systems of only 100 particles. More elaborate controls, either instantaneous or time-averaged, over two or three force or velocity moments were illustrated for simple one-dimensional models. The ergodicity of some models for systems as simple as a single one-dimensional harmonic oscillator demonstrated the exact consistency of some of these new forms of mechanics with Gibbs' statistical mechanics. We introduced the ϕ^4 model, which harbors chaos and obeys Fourier's Law, in contrast to the truly time-reversible behavior of the harmonic chain. We reiterated that irreversible processes *can* be reversed precisely on an atomistic scale by using Levesque and Verlet's integer leapfrog algorithm.

References

1. Wm. G. Hoover, *Smooth Particle Applied Mechanics - the State of the Art* (World Scientific, Singapore, 2006).
2. Wm. G. Hoover and H. A. Posch, "Entropy Increase in Confined Free Expansions *via* Molecular Dynamics and Smooth-Particle Applied Mechanics", Physical Review E **59**, 1770-1776 (1999).
3. Wm. G. Hoover and H. A. Posch, "Numerical Heat Conductivity in Smooth Particle Applied Mechanics", Physical Review E **54**, 5142-5145 (1996).
4. H. A. Posch, Wm. G. Hoover, and O. Kum, "Steady-State Shear Flows *via* Nonequilibrium Molecular Dynamics and Smooth-Particle Applied Mechanics", Physical Review E **52**, 1711-1720 (1995).
5. O. Kum, Wm. G. Hoover, and C. G. Hoover, "Temperature Maxima in Stable Two-Dimensional Shock Waves", Physical Review E **56**, 462-465 (1997).
6. F. Ricci-Tersenghi, "The Solution to the Challenge in 'Time-Reversible Random Number Generators' by Wm. G. Hoover and Carol G. Hoover", arχiv.1305.1805.
7. Wm. G. Hoover and C.G. Hoover, "Time-Reversible Random Number Generators: Solution of Our Challenge by Federico Ricci-Tersenghi", arχiv.1305.0961.
8. Wm. G. Hoover, "Atomistic Nonequilibrium Computer Simulations", in the Proceedings of the Bureau of Standards Conference *Nonequilibrium Fluid Phenomena* at Boulder, Colorado, June, 1982, in Physica **118A**, 111-122 (1983).
9. K. P. Travis and C. Braga, "Configurational Temperature Control for Atomic and Molecular Systems", The Journal of Chemical Physics **128**, 014111 (2008).
10. A. Sergi and G. S. Ezra, "Bulgac-Kusnezov-Nosé-Hoover Thermostat", Physical Review E **81**, 036705 (2010).
11. P. K. Patra and B. Bhattacharya, "Improving the Ergodic Characteristics of Thermostats Using Higher-Order Temperatures", arχiv 1411.2194.
12. D. Kusnezov, A. Bulgac, and W. Bauer, "Canonical Ensembles from Chaos", Annals of Physics **204**, 155-185 (1990) and **214**, 180-218 (1992).
13. Wm. G. Hoover and B. L. Holian, "Kinetic Moments Method for the Canonical Ensemble Distribution", Physics Letters **211A**, 253-257 (1996).

14. G. J. Martyna, M. L. Klein, and M. Tuckerman, "Nosé-Hoover Chains: the Canonical Ensemble *via* Continuous Dynamics", The Journal of Chemical Physics **97**, 2635-2643 (1992).
15. P. K. Patra, Wm. G. Hoover, C. G. Hoover, and J. C. Sprott, "The Equivalence of Dissipation from Gibbs' Entropy Production with Phase-Volume Loss in Ergodic Heat-Conducting Oscillators", International Journal of Bifurcation and Chaos **26**, 1650089 (2016) = arχiv 1511.03201.
16. D. Tapias, A. Bravetti, and D. P. Sanders, "Ergodicity of One-Dimensional Systems Coupled to the Logistic Thermostat", Computational Methods in Science and Technology **23**, 11-18 (2017).
17. Wm. G. Hoover, K. Aoki, C. G. Hoover, and S. V. De Groot, "Time-Reversible Deterministic Thermostats", Physica **187D**, 253-267 (2004).
18. Wm. G. Hoover and C. G. Hoover, "Nonequilibrium Temperature and Thermometry in Heat-Conducting ϕ^4 Models", Physical Review E **77**, 041104 (2008) = arχiv 0801.3051.
19. Wm. G. Hoover and K. Aoki, "Order and Chaos in the One-Dimensional ϕ^4 Model : N-Dependence and the Second Law of Thermodynamics", Communications in Nonlinear Science and Numerical Simulation **49**, 192-201 (2017) = arχiv 1605.07721
20. Wm. G. Hoover, C. G. Hoover, and F. Uribe, "Flexible Macroscopic Models For Dense-Fluid Shockwaves: Partitioning Heat and Work; Delaying Stress and Heat Flux; Two-Temperature Thermal Relaxation", Proceedings of the 38th Summer School *Advanced Problems in Mechanics*, Saint Petersburg, Russia, 2010 = arχiv 1005.1525.
21. D. Levesque and L. Verlet, "Molecular Dynamics and Time Reversibility", Journal of Statistical Physics **72**, 519-537 (1993).
22. O. Kum and Wm. G. Hoover, "Time-Reversible Continuum Mechanics", Journal of Statistical Physics **76**, 1075-1081 (1993).

Chapter 5

Ergodicity and Its Importance in Small Systems

/ Ergodicity in Small Systems / Nosé's Time-Scaling Thermostat / Dettmann's Hamiltonian / Gibbs' Ensembles and Linear-Response Theory / Ergodicity of Hard-Disk and Soft-Disk Cell Models / Nosé and Nosé-Hoover Oscillator and the Chaotic Sea / General Treatment of Thermostats / Liouville Approach to Thermostatting / Bhattacharya-Ezra-Patra-Sergi Model / Bauer, Bulgac, Ju, and Kusnezov / Weak Control / Ergodicity from 1984-2016 / Singly-Thermostatted Ergodicity and the Logistic Thermostat / Wang and Yang's Discovery of Knotted Orbits / Summary of Lecture 5 /

5.1 Ergodicity and its Importance for Small Systems

There is a Merriam-Webster dictionary definition of ergodicity: "Having zero probability that any state will never recur." This brief, but somewhat negative, definition is itelf not directly relevant to our use of fast computers in simulation. Our work is necessarily deterministic rather than probabilistic in the sense that all of our "problems" have definite answers, typically with double-precision detail. The Ehrenfests' dog-flea model illustrates a hard fact of simulation: there is neither time nor space to sample *all* the states. The exponential growth of complexity and chaos outpace and outreach our linear ability to create and explore computational models.

Gibbs' work[1] was abstract. He imagined a phase space in which *all* the (q, p) states were accessible. Hamiltonian mechanics was to provide the mechanism to travel from one state to another. In theory any Hamiltonian system, be it a single oscillator or a manybody fluid, has a definite state which changes with time according to the motion equations $\dot{q} = +(\partial \mathcal{H}/\partial p)$ and $\dot{p} = -(\partial \mathcal{H}/\partial q)$. Our attempts to capture this motion computationally are necessarily approximate as we have no time for more than a quick double-precision look at a few trillion timesteps. We hope these relatively few states are good representatives of the infinitely-many states we have no time to explore. We strive to make our work reproducible so that others can rely upon it.

Gibbs envisioned "ensembles", made up of states sharing the same energy or pressure, as a concept supporting instantaneous ensemble averages rather than long-time averages over the states that a typical dynamical system would traverse. He discovered what he called the "microcanonical ensemble". This collection of states, equivalent to a hypervolume in phase space, is a measure of the thermodynamic entropy $S(E, V)$:

$$(S/k) = \ln \Omega \simeq \ln[\int \cdots \int \prod dq dp \,] \text{ with } [\text{ E and V fixed }] \,.$$

The possibility of reducing thermodynamics to a sum or to an integral over all states consistent with system energy and volume provided a seemingly-welcome alternative to time averages based on dynamical trajectory calculations, the "many body problem". But it is only for small systems that an exhaustive sampling is possible.

Liouville's Theorem, the phase-space continuity equation, bridges the divide between Gibbs' many-dimensional ensembles and Newton's or Hamilton's one-dimensional trajectories. By demonstrating the conservation of comoving phase volume the Theorem shows that all accessible states must be equally weighted. It is assumed that this stationary choice of phase volume represents the properties of "equilibrium". This assumed correspondence linking the microcanonical ensemble to dynamics is often false – typically Hamiltonian phase space has separate toroidal and chaotic regions. A system starting in either place can never leave it. Such a dynamical system is not ergodic so that Gibbs' ensembles do not describe it.

Gibbs' most useful ensemble is the "canonical" one, where canonical means "fundamental" or "prototypical". Gibbs imagined that a weak thermal coupling linking a system to a heat reservoir would make the system take on isothermal rather than isoenergetic properties. All states would become available, with a probability or frequency now proportional to $e^{-\mathcal{H}/kT}$. The weakness in this correspondence between the canonical ensemble and dynamics is that the "weak coupling" which would enable access to "all" the states is left vague. Without an operational definition of the coupling one cannot proceed.

A way to cheat suggests itself. The "Monte Carlo" algorithm[2] in the phase space proceeds through a sequence of state-to-state "jumps", either

jumps at constant energy [microcanonical ensemble] or jumps with probability $e^{[-\Delta E/kT]}$ [canonical ensemble]. Provided that the jumping satisfies
reversibility rules and makes all states accessible this Monte Carlo method
is just an awkward tour of one of Gibbs' ensembles. The algorithm can
jump from one of two potential minima to the other even when Hamilton's
dynamics cannot possibly do so.

We reject stochastic algorithms (although we are happy with our random number generator RUND(INTX,INTY) which is quite deterministic and
portable and reproducible). In this Lecture we will explore ways to simulate Gibbs' ensembles with deterministic thermostats. **Figure 5.1** suggests
the difficulty of the task. It represents parts of a $p = 0$ section which slices
through a trajectory in the three-dimensional (q, p, ζ) Nosé-Hoover phase
space. The equations of motion[3, 4] are familiar :

$$\dot{q} = p \; ; \; \dot{p} = -q - \zeta p \; ; \; \dot{\zeta} = p^2 - 1 \; [\text{Nosé} - \text{Hoover Oscillator}] \; .$$

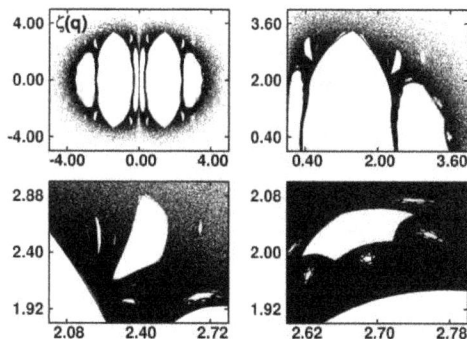

Fig. 5.1: This $(p = 0)$ section can be obtained with initial conditions
$(q, p, \zeta) = (3, 3, 0)$ or $(0, 5, 0)$, or $(0.0001, 0.0001, 0.0001)$. It is unar-
guable that these initial conditions generate trajectories occupying the
same chaotic sea and that they never do exit into the "holes" indicating
the penetration of the $(p = 0)$ plane by stable tori. The successive en-
largements up to fifty-fold in q and ζ suggest a chaotic "fat fractal" sea,
having the full dimensionality of (q, p, ζ) space, three. 10^{12} timesteps
were used for the lower righthand enlargement with $dt = 0.00025$.

A quick look suggests that there are an infinite number of tori surrounding stable periodic orbits in the phase space. The tori thread through the section an even number of times and conform to a few of the infinitely many holes that perforate the cross section on all scales.

Let us look more closely into Nosé's approach to simulating Gibbs' canonical ensemble. We will mainly stick with the one-dimensional harmonic oscillator of **Figure 5.1**, with unit mass, unit force constant, unit Boltzmann's constant, and unit temperature. These simplifications moderate both the notational burden and the conceptual difficulty of our task, bridging the gap between Gibbs' many-dimensional ensembles and one-dimensional dynamical trajectories.

5.1.1   *Nosé's Time-Scaling Thermostat*

Shuichi Nosé developed and described a completely new approach to atomistic dynamics in two 1984 papers, "A Unified Formulation of the Constant Temperature Molecular Dynamics Methods"and "A Molecular Dynamics Method for Simulations in the Canonical Ensemble".[5,6] Perhaps the rationale for two papers rather than just one was to show both [1] that he had surveyed what was already known and [2] that he had something "new" to contribute. Until his work the question of dynamical ergodicity for harmonic oscillator dynamics had not been raised and ergodicity (coming arbitrarily close to all the states in a well-defined statistical ensemble) was usually assumed rather than studied. It was generally accepted that all would be well in a vaguely-defined "thermodynamic limit" where the number of particles and the volume both diverged while the boundary and initial conditions became unimportant.

In those days conservative equilibrium dynamical systems could be divided into three classes: [1] regular systems, with toroidal solutions like the periodic oscillator orbit; [2] chaotic systems like hard disks and spheres; and [3] systems exhibiting "Hamiltonian Chaos". This last class of systems included a wide variety of systems with messy phase-space distributions such as those associated with the springy pendulum and Nosé's new version of the harmonic oscillator, shown in **Figure 5.1**. Through Nosé's work a new class of dynamical systems began to emerge, time-reversible, deterministic, *and* by now even ergodic, with unbounded phase-space distributions.

The oscillator cross sections show more and more structure with enlargement. They show fourfold symmetry $(\pm q, \pm \zeta)$ so that we can restrict our attention to the first quadrant $(+q, +\zeta)$. Having a look at the small hole near $(q, \zeta) = (2.68, 0, 2.03)$ reveals five much smaller surrounding holes. This complexity is characteristic of the transitioning from regular smooth tori (in the largest holes) to the chaotic sea which surrounds all of them. This complexity is the enemy which we must defeat in order to connect small-system dynamics with Gibbs' ensemble theory.

With Nosé's mechanics even the simple oscillator generates incredible complexity. In **Figure 5.1** the full-sized and three enlarged views show 300,000 successive penetrations of a selected part of the $p = 0$ Poincaré plane. Notice the pentagonal hole in the bottom right enlargement of the plane. The stable periodic orbit at the center of that hole penetrates the plane 34 times! Although we show no more detail here it is "well understood" that in the immediate vicinity of this periodic orbit there is more and more complexity, at all spatial scales, well beyond the present or future capabilities of computation. We will come to see how complex systems based on the simple harmonic oscillator furthered the development of new dynamics tracing out *all* of Gibbs' ensembles not just a part. These ergodic systems, making possible isothermal and isobaric simulations of Gibbs' ensembles, give a new category of dynamical systems, deterministic, time-reversible, ergodic, and unbounded. Their nonequilibrium cousins are stranger yet, converting Gibbs' smooth distributions to multifractals with *fractional* dimensionality.

The one-dimensional oscillator's constant-energy circular phase-space orbit is the best known example problem in statistical mechanics. In its period of 2π the harmonic oscillator cycles through all of its constant-energy states as dictated by the Hamiltonian,

$$\mathcal{H} = (q^2/2) + (p^2/2) \longrightarrow \dot{q} = +p \; ; \; \dot{p} = -q \rightarrow q = \cos(t) \; ; \; p = -\sin(t) \; .$$

Because it samples all of its accessible states this microcanonical oscillator could be termed ergodic. The situation is totally different if instead of a single oscillator energy we wish for *all* energies to become involved, covering the entire canonical-ensemble phase space, all of the (q, p) plane. Statistical mechanics addresses ensembles of such systems with all values of their energies. Gibbs' canonical distribution for a harmonic oscillator in contact with a heat bath at temperature T is the Gaussian probability density in "phase space" $f(q, p)$ which includes *all* oscillator energies, $0 < \mathcal{H} < \infty$:

$$(2\pi)f(q, p) = e^{-\mathcal{H}/kT} = e^{-(q^2/2)-(p^2/2)} \longrightarrow \int_{-\infty}^{+\infty} \int_{-\infty}^{+\infty} f(q, p)dqdp \equiv 1 \; .$$

Here we make our usual simplifying assumptions that Boltzmann's constant k, the kinetic temperature T, the oscillator mass m and force constant

κ are all of them set equal to unity. The canonical distribution $f(q,p)$ is normalized to unity in its two-dimensional (q,p) phase space and stretches all the way to infinity. The classical view was that coupling the oscillator to an infinitely-massive and vaguely-described heat reservoir would allow it to sample *all* of its energy states. From the computational standpoint Metropolis, the Rosenbluths, and the Tellers formalized a qualitative simplification, replacing the reservoir interaction with its infinite number of degrees of freedom by a sequence of random displacements in coordinate space (because the Maxwell-Boltzmann distribution in the momentum subspace $e^{-(p^2/2)}$ was already known). To accomplish this they formulated a simple rule: all decreases in potential energy were acceptable while any energy increase was accepted with a probability $e^{-\Delta E/kT}$ less than unity.[2]

Exercise: Write a Monte Carlo harmonic-oscillator program with random coordinate moves in the interval $-\delta < dx < +\delta$ and determine the best choice of the maximum jump length δ from the standpoint of the rapidity of convergence of the potential energy $\langle\, q^2/2\,\rangle = (1/2)$ and its fluctuation, $\langle\,[\,(q^2/2)-\langle\,(q^2/2)\,\rangle\,]^2\,\rangle = (3/4)-(2/4)+(1/4) = (1/2)$.

Monte Carlo algorithms that involve moving particles, one at a time or in small groups so as to keep the center of mass or the moment of inertia fixed, could be proven ergodic so long as [1] all states are accessible and [2] the transition probabilities satisfy "detailed balance", with the probabilities of going from state I to J and from J to I equal :

$$N_I W_{I\to J} = N_J W_{J\to I} \longrightarrow$$
$$[\,N_I/N_J\,] = [\,W_{J\to I}/W_{I\to J}\,] = e^{-E_I/kT}/e^{-E_J/kT}\;.$$

A very similar statistical condition underlies the Boltzmann equation, and is called "molecular chaos". It means that knowing the number of particles with velocities v_i and v_j one can calculate the rate at which such particles would collide resulting in the velocities v_i' and v_j'. The rate is proportional to the relative speed $|v_i - v_j|$ with the collision parameters (one of them in two dimensions and two of them in three) chosen "randomly". By its geometric construction the "reversed inverse" collision in which v_i' and v_j' collide to yield v_i and v_j has the same detailed mechanism as the original collision. See again **Figure 3.5**, of **Lecture 3**, page 114, for an illustration.

Molecular dynamics with velocity rescaling generates an ensemble intermediate between the microcanonical and the canonical. Velocity rescaling keeps the kinetic energy constant while the motion is consistent with a canonical potential-energy distribution $e^{-\Phi/kT}$. Shuichi Nosé was the first person to make a serious effort toward generating the *entire* canonical ensemble by the addition of just one degree of freedom to an underlying Hamiltonian description. His new variable, called s, "scales the time". Its conjugate momentum plays the role of a friction coefficient, so that we denote it ζ rather than p_s in anticipation of its ultimate use. Through s, with its time-scaling of Nosé's equations of motion, the kinetic energy is controlled. In Nosé's formulation the kinetic energy is $\sum(1/2)(p/s)^2$ rather than $\sum(p^2/2)$. Nosé also formulated a logarithmic potential for s, $(\#kT/2)\ln(s^2)$. This potential can be arbitrarily large and negative for s near zero, making arbitrarily large energies available for the conventional Hamiltonian variables $\{q,p\}$. Here $\#$ is the number of degrees of freedom in the Hamiltonian, among which Nosé included the (s,ζ) pair. Here we will instead set $\# = 1$ for the one-dimensional oscillator problem rather than $\# = 2$. This choice matches that modification of Nosé's dynamics to the nonHamiltonian one which I greatly prefer, Nosé-Hoover dynamics.

In the harmonic oscillator case the Nosé Hamiltonian and its equations of motion are as follows :

$$\mathcal{H}_{\text{Nosé}} = (1/2)[\ (p/s)^2 + q^2 + \ln(s^2) + \zeta^2\ ] \longrightarrow$$

$$\dot{q} = (p/s^2)\ ;\ \dot{p} = -q\ ;\ \dot{\zeta} = (p^2/s^3) - (1/s)\ ;\ \dot{s} = \zeta\ [\ \text{Nosé}\ ]\ .$$

Nosé arbitrarily multiplied these equations of motion by his time-scaling variable s in order to obtain motion equations consistent with the canonical ensemble. That scaling step was crucial to developing a useful algorithm. The original Nosé equations, without scaling, are extremely "stiff". Their numerical solution can be "difficult", and even impractical unless adaptive integration is used, as we mentioned in **Lectures 1 and 2**. See page 23.

5.1.2 *Carl Dettmann Derives the Oscillator Equations*

One evening in 1996, at a CECAM meeting in Lyon [Centre Européen de Calcul Atomique et Moléculaire], I asked Carl Dettmann if he could think of a Hamiltonian giving Nosé's equations of motion without the need for time scaling.[7] The next morning he had the answer,[8] obtained by using another novel idea, setting the Hamiltonian equal to zero! Multiplying Nosé's Hamiltonian by s gives new motion equations with all the rates, except $\dot{\zeta}$, multiplied by s, thus (almost) scaling the time automatically :

$$\mathcal{H}_{\text{Dettmann}} = s\mathcal{H}_{\text{Nosé}} \longrightarrow$$

$$(\dot{q}, \dot{p}, \dot{s})_{\text{Dettmann}} = s(\dot{q}, \dot{p}, \dot{s})_{\text{Nosé}} \ ; \ \dot{\zeta}_{\text{Dettmann}} = s\dot{\zeta}_{\text{Nosé}} - \mathcal{H}_{\text{Nosé}} \ .$$

Note here that setting $\mathcal{H}_{\text{Nosé}} = 0$ removes the extra contribution to $\dot{\zeta}$.

$$\mathcal{H}_{\text{Dettmann}} = (1/2)[\ (p^2/s) + sq^2 + s\ln(s^2) + s\zeta^2 \ ] \equiv 0 \longrightarrow$$

$$\dot{\zeta} = (1/2)[\ (p/s)^2 - q^2 - \ln(s^2) - \zeta^2 \ ] - 1 \equiv (p/s)^2 - 1 \ .$$

The "Dettmann" motion equations replicate the "Nosé-Hoover" form once the scaled momentum (p/s) is replaced by p :

$$\dot{q} = (p/s) \ ; \ \dot{p} = -qs \ ; \ \dot{\zeta} = (p^2/s^2) - 1 \ ; \ \dot{s} = s\zeta \ [\ \text{Dettmann} \ ] \ .$$

Notice that the (q, p, s, ζ) trajectory from Dettmann's Hamiltonian is identical to Nosé's. The last remaining simplification is to replace the scaled momentum (p/s) by the actual momentum (for which we still use the symbol p) and to discard the extraneous time-scaling variable s :

$$\dot{q} = p \ ; \ \dot{p} = -q - \zeta p \ ; \ \dot{\zeta} = p^2 - 1 \ ; \ [\ \text{Nosé} - \text{Hoover} \ ] \ .$$

I developed these last "Nosé-Hoover" equations in the summer of 1984 in an effort to understand Nosé's new approach to a thermostatted mechanics. This last form, augmented by a relaxation time, is the one which has proved extremely useful to those interested in simulating the mechanics of thermostatted many-body systems.

5.2 Ergodicity and Linear Response Theory

Statistical Mechanics provides all of equilibrium thermodynamics and even a little (linear) bit of hydrodynamics close to equilibrium. This information is valid whenever the dynamics is either ergodic or, in cases where the dynamics is too complex for ergodicity, is at least well-represented by equilibrium Gibbs' ensembles. The nonequilibrium link to hydrodynamics describes the mass, momentum, and energy currents which can be excited by external fields or driven by boundary or initial conditions. Provided that these perturbations are small a well-established "linear-response" theory (a first-order perturbation theory) due to Green and Kubo in the 1950s expresses the corresponding transport coefficients as time integrals of equilibrium correlation functions.

As an example consider the Hamiltonian perturbation describing a strain rate $\dot{\epsilon}$:

$$\Delta\mathcal{H} = \dot{\epsilon} \sum yp_x \longrightarrow \{ \ \Delta\dot{x} = \dot{\epsilon}y \ ; \ \Delta\dot{p}_y = -\dot{\epsilon}p_x \ \} \ .$$

This is the "Doll's Tensor" perturbation, a pun relating the child's "Kewpie Doll" of the early 20th century to the $(qp) \simeq (yp_x)$ form of the added shear-flow term.[9] This perturbation adds a small shear flow, $\dot{x} \simeq (du/dy) \propto y$, to the motion, and also changes the energy, which in turn changes the initially-canonical distribution :

$$\{ \, \Delta\phi_{ij} = -\dot{\epsilon}[(yF_x)_i - (yF_x)_j] \; ; \; \Delta(p_y^2)/2 = -\dot{\epsilon}p_x p_y \, \} \longrightarrow \Delta\dot{E} \equiv -\dot{\epsilon}P_{xy}V \, .$$

Evidently the linear perturbation provides a mean value of $P_{xy} = -\eta\dot{\epsilon}$ at time t in terms of Gibbs' canonical equilibrium distribution function :

$$\eta = (V/kT) \langle \, \textstyle\int_0^t P_{xy}(t')P_{xy}(t)dt' \, \rangle_{eq} \, .$$

Thus Green and Kubo's linear perturbation theory provides a formulation of the Newtonian shear viscosity η in terms of equilibrium stress fluctuations. The equilibrium time integral of the large-system limiting correlation function runs from zero to infinity. There are similar results for the thermal and electrical conductivities as well as for the self-diffusion coefficient.

5.3 Ergodicity of Hard-Disk and Soft-Disk Cell Models

Before computer simulations the structure of dense fluids was not well understood. Johannes van der Waals (1837-1923) had the idea that the particles took up space (van der Waals' b) and that configurational probabilities were related to their mutual attraction (van der Waals' a). These are the two nonideal features in van der Waals' equation of state for fluids:

$$PV/NkT = [\, (V/(V - Nb) \,] - (Na/VkT) \, .$$

These simple ideas were enough to formulate his equation of state which nicely fits the experimental fact of the coexistence of two fluid phases, "gas" and "liquid" below a "critical temperature" above which the two phases become indistinguishable. John Bernal (1901-1971) had the idea that liquid structure could be modelled by a tightly packed collection of ball bearings. Joel Hildebrand (1881-1983) pursued a similar idea, photographing a collection of gelatin spheres in an aquarium-like tank.

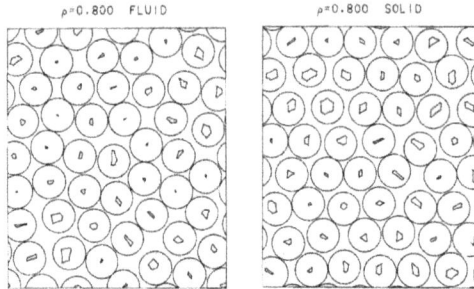

Fig. 5.2: Here we see the "free volumes" [areas in this two-dimensional case] available to hard disks in the fluid phase and the solid phase. Here the density relative to close packing is (4/5). It is interesting to see that the solid-phase free volume is considerably larger than that of the fluid. The link between the free volumes and thermodynamics motivates the use of cell models as illustrated in **Figures 5.3 and 5.4**.

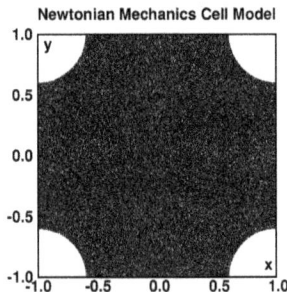

Fig. 5.3: Here 300,000 points from a 30,000,000-point cell-model trajectory are shown for a soft-disk potential $\phi(r < 1) = (1 - r^2)^4$. The forces on the wanderer particle are exerted by four scatterers located at the vertices of the 2×2 periodic cell. The initial condition, (0,0) for the coordinates and (0.6,0.8) for the velocity produces this trajectory which looks as if it could be ergodic. The following **Figure 5.4** shows that this one-body problem is not at all ergodic. The timestep is 0.001.

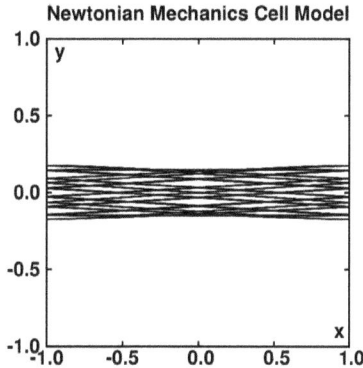

Newtonian Mechanics Cell Model

Fig. 5.4: 300,000 points from a simulation like that in **Figure 5.3** but with initial velocity $(0.998, \sqrt{1 - 0.998^2})$. This "regular" trajectory shows that the energy surface for this one-body problem has not only a relatively large chaotic sea but also stable tori and periodic orbits. This complexity typifies small-system Hamiltonian chaos and results in time averages which are not the same as Gibbs' ensemble averages.

A computational analog, pursued by those seeking to understand liquid structure during the period from the 1920s to the 1960s, was to create structural models of a single typical particle's environment in an N-body system.[10] The idea was to carry out single-particle phase-space integrations as a replacement for the (N-th root of the) coordinate-dependent part of the statistical-mechanical partition function. The one-particle "configurational integral", using the canonical probability density $e^{-\mathcal{H}/kT}$, was termed the "free volume" v_f :

$$e^{-A/kT} \equiv e^{-E/kT} e^{S/k} = \int \cdots \int e^{-\mathcal{H}/kT} \prod dq dp \simeq$$

$$e^{-\Phi_0/kT} [\int e^{-p^2/2mkT} dp\,]^{DN} v_f^N \; ; \; v_f = \int e^{-\delta\phi/kT} dq \; .$$

Here Φ_0 represents the system potential energy with all particles at their cell centers and $\delta\phi$ represents the change in potential when the "free-volume" integration over a cell is carried out. The momentum integral is just a Gaussian integral, $(2\pi mkT)^N$ for a two-dimensional system. The replacement of the configurational integral by the Nth power of a single particle integral is usually viewed as a "reasonable approximation". An alternative way to visualize the free volume is to imagine the isothermal dynamics of a very light particle, otherwise identical to its much heavier neighbors. Such a particle would explore its "cell" while all the others would be nearly motionless, providing a physical picture for the states accessible to a particle in terms of a cell integral.

This imagery is quite compelling for hard disks and hard spheres, for which free-volume calculations can provide an *exact* description of the thermodynamic equation of state. This fortunate circumstance is a consequence of the mass-independence of the potential energy. Light particles or heavy particles have exactly the same spatial distributions and pair correlation function because these only depend upon the (mass-independent) potential energy function. For hard disks of diameter σ the exact relation comes from the virial theorem and is

$$(PV/NkT) = 1 + (\sigma/4)\langle\, s_f/v_f \,\rangle \ .$$

Here s_f is the surface of the free volume v_f. **Figure 5.2** showed two 48-disk periodic systems in both the fluid and in the solid phases 25% expanded from the closest-packed number density $(N/V) = \sqrt{(4/3)}/\sigma^2$.

It is interesting that the more-nearly-regular structure of the solid phase gives a larger free volume than that of the fluid. At high densities near close packing it is apparent that the mean value of the free volume is much less than the volume per particle, (V/N). At low density the free volume is $(V - N\pi\sigma^2)$ with a surface $2\pi N\sigma$, reproducing the second virial coefficient for the hard disks. The change from an intensive free volume $\simeq (V/N)$ to an extensive one $\simeq V$ behaves like a phase transition, but this "percolation transition" occurs at a density near one-fourth of the close-packed density, while the actual melting transition density is about three times greater than that.

For hard disks there is evidently a theoretical proof of ergodicity which would apply to this system as well as to the prototypical smallest one, a single disk bound to its cell by four neighbors. In 1997 with Harald Posch and Franz Vesely, I published cell-model results for this geometry using a soft-disk[4] potential. We found good agreement with Gibbs' isokinetic distribution. With the advantage of 30 years' experience and faster computers we now know that trajectories close to horizontal or vertical, nearly bisecting the cell, exhibit the same sorts of complexity found with smooth Hamiltonian chaos. **Figures 5.3 and 5.4** compare soft-disk trajectories for a time of 500 with two different initial velocities. The qualitative difference shows that chaotic and regular trajectories coexist. Regular trajectories, close-to-horizontal or close-to-vertical, show a lack of ergodicity for soft disks, qualitatively different to the behavior of hard disks. Though we do not display them here, trajectories for isokinetic Gaussian dynamics show qualitative behavior similar to that of their isoenergetic analogs.

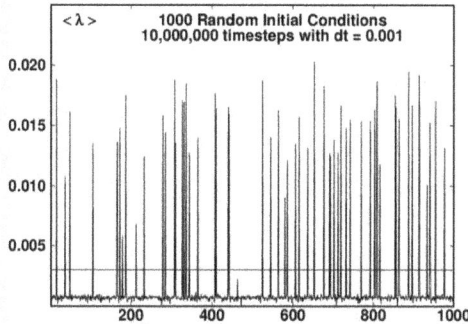

Fig. 5.5: Time-averaged Lyapunov exponents are shown here for 1000 randomly-chosen initial conditions selected from the stationary Nosé-Hoover oscillator distribution, $e^{-q^2/2}e^{-p^2/2}e^{-\zeta^2/2}/(2\pi)^{(3/2)}$. The abscissa numbers the sequences of the 1000 separate simulations reaching a time of 10,000. Sixty of the trials had average Lyapunov exponents greater than 0.003, the horizontal line separating likely chaotic solutions from likely regular solutions. The results cluster about 0 and 0.0139, the two long-time-averaged possibilities for λ_1. The 1000 cumulative Lyapunov exponents, $(1/t)\int_0^t \lambda_1(t')dt'$, are shown.

5.4 The Nosé and Nosé-Hoover Oscillators' Chaotic Sea

We pointed out, in connection with **Figure 5.1**, that the Nosé-Hoover oscillator's chaotic sea includes about 6% of its Gaussian measure, as shown in **Figure 5.5**. Nosé's Hamiltonian should have a canonical measure

$$f(q, p, s, \zeta) \propto e^{-(p/s)^2/2}e^{-q^2/2}e^{-\zeta^2/2}d(p/s)dqd\zeta ds .$$

We emphasize that both models share exactly the same cross section because any three-dimensional state with $p = 0$ corresponds to a family of similar four-dimensional states with $(p/s) = 0$ regardless of the value of s.

For both mechanics this model nicely illustrates typical aspects of Hamiltonian chaos where stable tori, with a vanishing largest Lyapunov exponent, neighbor a unique chaotic sea made up of exponentially unstable trajectories tracing out a "holy sea", perforated with toroidal holes but generating, as time goes on, three- or four-dimensional "fat-fractal" measures.

We saw in **Lecture 3** that Wang and Yang[11] were able to find, easily, fifteen separate varieties of knotted orbits for the Nosé-Hoover model in the slightly more complex case that results with a stiffer friction coefficient: $\dot\zeta = 10(p^2 - 1)$. In addition they found "trefoil" knots tied together by a single periodic orbit. These are the simplest knots that can be tied in a single length of rope. To tie a trefoil imagine an overhand knot in a short length of rope. Joining the free ends creates the trefoil knot.

No doubt many more topological examples are present in the phase space of this harmonic oscillator problem where the dynamics is regular rather than ergodic. In addition to the stable periodic orbits which are surrounded by concentric tori there are (obviously) infinitely-many unstable periodic orbits which are not. These latter orbits no doubt conceal even more complex knots. For their exploration good computer graphics software, which can expand and symmetrize such structures is a necessary aid. With this pervasive lack of ergodicity fresh in our minds let us review the thirty years' progress which was required in order to overcome this difficulty. We begin with simple diagnostic tests for ergodicity.

5.5   General Treatment of Bauer, Bulgac, Ju, and Kusnezov

Bauer, Bulgac, Ju, and Kusnezov (1990-1993)[12-14] took a much more wide-ranging approach to thermostatting using two or more control variables with a variety of functional forms. They discovered and emphasized the usefulness of cubic forces in promoting chaos and ergodicity. An example set of oscillator equations uses cubic friction as well as a cubic dependence on momentum. Bulgac and Ju found such a combination useful in modelling the chaotic dynamics of few-particle droplets.

$$\dot q = p \; ; \; \dot p = -q - \zeta^3 p - \xi p^3 \; ; \; \dot\zeta = p^2 - 1 \; ; \; \dot\xi = p^4 - 3p^2 \; [\text{Ju} - \text{Bulgac}].$$

They emphasized the usefulness of the phase-space continuity equation in finding thermostat motion equations consistent with given generalized Gaussian distributions. The Ju-Bulgac oscillator equations are consistent with this example:

$$f_{JB}(q,p,\zeta,\xi) \propto e^{-q^2/2}e^{-p^2/2}e^{-\zeta^4/4}e^{-\xi^2/2} ,$$

Gaussian in (q,p,ζ^2,ξ) . Such a consistency test must necessarily be passed by *any* ergodic set of motion equations. This is the gist of their work.

5.6   Ergodicity of Oscillator Models from 1984 to 2016

Many thermostat models have been developed since Nosé's 1984 efforts. When the Nosé-Hoover oscillator system failed ergodicity, three improved models followed, each based on four differential equations rather than three. The three models were successful in generating ergodic distributions consistent with Gibbs' canonical distribution. Two of them were consistent with the same four-dimensional Gaussian phase-space distribution :

$$(2\pi)^2 f(q, p, \zeta, \xi) = e^{-q^2/2} e^{-p^2/2} e^{-\zeta^2/2} e^{-\xi^2/2}$$

[(Martyna − Klein − Tuckerman) and (Hoover − Holian)] .

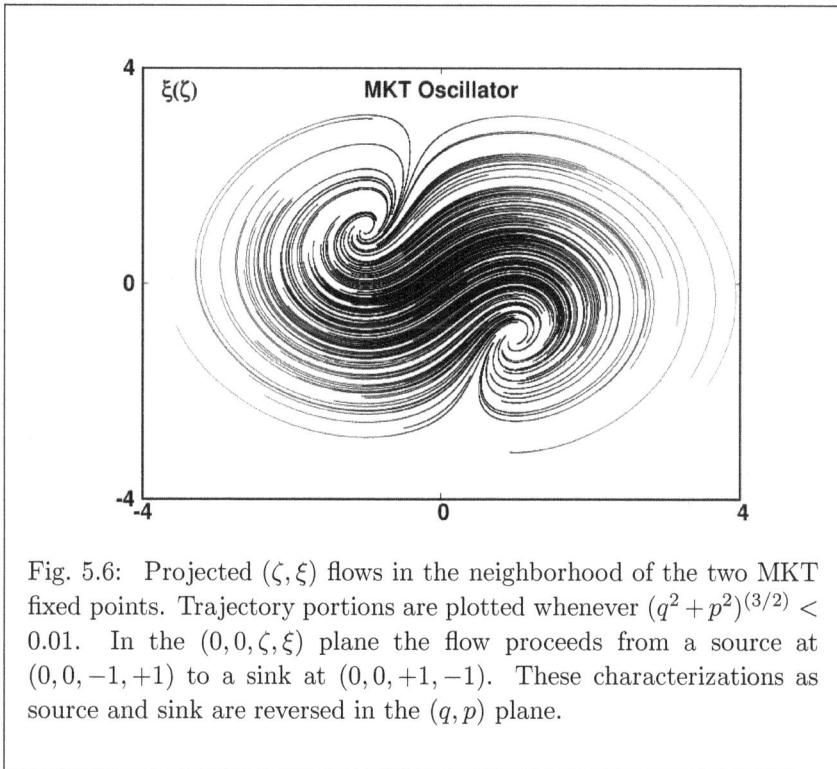

Fig. 5.6: Projected (ζ, ξ) flows in the neighborhood of the two MKT fixed points. Trajectory portions are plotted whenever $(q^2 + p^2)^{(3/2)} <$ 0.01. In the $(0, 0, \zeta, \xi)$ plane the flow proceeds from a source at $(0, 0, -1, +1)$ to a sink at $(0, 0, +1, -1)$. These characterizations as source and sink are reversed in the (q, p) plane.

In their work Martyna, Klein, and Tuckerman[15] add the friction coefficient ξ as a control variable governing the first control variable ζ :

$$\dot{q} = p \; ; \; \dot{p} = -q - \zeta p \; ; \; \dot{\zeta} = p^2 - 1 - \xi\zeta \; ; \; \dot{\xi} = \zeta^2 - 1$$

[Martyna − Klein − Tuckerman] .

Brad Holian and I had the idea of controlling the fourth moment of the momentum in addition to the second, using the new control variable ξ for that purpose.[16] The motion equations were no longer quadratic, but the stationary distribution function remained Gaussian in all four variables :

$$\dot{q} = p \; ; \; \dot{p} = -q - \zeta p - \xi p^3 \; ; \; \dot{\zeta} = p^2 - 1 \; ; \; \dot{\xi} = p^4 - 3p^2$$

[Hoover − Holian] .

Looking for "holes" in a four-dimensional distribution is more demanding than examining the three-dimensional sections of the Nosé and Nosé-Hoover thermostats. Double cross sections, confining $q^2 + p^2$ or $\zeta^2 + \xi^2$ to a small range are feasible. A simpler test is to measure the nonvanishing time-averaged moments. There are 4 second moments, 10 fourth moments, and 20 sixth moments. All of them can be evaluated by integrating new ordinary differential equations. For example, by integrating the equation $d(xxxxyy)/dt = x^4 y^2$ from 0 to t and dividing by t we get the cumulative average $\langle\, x^4 y^2 \,\rangle$. In this way it is feasible to determine all these moments to three- or four-figure accuracy.

5.6.1   *Stability Analysis Around the MKT Fixed Points*

The two fixed points of the MKT equations, $(q, p, \zeta, \xi) = (0, 0, \pm 1, \mp 1)$, are both of them *unstable*. Both act as simultaneous sources and sinks. See **Figure 5.6** as well as the Falling-Particle analog of **Figure 5.7**. It is worthwhile to linearize the equations about the fixed points. The reader should be able to show that introducing the infinitesimal deviations { δ } from the first fixed point, $(0, 0, +1, -1)$, gives the following set of four ordinary differential equations :

$$\dot{\delta}_q = \delta_p \; ; \; \dot{\delta}_p = -\delta_q - \delta_p \; ; \; \dot{\delta}_\zeta = \delta_\zeta - \delta_\xi \; ; \; \dot{\delta}_\xi = 2\delta_\zeta \; .$$

Another time differentiation of $\dot{\delta}_q$ and $\dot{\delta}_\zeta$ gives two second-order differential equations. They both have the form of damped oscillator equations with either stable or unstable damping, $\ddot{\delta} \simeq \mp \dot{\delta}$:

$$\ddot{\delta}_q = -\delta_q - \dot{\delta}_q \; ; \; \ddot{\delta}_\zeta = \dot{\delta}_\zeta - 2\delta_\zeta \; .$$

Perhaps coincidentally, perhaps not, the $\ddot{\delta}_p$ and $\ddot{\delta}_\xi$ equations have these same two forms so that the motion in the qp plane is stable and in the $\zeta\xi$ plane is unstable. It is remarkable that the motion around the other fixed point $(0, 0, -1, 1)$ is exactly opposite, unstable in the qp plane and stable in the $\zeta\xi$ plane. The upshot is that both of the fixed points are unstable. This instability is consistent with ergodicity though it does not show it.

In connection with the fixed points it is interesting to consider the simple problem of a one-dimensional point mass in a constant negative field with a Nosé-Hoover temperature equal to 1. The three motion equations,

$$\dot{x} = v \; ; \; \dot{v} = -1 - \zeta v \; ; \; \dot{\zeta} = v^2 - 1 \; ,$$

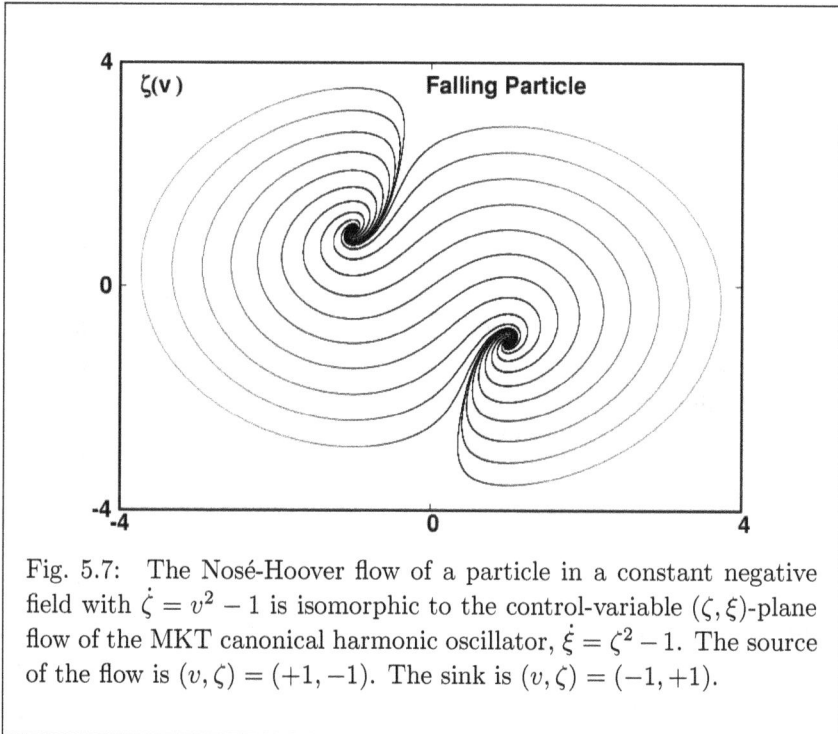

Fig. 5.7: The Nosé-Hoover flow of a particle in a constant negative field with $\dot{\zeta} = v^2 - 1$ is isomorphic to the control-variable (ζ, ξ)-plane flow of the MKT canonical harmonic oscillator, $\dot{\xi} = \zeta^2 - 1$. The source of the flow is $(v, \zeta) = (+1, -1)$. The sink is $(v, \zeta) = (-1, +1)$.

have two fixed points in the (v, ζ) plane, $(+1, -1)$ and $(-1, +1)$. Here the $\dot{x}$ equation is irrelevant so that we can and do ignore it. Linearization around these two fixed points gives two different equations for $\ddot{\delta}_v$:

$$(+1, -1) \rightarrow \ddot{\delta}_v = -2\delta_v + \dot{\delta}_v \; ; \; (-1, +1) \rightarrow \ddot{\delta}_v = -2\delta_v - \dot{\delta}_v \; .$$

The first (unstable) point acts as a source for the flow while the second acts as a sink. These equations are precisely isomorphic to the MKT flow in the (ζ, ξ) plane. See again **Figure 5.7** for representative streamlines of this simple two-dimensional flow.

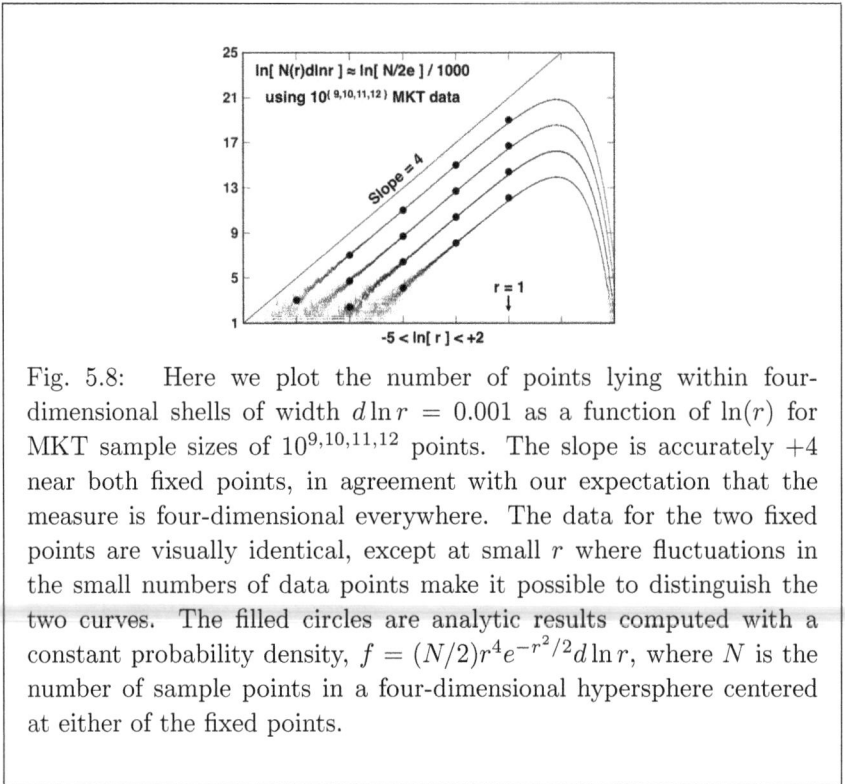

Fig. 5.8: Here we plot the number of points lying within four-dimensional shells of width $d\ln r = 0.001$ as a function of $\ln(r)$ for MKT sample sizes of $10^{9,10,11,12}$ points. The slope is accurately $+4$ near both fixed points, in agreement with our expectation that the measure is four-dimensional everywhere. The data for the two fixed points are visually identical, except at small r where fluctuations in the small numbers of data points make it possible to distinguish the two curves. The filled circles are analytic results computed with a constant probability density, $f = (N/2)r^4 e^{-r^2/2} d\ln r$, where N is the number of sample points in a four-dimensional hypersphere centered at either of the fixed points.

5.6.2 *Liouville Helps to Confirm the Ergodic Measure*

Any *steady* phase-space flow is unchanged by the passage of time, $(\partial f/\partial t) = 0$. Equivalently the flows into and out of any fixed phase volume must balance, $-\nabla_r \cdot (fv) = 0$. This is just a many-dimensional example of the continuity equation for the steady flow of mass density.

$$0 = (\partial\rho/\partial t)_r = -\nabla_r \cdot (\rho v) = 0 \ [\text{ Steady} - \text{State Continuity Equation }].$$

For consistency with any steady Gibbs' distribution the equations of motion, $v(r)$, must satisfy this steady-state Continuity Equation.

When the motion is ergodic and the distribution is Gaussian, as in the HH and MKT models, then the measure *at any point* is given by the four-dimensional Gaussian function

$$f(q,p,\zeta,\xi) = e^{-q^2/2} e^{-p^2/2} e^{-\zeta^2/2} e^{-\xi^2/2} dqdpd\zeta d\xi /(2\pi)^2 .$$

N points from this distribution can be sampled using ergodic dynamics, solving the MKT equations, or by Monte Carlo, using the same procedure we developed for the Maxwell-Boltzmann velocity distribution :

```
10    TRY  = 10*(RUND(INTX,INTY)-0.5D00)
      BOLTZ = DEXP(-TRY*TRY/2)
      IF(BOLTZ.LT.RUND(INTX,INTY) GO TO 10
      Q = TRY
```

The two random numbers appearing here are different as each call of the generator changes the two seeds `INTX` and `INTY` . Four such sequences give a sampled (`Q,P,Z,X`) point from the Gaussian distribution. We can then measure the number of points in a hyperspherical shell of radius r and thickness $d\ln(r)$. It is straightforward to show that the hypervolume of a four-dimensional spherical shell of thickness $d\ln r$ is $2\pi^2 r^4 d\ln r$ so that measuring the number of points in such a four-dimensional shell provides a quantitative check for ergodicity. At a distance r from the origin the number of points is $(N/2)r^4 e^{-r^2/2} d\ln r$. **Figure 5.8** shows the distributions of $10^{9,10,11,12}$ points from a simulation of the MKT equations where r is the distance from the two fixed points $(q, p, \zeta, \xi) = (0, 0, \pm 1, \mp 1)$. The individual points shown here are $(N/2e)r^4 d\ln(r)$ where the "e" in the denominator takes into account the distance $\sqrt{2}$ of the two fixed points from the origin.

In the MKT case Carol and I computed the measures at all the integer points on and inside a $2 \times 2 \times 2 \times 2$ tesseract of sidelength 2 and found that all 81 combinations of $(-1, 0, +1)$ agreed, within three-figure accuracy, with the analytic values, $\{e^{-n/2} = 0.60653^n\}$, where n varies from 0 to 4.

The Hoover-Holian equations have no fixed point, but there is a stationary flow along the ζ axis if all of the other variables vanish. This flow is unstable transverse to the axis and the HH equations, like the MKT and BBJK equations, are ergodic for the oscillator. In the end all three of the four-equation models we have detailed, MKT, HH, and JB are ergodic.

The only useful ergodicity test we have omitted here for these four-dimensional models is the calculation of the largest Lyapunov exponent, which has a value independent of the initial condition in the ergodic case. We omit the details of that diagnostic here as we discussed the diagnostic algorithm earlier in connection with the Nosé and Nosé-Hoover oscillator problems, illustrated in **Figure 5.5**. In the latter nonergodic three-dimensional flow only a small fraction of initial conditions, about six percent of those chosen from the stationary measure, was chaotic.

In 2016 I discovered that it was possible, and in fact relatively easy, to thermostat a one-dimensional oscillator with a single control variable rather than two. Tapias, Bravetti, and Sanders were able to extend that idea to the more complicated Mexican Hat problem. Let us chronicle some of the high spots along the road to those successful efforts.

5.6.3 *2001: Sergi and Ferrario's Thermostatted Oscillator*

In 2001 Sergi and Ferrario introduced an oscillator model with a single thermostat variable.[17] The simplest version of their equations is as follows:

$$\dot{q} = p + \zeta p \; ; \; \dot{p} = -q - \zeta p \; ; \; \dot{\zeta} = p^2 - 1 - qp \; [\; \text{Sergi} - \text{Ferrario} \;] .$$

Their claim that this model is ergodic turned out to be false.[18] In fact an initial condition $(q, p, \zeta) = (1, 1, 1)$ yields a simple torus. See **Figure 5.9**. It seems likely that the mistaken identification of this model as ergodic was due to the authors' solving four differential equations (one of them irrelevant, like Nosé's $\dot{s}$ equation) rather than three, complicating the diagnosis of the many tori that infest the chaotic sea generated by their "ergodic" model. With Puneet Patra and Clint Sprott I carried out a detailed investigation of this model and one of its cubic generalizations, but we found no easy route to ergodicity.

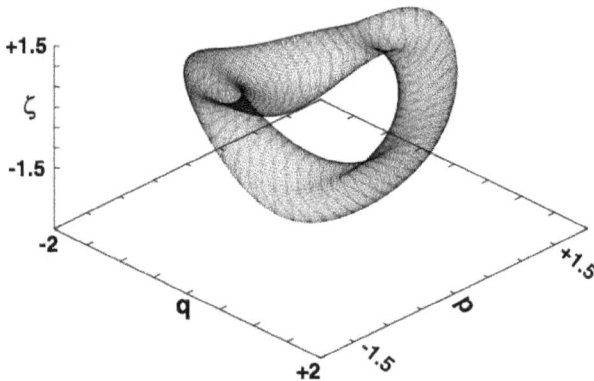

Fig. 5.9: Here is the torus generated by the Sergi-Ferrario oscillator equations[18] with the initial conditions $(q, p, \zeta) = (1, 1, 1)$. 50,000 out of 5,000,000 points generated with $dt = 0.001$ are shown here.

5.6.4   *2010: Bhattacharya-Ezra-Patra-Sergi Ergodicity*

Sergi and Ezra (2010) and Patra and Bhattacharya (2014) independently had the same idea,[19, 20] thermostatting both the coordinate (or force) and the momentum with separate control variables, ξ and ζ respectively:

$$\dot{q} = p - \xi q \; ; \; \dot{p} = -q - \zeta p \; ; \; \dot{\zeta} = p^2 - 1 \; ; \; \dot{\xi} = q^2 - 1 \; [\text{ BEPS }] \; .$$

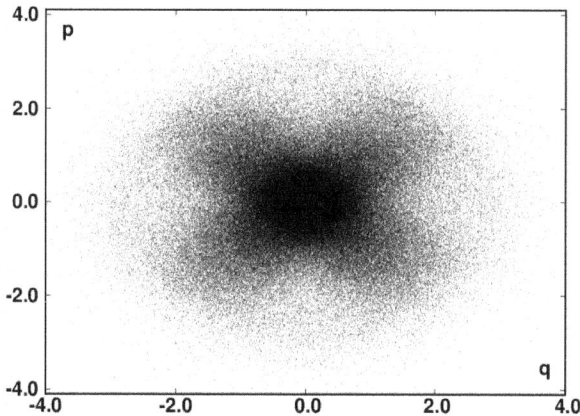

Fig. 5.10:   Here is a projected (q, p) time exposure of the chaotic sea resulting from the Bhattacharya-Ezra-Patra-Sergi thermostatted oscillator with the initial conditions $(q, p, \zeta, \xi) = (1, 1, 1, 1)$. The lack of circular symmetry is striking, despite the analytic steady-state distribution $f(q, p, \zeta, \xi) = e^{-q^2/2}e^{-p^2/2}e^{-\zeta^2/2}e^{-\xi^2/2}/(2\pi)^2$. Compare the two figures 2 plotted by BEPS in References 19 and 20. The fluctuations in this projection are quite slow, such that two successive time exposures of this length (300,000 points spaced equally at $\Delta t = 1000dt = 1$) can look very different, though the one shown here is preponderant over the others.[18] **Figure 5.10** shows 300,000 out of a batch of 300,000,000 points. This batch was selected from ten as a "typical" one.

Just as in the Sergi-Ferrario model both variables (q, p) are thermostatted, though here with two different control variables rather than one. These equations do not provide ergodicity. They do not even provide circular symmetry in the (q, p) plane. They are consistent with the four-dimensional Gaussian distribution, but only about half the measure is chaotic. Because both groups of researchers plotted the (q, p) projection of the chaotic sea in their publications the visual resemblance triggered a *déjà vu* moment in me. I realized, on seeing the Sergi-Ezra paper, that the two models were *identical*. See **Figure 5.10** for the (q, p) projection of the BEPS chaotic sea.

Exercise: There is a cubic variation of the BEPS model which *is* apparently ergodic (with all the parameters equal to unity). See if you can find it.

5.6.5 *Ergodicity and Weak Control*

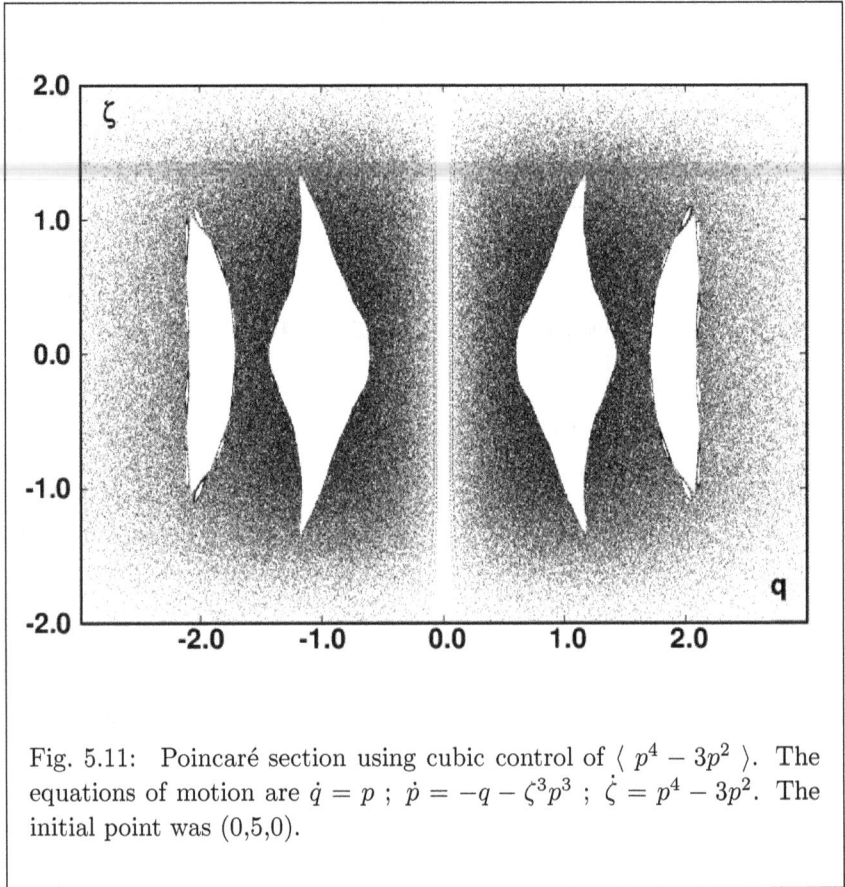

Fig. 5.11: Poincaré section using cubic control of $\langle\, p^4 - 3p^2 \,\rangle$. The equations of motion are $\dot{q} = p$; $\dot{p} = -q - \zeta^3 p^3$; $\dot{\zeta} = p^4 - 3p^2$. The initial point was $(0,5,0)$.

The work in the previous two Sections suggested the possibility of controlling two oscillator moments "weakly" as in this example :[21]

$$\dot{q} = p - 0.827\zeta^3 q \; ; \; \dot{p} = -q - 0.273\zeta^3 p^3 \; ; \; \dot{\zeta} = 0.827(q^2 - 1) + 0.273(p^4 - 3p^2).$$

The parameters 0.273 and 0.827 were found by a computerized search, using enough trajectory points so that the generation time took the same number of seconds required for a quick look at a cross section. "Interesting" cross sections were then examined in more detail in the search for ergodicity. The example just given was found in this way. See **Figure 5.11** for the special case where $\dot{q} = p$. The resulting section is clearly not ergodic.

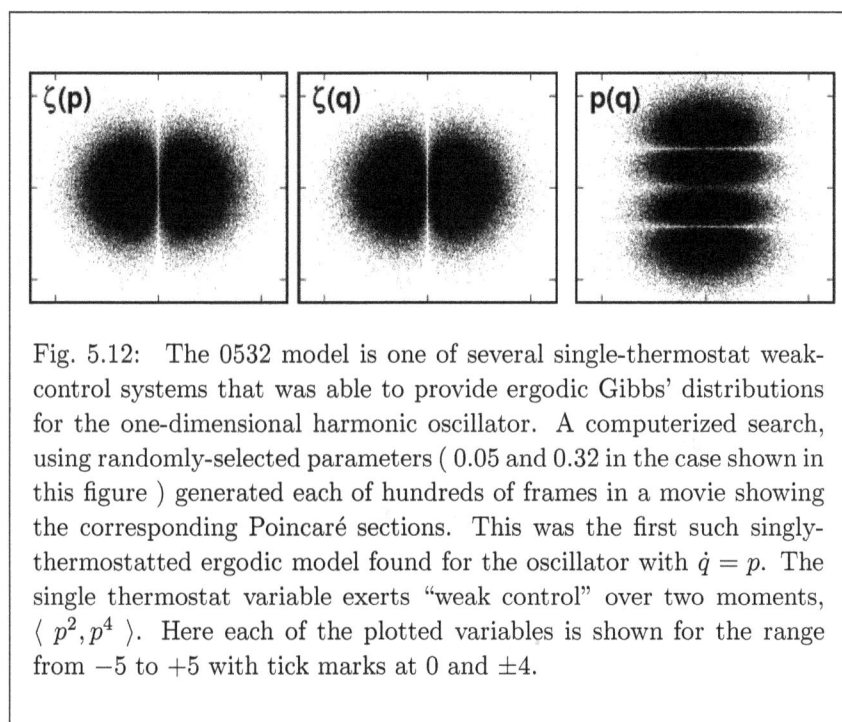

Fig. 5.12: The 0532 model is one of several single-thermostat weak-control systems that was able to provide ergodic Gibbs' distributions for the one-dimensional harmonic oscillator. A computerized search, using randomly-selected parameters (0.05 and 0.32 in the case shown in this figure) generated each of hundreds of frames in a movie showing the corresponding Poincaré sections. This was the first such singly-thermostatted ergodic model found for the oscillator with $\dot{q} = p$. The single thermostat variable exerts "weak control" over two moments, $\langle p^2, p^4 \rangle$. Here each of the plotted variables is shown for the range from -5 to $+5$ with tick marks at 0 and ± 4.

Because it is highly desirable to retain unchanged the equation of motion defining velocity, $\dot{q} = p$, it was natural after successful weak joint control of a coordinate moment and a velocity moment to direct additional weak-control searches toward two velocity moments. Because the velocity distribution is, at equilibrium, always the same, the simple Maxwell-Boltzmann Gaussian, it is sensible to seek to reproduce that Gaussian rather than to consider the case-by-case treatments of different potential functions. Sure enough, a search for simple combinations consistent with the even moments from Gibbs' distribution :

$$\langle\, p^{\{\,2,\,4,\,6\,\}}\,\rangle = \{\,1,\,3,\,15\,\}$$

turned up a variety of promising candidates. A relatively simple choice, which we call the 0532 Model, passes all the tests for ergodicity.

$$\dot{q} = p\ ;\ \dot{p} = -q - \zeta p(0.05 + 0.32p^2)\ ;\ \dot{\zeta} = 0.05(p^2 - 1) + 0.32(p^4 - 3p^2)$$

[0532 Model] .

See **Figure 5.12** for the Poincaré sections of the 0532 Model. The same ideas can be pursued for other simple systems like the pendulum and can also be applied to nonequilibrium systems. My attempts to extend the same idea to the quartic and Mexican Hat potentials failed, and motivated related work carried out by Tapias, Bravetti and Sanders.[22]

5.6.6 *2016: Tapias-Bravetti-Sanders' Logistic Thermostat*

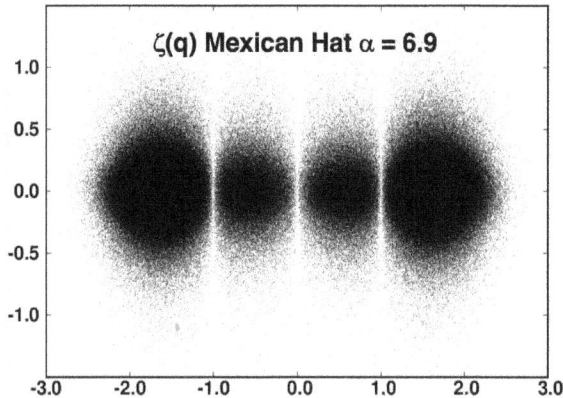

Fig. 5.13:   A Poincaré section for the Mexican Hat potential found using Tapias, Bravetti, and Sanders' hyperbolic-tangent thermostat. The fat tail of the sech($\alpha\zeta$) distribution, where ζ is the thermostat variable and α is a parameter, provides larger moments of ζ. The ratio of the fourth to the square of the second moment is about five rather than precisely three found with a Gaussian distribution.

Our feeling of success in finding a variety of ergodic thermostats, all based on the weak control of two or more moments, was dampened when we attempted the extension of these same models to the quartic oscillator and the Mexican Hat potential,

$$\phi_{\text{quartic}}(x) = (x^4/4) \ ; \ \phi_{\text{MH}}(x) = -(x^2/2) + (x^4/4) \ .$$

I looked at hundreds, perhaps even thousands, of randomly-generated weak-control recipes applied to the Mexican Hat problem. None of them was successful. In the end Carol and I chose this "hard problem" as the Ian Snook Prize Problem of 2016, hoping it would catch the eye of an imaginative physicist or mathematician. We were very pleased to have a prize-winning three-man entry solve the problem.[22] The key idea was to use a different hyperbolic-function form for a thermostat variable rather than the linear and cubic ideas which we were unable to apply successfully.

Consider using a hyperbolic-tangent thermostat for the oscillator :

$$\dot{q} = p \ ; \ \dot{p} = -q - \alpha p \tanh(\alpha\zeta) \ ; \ \dot{\zeta} = p^2 - 1 \ .$$

$$\longleftrightarrow (2\pi^2/\alpha)f(q,p,\zeta) = e^{-q^2/2}e^{-p^2/2}/\cosh(\alpha\zeta) \ [\text{ TBS }] \ .$$

We have included a parameter α here because this thermostat with a coefficient of 1 or 2 does not provide ergodicity for the simple harmonic oscillator. Three is enough ! The hyperbolic function is a handy probability density with the simple normalization :

$$\int_{-\infty}^{+\infty} \text{sech}(\alpha\zeta)d\zeta \equiv (\pi/\alpha) \ .$$

The "sech" distribution has a "fatter tail" than the Gaussian so that the fourth moment divided by the square of the second is about five rather than exactly three. The analog of Liouville's Theorem using the hyperbolic tangent friction with the $\text{sech}(\alpha\zeta)$ momentum distribution satisfies the steady-state phase-space continuity equation :

$$(\partial \ln f/\partial t) = 0 =$$

$$-(\partial\dot{p}/\partial p) - (\partial \ln f/\partial q)\dot{q} - (\partial \ln f/\partial p)\dot{p} - (\partial \ln f/\partial \zeta)\dot{\zeta} =$$

$$\alpha\tanh(\alpha\zeta) + qp + p[-q - \alpha p \tanh(\alpha\zeta)] + \alpha\tanh(\alpha\zeta)(p^2-1) \equiv 0 \ .$$

I found that Liouville's continuity equation was the most useful tool for analyzing chaos in nonHamiltonian systems. Any smooth and stationary distribution, where $(\partial f/\partial t)$ necessarily vanishes, must simultaneously follow the flow rule describing the flux of probability $f(r,t)$ into and out of a small Cartesian volume element.

$$0 = (\partial f/\partial t) \equiv -\nabla_r \cdot [\, f(r)v(r)\,] = 0 \;;\; f(r) \propto f_{\text{Gibbs}} \;.$$

Here both the phase-space variables $\{\,r\,\}$ and their time derivatives $\{\,v = \dot{r}\,\}$ include any time-scaling or thermostatting variables required by an algorithm leading to the canonical distribution. I applied this rule to show that the distribution of the friction coefficient ζ is Gaussian for the Nosé-Hoover motion equations. Tapias, Bravetti, and Sanders introduced their "Logistic" probability density using this same idea. The Logistic name comes from two of the simplest evolutionary models, the Logistic Map and the Logistic [differential] Equation :

$$y(t+dt) \propto y(t)[\,1 - y(t)\,] \;;\; \dot{y} \propto y(1-y) \;,$$

as described further in Wikipedia.

Nosé was fully aware of the need for ergodicity in his approach to the canonical ensemble. Because there is no foolproof method for establishing the ergodicity of a particular set of ordinary differential equations, he simply assumed it, reasoning that ergodicity would not be a problem in simulating the molecular dynamics of many-body systems.

When I checked numerical solutions of Nosé's motion equations for the harmonic oscillator I immediately found tori rather than the expected canonical Gaussian distributions. Within about a year Harald Posch, Franz Vesely and I believed we had fully documented the complexity of the oscillator problem.[4] In the process the three of us decorated all the accessible horizontal surfaces in Harald's Viennese laser lab with hundreds of computer-generated plots of thermostatted oscillator dynamics.

Figure 5.14 shows some interesting detail of which we were blissfully unaware in 1985. Choosing 71 initial conditions equally spaced along the q axis $(0.50, 0, 0) \leq (q, p, \zeta) \leq (1.20, 0, 0)$ I plotted 5000 penetrations of the $(p = 0)$ plane for each. Apparent tori resulted most of the time. The rings with "holes" correspond to "tiny" islands in the plane indicating the presence of tiny thin tori perpendicular to the plane with a huge number of penetrations for the periodic orbits at the centers of those islands.

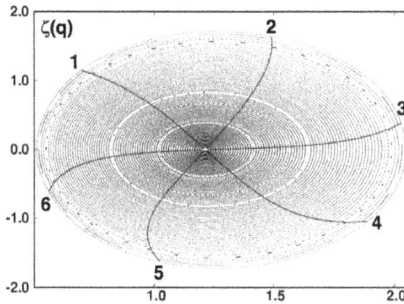

Fig. 5.14: Successive penetrations of the $p = 0$ plane from 71 equally-spaced initial conditions along the q axis from 0.50 to 2.20. At the central periodic orbit the winding number is apparently six, with twelve penetrations of the plane necessary to return to the initial point. The winding slows away from the center as shown by the six lines of even-numbered penetrations, which fail to return to the q axis except very near the central periodic orbit. Notice the irregular toroidal solutions for some of the initial points indicating that this Hamiltonian problem (Nosé's Hamiltonian) has the usual regular islands interspersed in the chaotic sea and separating some of the tori from their neighbors.

Shown also in the figure are nearly radial sets of 71 points each at the locations of Poincaré penetrations after 1, 2, 3, 4, 5, and 6 trips the long way around the torus. Evidently the large torus is undergoing shear as the angular velocity varies with radius and the sixth penetration is only coincident with the initial condition (the q axis) near the central periodic orbit. It appears that the winding number (times around the torus the "long" way for the central periodic orbit) is six.

In **Figure 5.15** we show a torus as well as the track of a (very-nearly) periodic orbit on the torus' surface. Also shown are the 12 penetrations, 6 with positive $\dot{p} > 0$ and six with negative $\dot{p} < 0$. It would take us too far afield to discuss the complex geometry of shearing tori with a mulitiplicity of thin tori and very long periodic orbits wound around them.

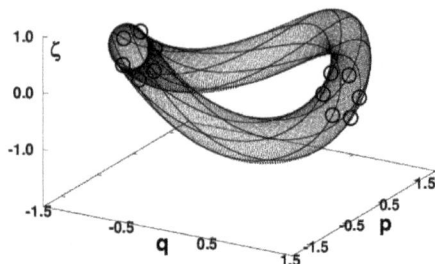

Fig. 5.15: Here a torus centered on the periodic orbit of **Figure 5.14** shows on its surface an orbit with a winding number close to six. The intersections of that orbit with the Poincaré plane are shown as the 12 open circles. The shear resulting from the dependence of the winding number on radius is a consequence of the complex nature of Hamiltonian chaos. The initial condition (1,0,0) was followed for 33,696 timesteps with $dt = 0.001$.

5.7 The Knotted Complexity of the Nosé-Hoover Tori

Recent work on the Nosé-Hoover oscillator, 30 years after my work with Harald and Franz,[4] revealed a new feature, *interlocking tori*.[11] And soon after that it was discovered that many pairs of the oscillator tori are tied together in knots ! Thus the simple oscillator is not so simple.[23, 24] In fact it is a computational challenge to determine whether or not tori are knotted (or are not knotted). Specialized software treating surfaces, such as the surface of a torus, can be used to render groups of tori, showing which are knotted and which are not. As trajectories only approximate surfaces a solution of this problem using points rather than surfaces is desirable.

Figure 5.16 shows the result of a pointwise algorithm designed to remove "hidden lines". To generate it the two tori shown in the Figure were represented by 50,000 points each of the tori. Then each pair of points, one point from each of the tori, was checked to determine its separation projected into the (q, p) plane. Whenever this distance was less than $10^{-3/2}$ the point corresponding to the smaller value of ζ was discarded. In this operation one torus lost about 500 points and the other 1000, providing the two gaps which can be seen in the Figure. Because this is a pairwise operation, which can be applied to any pair of torus' point sets this algorithm can be applied to more complex situations such as those discussed by Wang and Yang.[11]

Piotr Pieranski has devoted much of his life to both artistic and scientific studies of knots and to the optimization of their representations. The reader is urged to look for his innovative work on the internet. The study of knots is a classical field of mathematics in which noted men, Gauss, Kelvin, and Tait among them, have made contributions. I was most surprised to find knots appearing in familiar phase-space orbits, underlining the complexity of the thermostatted oscillator problem. Although the knotted tori in Wang and Yang's Figure 6[11] are stable it is (almost) obvious that the chaotic sea contains infinitely-many knots comprised of Lyapunov-unstable periodic orbits. Ergodicity does not rule out knots !

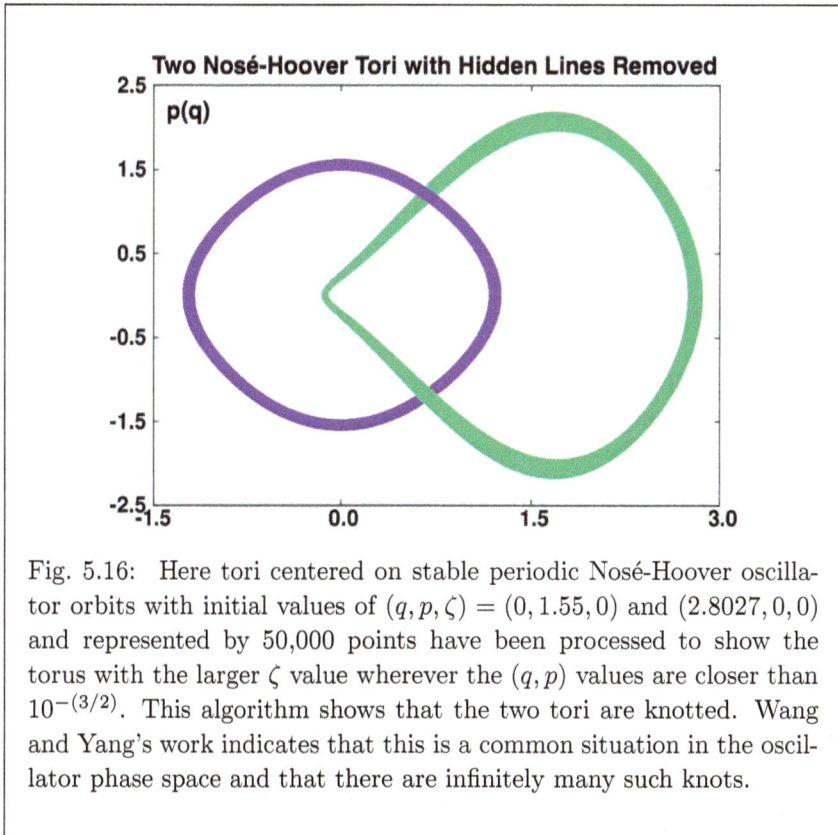

Fig. 5.16: Here tori centered on stable periodic Nosé-Hoover oscillator orbits with initial values of $(q, p, \zeta) = (0, 1.55, 0)$ and $(2.8027, 0, 0)$ and represented by 50,000 points have been processed to show the torus with the larger ζ value wherever the (q, p) values are closer than $10^{-(3/2)}$. This algorithm shows that the two tori are knotted. Wang and Yang's work indicates that this is a common situation in the oscillator phase space and that there are infinitely many such knots.

Finding an ergodic algorithm free of holes and tori took a few decades. Four differential equations with Gaussian distributions for two control variables, sometimes applied to forces and sometimes to momenta eventually led to three differential equations with both Gaussian and hyperbolic distributions using weak control rather than the usual differential or integral feedback. No doubt these ideas will prove useful in a variety of applications.

5.8   Summary of Lecture 5

An ergodic dynamics includes all the states in an appropriate ensemble where that ensemble is the stationary solution of the phase-space continuity equation :

$$0 = (\partial f/\partial t) = -\nabla_r \cdot (fv) = -v \cdot \nabla_r f - f\nabla_r \cdot v = 0 \ .$$

We will see that isothermal dynamics can provide realistic nonequilibrium simulations in situations which Hamiltonian mechanics cannot address. The Hamiltonian mechanics of small systems is typically complex, with structure on all spatial scales, while ergodic dynamics can provide smooth and accurate equilibrium ensemble averages with only a little extra work.

Nosé's time-scaling idea, introducing energy fluctuations into a time-scaled Hamiltonian system, does not reproduce Gibbs' canonical distribution for simple systems like the harmonic oscillator and the Mexican Hat potential. Until recently it was thought that at least two control variables were required for successful canonical averages. Recent work has shown that "weak control" of two or more dynamical averages or hyperbolic-tangent "logistic thermostat" control (rather than linear or cubic control) can replicate Gibbs' canonical distribution for a variety of one-dimensional problems.

The phase-space continuity equation has proved its worth over more than 30 years. It is a valuable generalization of Liouville's flow equation. *Any* dynamical model necessarily follows this rule, whether or not the co-moving phase volume varies along a trajectory. Here we have explored other tools, Poincaré sections, moments, Lyapunov-instability analysis, and measure theory. With these aids we are able to control and validate chaos so as to obtain ergodicity. Gibbs' "weak coupling" from 1902 has finally been made explicit though useful couplings have proved not to be particularly weak. The new ideas have already proved to be valuable in a variety of equilibrium and nonequilibrium settings.[23, 24]

References

1. J. W. Gibbs, *Elementary Principles in Statistical Mechanics* (Chas. Scribner's, 1902).
2. N. Metropolis, A. W. Rosenbluth, M. N. Rosenbluth, A. H. Teller, and E. Teller, "Equation of State Calculations by Fast Computing Machines", The Journal of Chemical Physics **12**, 1087-1092 (1953).
3. Wm. G. Hoover, "Canonical Dynamics: Equilibrium Phase-Space Distributions", Physical Review A **31**, 1695-1697 (1985).
4. H. A. Posch, Wm. G. Hoover, and F. J. Vesely, "Canonical Dynamics of the Nosé Oscillator: Stability, Order, and Chaos", Physical Review A **33**, 4253-4265 (1986).
5. S. Nosé, "A Molecular Dynamics Method for Simulations in the Canonical Ensemble", Molecular Physics **52**, 255-268 (1984).
6. S. Nosé, "A Unified Formulation of the Constant Temperature Molecular Dynamics Methods", The Journal of Chemical Physics **81**, 511-519 (1984).
7. Wm. G. Hoover, "Mécanique de Nonéquilibre à la Californienne", Physica **240A**, 1-11 (1997).
8. C. P. Dettmann and G. P. Morriss, "Hamiltonian Formulation of the Gaussian Isokinetic Thermostat", Physical Review E **54**, 2495-2500 (1996).
9. Wm. G. Hoover, C. G. Hoover, and J. Petravic, "Simulation of Two- and Three-Dimensional Dense-Fluid Shear Flows *via* Nonequilibrium Molecular Dynamics: Comparison of Time-and-Space-Averaged Stresses from Homogeneous Doll's and Sllod Shear Algorithms with Those from Boundary-Driven Shear", Physical Review E **78**, 046701 (2008).
10. J. O. Hirschfelder, C. F. Curtiss, and R. B. Bird, *The Molecular Theory of Gases and Liquids* (Wiley, New York, 1954).
11. L. Wang and X-S. Yang, "The Invariant Tori of Knot Type and the Interlinked Invariant Tori in the Nosé-Hoover System", The European Physical Journal **88B**, 78 (2015) = arχiv 1501.03375.
12. A. Bulgac and D. Kusnezov, "Canonical Ensemble Averages from Pseudomicrocanonical Dynamics", Physical Review A **42**, 5045-5048 (1990).
13. D. Kusnezov, A. Bulgac, and W. Bauer, "Canonical Ensembles from Chaos", Annals of Physics **204**, 155-185 (1990) and **214**, 180-218 (1992).
14. N. Ju and A. Bulgac, "Finite-Temperature Properties of Sodium Clusters", Physical Review B **48**, 2721-2732 (1993).
15. G. J. Martyna, M. L. Klein, and M. Tuckerman, "Nose-Hoover Chains: the Canonical Ensemble *via* Continuous Dynamics", The Journal of Chemical Physics **97**, 2635-2643 (1992).
16. W. G. Hoover and B. L. Holian, "Kinetic Moments Method for the Canonical Ensemble Distribution", Physics Letters A **211**, 253-257 (1996).
17. A. Sergi and M. Ferrario, "Non-Hamiltonian Equations of Motion with a Conserved Energy", Physical Review E **64**, 056125 (2001).
18. Wm. G. Hoover, J. C. Sprott, and P. K. Patra, "Ergodic Time-Reversible Chaos for Gibbs' Canonical Oscillator", Physics Letters A, **379**, 2935-2940 (2015) = arχiv 1503.06749.

19. A. Sergi and G. S. Ezra, "Bulgac-Kusnezov-Nosé-Hoover Thermostats", Physical Review E **81**, 036705 (2010).
20. P. K. Patra and B. Bhattacharya, "Improving the Ergodic Characteristics of Thermostats Using Higher Order Temperatures" = arχiv 1411:2194.
21. Wm. G. Hoover, J. C. Sprott, and C. G. Hoover, "Ergodicity of a Singly-Thermostatted Harmonic Oscillator", Communications in Nonlinear Science and Numerical Simulation **32**, 234-240 (2016) = arχiv 1504.07654.
22. D. Tapias, A. Bravetti, and D. P. Sanders, "Ergodicity of One-Dimensional Systems Coupled to the Logistic Thermostat", Computational Methods in Science and Technology **23**, 11-18 (2017) = arχiv 1611.05090.
23. J. C. Sprott, W. G. Hoover, and C. G. Hoover, "Heat Conduction and the Lack Thereof in Time-Reversible Dynamical Systems : Generalized Nosé-Hoover Oscillators with a Temperature Gradient", Physical Review E **89**, 042914 (2014) = arχiv 1401.1762.
24. L. Wang and X-S. Yang, "A Vast Amount of Various Invariant Tori in the Nosé-Hoover Oscillator", Chaos **25**, 123110 (2015).

Chapter 6

Equilibrium Thermodynamics + Nonequilibrium Hydrodynamics

/ Fundamentals of Thermodynamics / Carnot Cycles and Entropy / Fundamentals of Hydrodynamics / Pressure Tensor, Stress, Heat Flux / Continuity Equations / Equation of Motion / Energy Equation / Atomistic *versus* Continuum Theories / Gibbs' Thermodynamics / Gibbs' Paradox / Jaynes' Maximum Entropy / Boltzmann Entropy / Mayers' Virial Expansion / Rods, Squares, Disks, Cubes, and Spheres / Harmonic Entropy / Stress with Gravity and Rotation / Nonequilibrium Entropy Production / Baker Maps / Summary of Lecture 6 /

6.1 Thermodynamics and Hydrodynamics

To begin, let us step back a moment from the microscopic dynamics of small systems and consider the alternative macroscopic view of continuum mechanics. Macroscopic descriptions treat systems of interest as continua within which constitutive equation-of-state and dynamical variables can depend upon both space and time. Isolated systems, at equilibrium, are described by thermodynamics. Descriptions of thermodynamic systems, in addition to composition, include two "state variables". That two are required, but not three or more, is phenomenological. The possibilities for these state variables include not only the mechanical variables : pressure, volume, and energy ; but also the thermal variables which include temperature, and entropy.

Hydrodynamics or "Fluid Mechanics" deals with matter in motion. Typical systems include evolving gradients in velocity, pressure, and temperature. Typical hydrodynamic variables are continuous functions of space and time, including the density $\rho(r, t)$, velocity $v(r, t)$, specific energy $e(r, t)$, temperature $T(r, t)$, stress $\sigma(r, t)$, and heat flux $Q(r, t)$. The presence of time derivatives along with space derivatives implies that the hydrodynamic evolution equations for $(\dot{\rho}, \dot{v}, \dot{e})$ can be expressed in terms of the various thermodynamic variables and their gradients, $(\nabla v, \nabla \cdot v, \nabla \cdot P, \nabla \cdot Q, \nabla T)$. The resulting evolution equations for the density, velocity, and energy are *partial* differential equations, with algorithms for their numerical solution typically involving grids of fixed or moving "zones" or "cells".

Gibbs' statistical mechanics[1] and Boltzmann's dilute-gas hydrodynamics[2] connect the macroscopic description to atomistic mechanics. The microscopic and macroscopic descriptions necessarily disagree, and for two important reasons : [1] atomistic properties are point properties while continuum properties are continuous functions of both space and time ; [2] atomistic evolution equations are (nearly always) time-reversible while the phenomenological constitutive equations required by the continuum evolution equations are (nearly always) *irreversible*, obeying the Second Law of Thermodynamics.

Our goal here is to understand the connections between the microscopic and macroscopic descriptions. This understanding cannot be complete for the two reasons just cited, granularity and reversibility. I will sketch the structures of thermodynamics, hydrodynamics, and statistical mechanics as they are today. There is plenty still to be learned by simulating simple problems elucidating the differences among these three approaches. Let us begin by reviewing macroscopic thermodynamics. For simplicity we will restrict our discussion here to "simple fluids". Extending the basic ideas to solids, mixtures, and multiphase systems is relatively straightforward and opens up a host of interesting research ideas. Let us begin with a short exposition of the equilibrium thermodynamics of simple fluids.

6.2   Fundamentals of Thermodynamics

Thermodynamics began as an empirical subject. The main goal in developing this subject was adding in "thermal processes" described by temperature, heat, and entropy, to mechanics. Thermodynamics deals primarily with "equilibrium states" of matter, which are unchanged so long as the system is isolated from sources of energy. It is assumed that these states can be characterized by two state variables, for instance energy and volume, (E, V) or pressure and temperature, (P, T). The two thermal variables, temperature and entropy, not appearing in mechanics, both need to be defined and analyzed. For this purpose the fundamentally important thermodynamic system is the equilibrium ideal gas, which can serve (by definition) as a thermometer or as a heat reservoir. Temperature is introduced as the pressure × volume product of such a gas, requiring a new constant—either Boltzmann's constant k or the "gas constant" $R = kN_{\text{Avogadro}}$ relating thermal energies to mechanical energies, as in the ideal gas law, $PV = NkT$. For simplicity, we will mostly choose to set k or R equal to unity.

With the concepts of state and ideal-gas temperature in mind the subject can be summarized concisely : [1] energy is a "state function" so that *changes* in energy (through heat and work) in any cyclic process sum to zero ; [2] entropy is likewise a state function so that the contributions to $\oint dQ/T$ in a cyclic "reversible" process likewise sum to zero ; [3] there are "irreversible" processes, where some of states encountered are not at equilibrium – in such a case the entropy always increases. In addition to the first law of thermodynamics (energy is conserved) and the second law (entropy cannot decrease) the "zeroth law" is assumed (two bodies in thermal equilibrium with a third are necessarily in thermal equilibrium with each other). Substituting "mechanical" for "thermal" gives another tacit assumption, a "Zeroth Law" for mechanical processes.

Thermodynamics is primarily concerned with "reversible processes". These are processes which evolve through a sequence of equilibrium states. Volume changes and temperature changes occur sufficiently slowly that an equilibrium description applies thoughout the process. If the rates of change are too fast (so that entropy increases) the process is "irreversible". Let us consider how the new state variable associated with heat and temperature is introduced into thermodynamics. The concept arises from an analysis of an ideal gas' Carnot cycle, a thought experiment involving three gases, a "cold reservoir", a "hot reservoir", and a "working fluid".

6.2.1   *Carnot Cycle Proves Entropy is a State Function*

The Carnot cycle is a sequence of four thermodynamically reversible processes : a pair of expansions with the first isothermal and the second adiabatic, followed by a pair of compressions, the first isothermal and the next adiabatic. From the First Law of thermodynamics the energy change for each process is equal to the heat taken in less the work done, $dE = dQ - dW$. For an ideal gas, and hence also for the "working fluid" the isothermal processes "1" and "3" obey $dE = 0 = dQ - dW$; the adiabatic processes "2" and "4" obey $dQ = 0 = dE + dW$. The four-part cycle returns the working fluid to its starting point with the effect of converting the net heat taken in $\oint dQ$ to the net work done $\oint PdV$. So long as the cycle is thermodynamically reversible the state function nature of energy guarantees this equivalence :

$$\oint dE \equiv 0 \rightarrow \oint dQ_{\mathrm{rev}} - \oint dW_{\mathrm{rev}} = \oint TdS - \oint PdV = 0 \,.$$

This cycle also provides a thermodynamic definition of the entropy S .

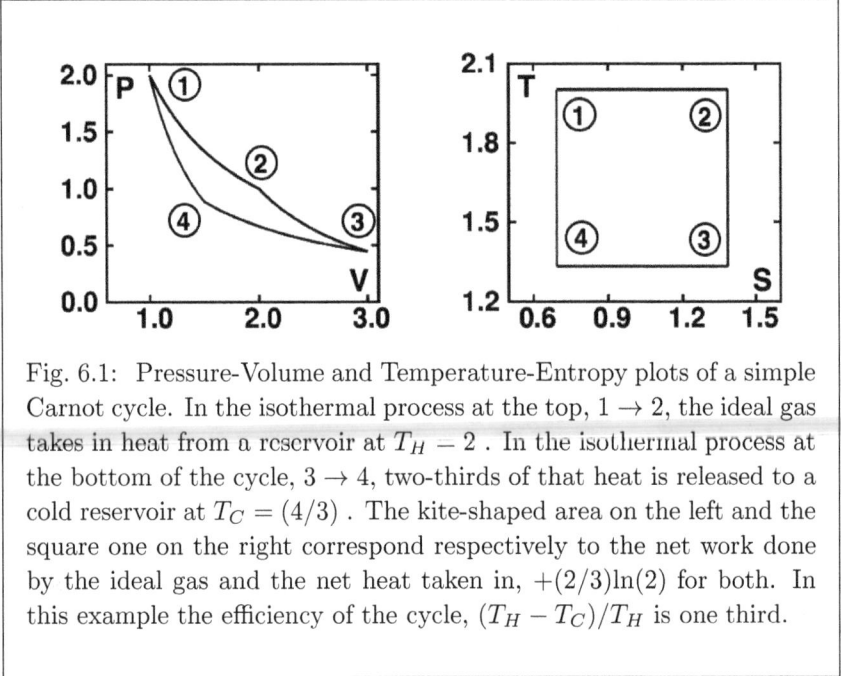

Fig. 6.1: Pressure-Volume and Temperature-Entropy plots of a simple Carnot cycle. In the isothermal process at the top, $1 \rightarrow 2$, the ideal gas takes in heat from a reservoir at $T_H - 2$. In the isothermal process at the bottom of the cycle, $3 \rightarrow 4$, two-thirds of that heat is released to a cold reservoir at $T_C = (4/3)$. The kite-shaped area on the left and the square one on the right correspond respectively to the net work done by the ideal gas and the net heat taken in, $+(2/3)\ln(2)$ for both. In this example the efficiency of the cycle, $(T_H - T_C)/T_H$ is one third.

A particular example for a two-dimensional ideal gas with the equation of state $PV = T = E$ and $S = \ln(V) + \ln(T)$ is illustrated in **Figure 6.1**, in both the (P, V) and (S, T) state planes. The Carnot Cycle is composed of the four reversible processes illustrated in **Figure 6.1** connecting the following four states of thermodynamic equilibrium :

1	$P = 2$	$V = 1$	$S = \ln(2)$	$T = 2$
2	$P = 1$	$V = 2$	$S = \ln(4)$	$T = 2$
3	$P = (4/9)$	$V = 3$	$S = \ln(4)$	$T = (4/3)$
4	$P = (8/9)$	$V = (3/2)$	$S = \ln(2)$	$T = (4/3)$

The four reversible processes making up the cycle are the following :

Isothermal	$1 \to 2$	$P = 2/V$	$dW = +dQ = +2\ln(2)$
Adiabatic	$2 \to 3$	$P = 4/V^2$	$dW = -dE = +(2/3)$
Isothermal	$3 \to 4$	$P = 4/(3V)$	$dW = +dQ = -(4/3)\ln(2)$
Adiabatic	$4 \to 1$	$P = 2/V^2$	$dW = -dE = -(2/3)$

The cycle is traversed clockwise so that the heat taken in and the work done are both positive and equal to $(2/3)\ln(2)$. Because the heat $T\Delta S$ taken in at the higher temperature is $2\ln(2)$ while two-thirds of that quantity, $(4/3)\ln(2)$, is released at the lower temperature the efficiency of the Carnot cycle is $(1/3) = (T_{hot} - T_{cold})/T_{hot}$. In the *general* case the efficiency has this same form $(T_{hot} - T_{cold})/T_{hot}$.

Although this Carnot cycle appears to be a very special case it should be noted that *any* cycle whatever can be composed of infinitesimal scale-model Carnot cycles. Because in each of the cycles the work done is equal to the heat taken in the same is true for the cycle as a whole. The fact that the entropy $S \equiv \oint dQ_{rev}/T$ is unchanged by the cycle shows that it is a "state function" for the ideal gas, just like P, V, E, and T. And because any fluid with a reversible equation of state $P(T,V)$ can be followed around such a cycle with $dW_{\text{fluid}} = -dW_{ideal}$ and $dQ_{\text{fluid}} = -dQ_{ideal}$ the entropy for such a fluid is also a state function.

To a large extent thermodynamics is concerned with tabulating the properties of materials. Such data make it possible to design processes obeying the first and second laws of thermodynamics. We will not touch on the details of chemical reactions here, but it is evident that energy conservation and the increase of entropy make it possible to distinguish the likely reaction pathways from the impossible ones in advance.

6.2.2   A Thermodynamic "Thought Experiment"

Before returning to our microscopic view based on the dynamics of many interacting particles we consider an example problem illustrating the predictive power of thermodynamics. Imagine a two-dimensional ideal gas, confined by a piston with mass M in a vertical gravitational field g. A second weight, in the form of a ball, also of mass M, is dropped from a height h above the first. See **Figure 6.2**. When equilibrium has ensued the weights are motionless, the volume is unchanged, and the energy lost by the falling ball has been transferred to the gas, doubling its temperature and pressure so as to support the additional weight. Energy conservation is enough to find the height from which the weight was dropped. Note that if the ball were applied gently the volume would be reduced to $V/\sqrt{2}$ as the process would be adiabatic so that PV^2 would be conserved. Alternatively, if the ball were dropped from a great height the energy and the final volume would become unbounded. Evidently the ball falls a net distance of (PV/Mg) so that the total energy of the gas is *doubled*. Although the details of equilibration are highly-complicated, involving repeated bounces of the ball upon the piston, and the dissipation of the piston motion into viscous heat, and the dissipation of thermal gradients to equilibrium through the mechanics of heat conductivity, there is no difficulty in finding the final state of the system. The details of several models of the nonequilibrium equilibration process can be found in Reference 3.

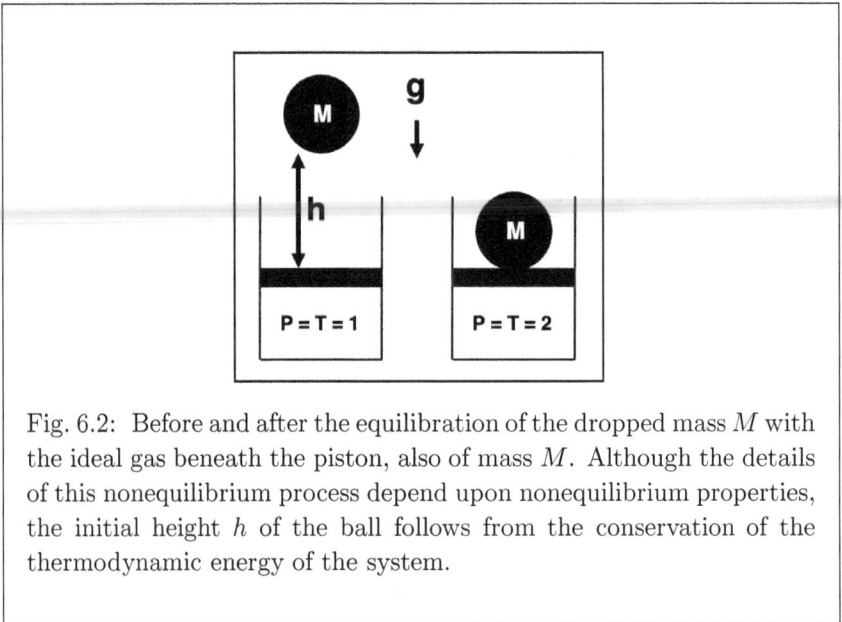

Fig. 6.2: Before and after the equilibration of the dropped mass M with the ideal gas beneath the piston, also of mass M. Although the details of this nonequilibrium process depend upon nonequilibrium properties, the initial height h of the ball follows from the conservation of the thermodynamic energy of the system.

Throughout the equilibration process the ball exerts occasional impulses to the piston below and that piston performs work on the ideal gas below. The time-integrated work done on the gas is precisely equal to the energy change of the gas and also to the lost ball energy :

$$Mgh = P_{\text{initial}}V \rightarrow E_{\text{final}} = 2E_{\text{initial}} \; ; \; S_{\text{final}} = S_{\text{initial}} + Nk\ln(2) \; .$$

The gas energy doubles and its entropy increases by $Nk\ln(2)$, so that thermodynamics can predict the final state *exactly*, though without any of the mechanistic details. These details necessarily depend upon nonequilibrium properties of the gas, such as the viscosity and heat conductivity. For details of such processes we need to solve the problem with molecular dynamics or with continuum fluid mechanics. We begin to develop these tools by discussing basic fluid dynamics (for which "hydrodynamics" is a usual synonym despite the subject's ability to treat fluids other than water) .

6.3 Fundamentals of Hydrodynamics

6.3.1 *Continuity Equations for Mass, Momentum, Energy*

Writing the conservation laws locally is reminiscent of Liouville's Theorem, which traces the flow of probability in phase space through Hamilton's equations of motion. To derive the continuum continuity equation in the one-dimensional case involves only the density and velocity. **Figure 6.3** shows the underlying "thought experiment" leading to the continuity equation. The change of density in a fixed length dx is due to the difference between the flows into and out of dx, $(\rho v)_{\mp dx/2}$. Provided that the variables are differentiable the small-dx limit gives the continuity equation, multiplied by dx :

$$(\partial \rho/\partial t)dx = -(\partial/\partial x)(\rho v_x)dx \longleftrightarrow (\partial \rho/\partial t) = -(\partial/\partial x)(\rho v_x) .$$

[Eulerian Continuity Equation with fixed grid]

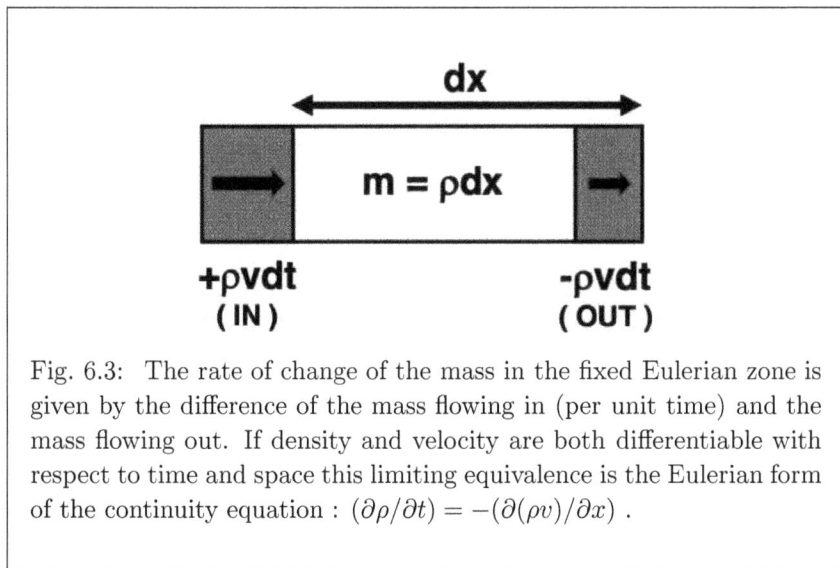

Fig. 6.3: The rate of change of the mass in the fixed Eulerian zone is given by the difference of the mass flowing in (per unit time) and the mass flowing out. If density and velocity are both differentiable with respect to time and space this limiting equivalence is the Eulerian form of the continuity equation : $(\partial \rho/\partial t) = -(\partial(\rho v)/\partial x)$.

The comoving time derivative of density, the time-rate-of-change of density following the flow, is denoted is denoted by a superior dot, $\dot{\rho}$:

$$\dot{\rho} = (\partial \rho / \partial t) + v_x(\partial \rho / \partial x) = -\rho(\partial v_x / \partial x) \ .$$

[Lagrangian Continuity Equation with comoving grid]

Evidently the two- and three-dimensional versions of the continuity equation follow from adding the y, or the y and the z, contributions to the mass flow parallel to the x axis. It is a useful exercise to show that the contributions of v_y and v_z flows make vanishing contributions to flows in the x direction so that the continuity equation can be written $\dot{\rho} = -\rho \nabla \cdot v$ in any number of dimensions. Notice also that the continuity equation is time-reversible, with both sides of the equation changing sign $dt \to -dt$ and $v \to -v$ if the motion is reversed.

Because momentum is conserved it satisfies a conservation equation, the "equation of motion" including convection as well as the comoving momentum flux P. Consider convection. The "convective" (meaning carried along with the flow) change of momentum $(d/dt)(\rho v)dx$ for a fixed dx is given by the gradient of $\rho v v_x$, multiplied by dx . In two or three dimensions v is a vector so that there can be a nonzero flow of y momentum in the x direction, as in a shear flow where $v_x \propto y$.

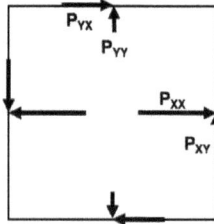

Fig. 6.4: The eight arrows shown within a small $L \times L$ comoving cell represent the directions and magnitudes of the momentum flux forces exerted *by* the element *on* its surroundings. The forces *on* the element are given by the products of pressure tensor components and the sidelength L. The shear components of the two-dimensional pressure tensor, P_{xy} and P_{yx} are necessarily equal, as is necessary for stability.

The continuum equation of motion is particularly easy to derive in the Lagrangian (comoving frame) because the convective flows of mass, momentum, and energy all vanish in that frame. The Lagrangian zone or "cell" or "element" moves with the fluid velocity. We need only follow the motion for a single timestep dt and are free to choose the "shape" of the zone to suit our convenience. Let us consider the two-dimensional $L \times L$ Lagrangian zone shown in **Figure 6.4**.

Now consider the pressure tensor P. Momentum *in* the zone changes due to the momentum flux $P = -\sigma$, where P is the (necessarily symmetric) pressure tensor and its negative σ is the stress tensor. The physical interpretation of pressure as a momentum flux is crucial to an understanding of the change of momentum $\rho v dx dy$ present in the fixed mass $\rho dx dy$ within its comoving frame. The zone takes in x momentum through the flux P_{xx} on its left side, $P_{xx} L dt$, during time dt. There is a corresponding loss, $-P_{xx} L dt$, on the right side so that the two contributions add to give $-(\partial P_{xx}/\partial x) L^2 dt$. In addition x momentum can also flow *vertically* into the comoving element, with $P_{yx} L dt$ entering at the bottom and leaving at the top making a net contribution $- (\partial P_{yx}/\partial y) L^2 dt$. Adding all four contributions to the comoving change of x momentum gives the x component of the vector Lagrangian equation of motion :

$$\rho \dot{v} \equiv \rho[\ (\partial v/\partial t) + v \cdot \nabla v\ ] = -\nabla \cdot P = \nabla \cdot \sigma\ .$$

[Lagrangian Equation of Motion]

The y component of the motion equation follows by symmetry, $\rho \dot{v}_y = -(\partial P_{xy}/\partial x) - (\partial P_{yy}/\partial y)$. The choice of a square cell (a cube in three dimensions) is perfectly general in that a cell of any shape can be constructed of infinitesimal square or cubic cells with the internal face contributions $\pm P dL dt$ cancelling precisely. Our derivation is valid for any smoothly differentiable stress distribution. An Eulerian derivation can be carried out in parallel to our treatment of the continuity equation, $(\partial \rho/\partial t) = -\nabla \cdot (\rho v)$, but the Lagrangian description is clearly simpler as well as more memorable, and hence preferable.

It is important to remember that the rotational forces (for simplicity in two dimensions) are proportional to the sidelength L of a small $L \times L$ volume element while the mass varies as L^2. This means that a small-L nonvanishing difference between P_{xy} and P_{yx} would provide a *divergent* angular acceleration. Necessarily the pressure tensor is *symmetric* with P_{xy} amd P_{yx} equal. The equations of motion in two dimensions are

$$\rho \dot{v}_x = -(\partial P_{xx}/\partial x) - (\partial P_{yx}/\partial y)\ ;\ \rho \dot{v}_y = -(\partial P_{yy}/\partial y) - (\partial P_{xy}/\partial x)\ .$$

"Newtonian" fluids have a linear dependence of stress on the velocity gradient. They are described by the viscous constitutive relation :

$$\sigma = \eta[\ (\nabla v) + (\nabla v)^t\ ] + [\ \lambda \nabla \cdot v + \sigma_{\text{eq}}\ ]\ I \ .$$

Here λ is the "second viscosity coefficient" and I is the unit tensor. This definition is *irreversible*, with (∇v) and its transpose $(\nabla v)^t$ changing sign when time is reversed while the atomistic equilibrium stress is unchanged. This is odd, as the motion equations for an atomistic fluid are reversible, with the stress unchanged if all the velocities are reversed. Despite this qualitative disagreement an atomistic fluid typically *does* have a shear viscosity η which is positive. Linear response theory provides an explanation : the time required for the stress to respond to a strain rate $[\ (\nabla v) + (\nabla v)^t\ ]$ is on the order of the time between collisions. Thus the continuum relation is an approximation, justified in a steady state of shear but exhibiting hysteresis if the motion is not steady.

Exercise : Show that rotating the (x, y) coordinate system by 45° converts the shear stress σ_{xy} to a difference in the diagonal components $\pm(1/2)(\sigma_{xx} - \sigma_{yy})$.

Exercise : Show that the bulk viscosity, which gives the additional stress for homogeneous compression or expansion is $\lambda + \eta$ in two dimensions and $\lambda + (2/3)\eta$ in three. The bulk viscosity gives the additional normal stress proportional to the strain-rate of the volume $\nabla \cdot v$.

6.3.2   The Energy Equation and Thermodynamics

The continuity equation for energy change applies both at and away from equilibrium :

$$\rho \dot{e} = -\nabla \cdot Q - P : (\nabla v) \ .$$

[Lagrangian Energy Equation]

The derivation is relatively tedious though straightforward in the Eulerian frame. The Lagrangian formulation just given is "almost obvious". It looks much like (and in fact is) the thermodynamic statement of energy conservation where the inequality in the thermodynamics is now taken up in the hydrodynamics by the irreversible parts of the Pressure tensor P and the Heat-flux vector Q. The positive entropy production in continuum mechanics is unchanged if all of the velocity components and the direction of time are reversed.

$$dE \equiv dQ - dW = TdS - P : (\nabla v) .$$

[Combined First and Second Laws of Thermodynamics] .

Here dS contains the contributions from viscosity and heat conductivity so that the energy equation is an equality both at and away from equilibrium.

The change of entropy dS includes the reversible transfer of heat dQ divided by the equilibrium temperature T. The irreversible work is included in the double dot product of the pressure tensor and (∇v). This product contains four terms in two dimensions and nine in three. In thermodynamics processes are typically "quasistatic" – the change in energy through heat transfer dQ and through work done (usually by pressure forces or by motion in a gravitational field) dW is so slow as to be unconnected to time. "Thermodynamics" always applies to slow processes connecting equilibrium states. A more descriptive name would be "thermostatics". Similarly "Hydrodynamics" is not usually a study of water. The word is usually thought of as synonymous with "fluid mechanics" and obeys the same conservation laws as solid mechanics though with different constitutive equations. Hydrodynamics simply replaces (dE, dQ, dW) with the corresponding rates $(\dot{E}, \dot{Q}, \dot{W})$. Here $\dot{Q}$ is the rate at which heat enters the system.

Caution : Note that Q can denote either a quantity of heat or the heat-flux vector. I trust readers to use context to distinguish the two.

There is also a new hydrodynamic variable, Q the heat-flux vector. Like P the heat flux Q needs to be measured in a "comoving" frame that moves with the fluid (including the sometimes controversial effects due to rotation[4]) and it describes the additional flux of energy by conduction rather than convection. The heat flux vector has units of energy per unit area and time. As an example of conductive heat flow consider a moving particle making a headon collision with a second motionless but otherwise identical particle. The moving particle transfers its energy across the space separating the two particles during their collision. Also, like P, the heat-flux vector Q takes a time on the order of a collision time to become established so that Fourier's Law, which is irreversible, $Q = -\kappa\nabla T$ is inconsistent with the time reversibility of the atomistic equations of motion. To see this notice that a right-to-left Fourier heat flow, responding to a positive temperature gradient (dT/dx) with a negative heat flux, is *unchanged* by time reversal while the atomistic Q does change sign. Apart from gravity "Action at a distance" is largely missing from continuum mechanics, where the constitutive relations, unlike atomistic forces, are strictly local.

6.4 How Atomistic and Continuum Theories are Different

We have noticed repeatedly differences in the reversibility of atomistic mechanics and the irreversibility of continuum mechanics. Loschmidt and Zermélo pointed this out in the 19th century.[2] Nosé provided a novel point of view in 1984, generalizing Hamiltonian mechanics to include temperature.[5,6] By adopting a nonHamiltonian version of his work I was able to extend his approach to irreversible processes.[7] That idea led to fractal phase-space distributions demonstrating the rarity of nonequilibrium states.[8]

The difference between the microscopic and macroscopic points of view is fundamental. It serves as a reminder that *both* these approaches are simply "models", approximate descriptions of parts of the real world around us. The atomistic approach provides mechanical recipes for the temperature, pressure, and heat flux using Newton's or Hamilton's time-reversible motion equations. Likewise, the Gauss, Hoover-Leete, and Nosé-Hoover motion equations are time-reversible. These generalizations of mechanics are quite capable of modelling nonequilibrium flows. All of the atomistic approaches, thermostatted or not, solve ordinary differential equations for the evolution of the particle coordinates and momenta. Energy is given as a function of the $\{q, p\}$ variables. The spatial dependence of temperature, pressure, and heat flux can be defined in a variety of ways for all these atomistic models and it is an æsthetic question as to which is "best" for a particular purpose.

My belief is that the smooth-particle approaches[9] provide good routes to spatial interpolation, making it possible to define continuum properties like temperature, pressure, and energy *everywhere* rather than just *at* or *between* the places where the particles are. Entropy is a challenge and seems like a fine research area from the standpoint of analyzing irreversible processes like the free expansion problem. These problems are ideally suited to small-scale simulations, often in one or two space dimensions, where new approaches can be tested readily and improved or discarded. We must keep the idea of modelling in mind, striving for simplicity. Let us consider thermodynamics and hydrodynamics from the standpoint of Gibbs' and Boltzmann's work, seeking a more-detailed microscopic way of thinking about thermodynamics, keeping the thought in mind that atomistic and continuum models are fundamentally different.

6.5   Thermodynamics from Gibbs' Canonical Ensemble

The simplest atomistic model for a fluid is a set of mass points interacting pairwise with a short-ranged repulsive potential $\phi(r)$. Hard disks and spheres fit that description and simplify it further as the particle momenta scale with temperature $\{\, p \propto \sqrt{T}\,\}$. Gibbs' treatment of the connection between particles and thermodynamics is elegant and the simplest possible. Gibbs used Liouville's Theorem to identify the thermodynamic entropy with the logarithm of the number of states or, equivalently, the hypervolume taken up by a continuous distribution of states in (q,p) phase space.

Maximizing the entropy to find the most-likely states returns the result that an ideal-gas heat reservoir and a fluid system with which it is in thermal contact equilibrate at the reservoir temperature. A thermal "reservoir" is imagined to have so many degrees of freedom that its temperature fluctuations vanish. If the fluid system has energy E so that the reservoir is left with $E_{\text{total}} - E$, the number of reservoir states is proportional to $e^{E_{\text{total}}/kT}e^{-E/kT}$.

$$\text{prob}(E) \propto e^{-E/kT} \longrightarrow \text{prob}(E) = e^{-E/kT}/Z \; ; \; Z = \sum e^{-E/kT} \; .$$

The sum over states ("Zustandsumme" in German) is just that. If the probability is sharply peaked the sum approaches the product

$$Z(V,T) = \sum e^{-E/kT} \longrightarrow e^{S/k}e^{-E/kT} = e^{-A/kT} \; .$$

Here A is Helmholtz' free energy, $E - TS$. This identification can be confirmed by using a thermodynamic identity :

$$d(-A/T) = d(-E/T) + dS =$$

$$(PdV - TdS)/T + (E/T^2)dT + dS = (PV/T)d\ln V + (E/T)d\ln T \; .$$

We see that the pressure and energy are given by the derivatives of the "canonical partition function" $Z(V,T)$:

$$(PV/kT) = (\partial \ln Z/\partial \ln V)_T \; ; \; (E/kT) = (\partial \ln Z/\partial \ln T)_V \; .$$

As Andy DeRocco [he was my thesis advisor at the University of Michigan] declared to his Statistical Mechanics class : "*If* you know Z you know *everything* !" From today's perspective this seems a bit exagerated but it is indeed an elegant connection between a sum or an integral, and all of thermodynamics. The sum over states is identified with Helmholtz' Free Energy. Helmholtz' free energy can be thought of as the total energy of a system linked to an ideal-gas heat reservoir, as shown in **Figure 6.5**. In the two-dimensional case the number of reservoir states is proportional to $(NkT - E_S)^N$, where N is the (huge) number of reservoir particles. Thus the probability of observing the system state is proportional to $[\, 1 - (E_S/NkT)\,]^N$, which approaches $e^{-E_S/kT}$ for large N .

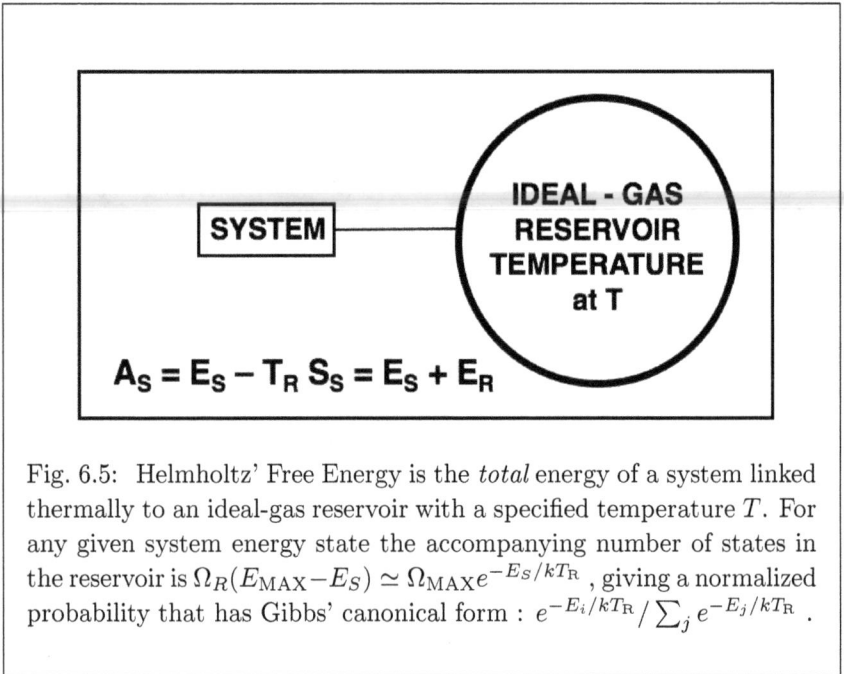

Fig. 6.5: Helmholtz' Free Energy is the *total* energy of a system linked thermally to an ideal-gas reservoir with a specified temperature T. For any given system energy state the accompanying number of states in the reservoir is $\Omega_R(E_{\mathrm{MAX}} - E_S) \simeq \Omega_{\mathrm{MAX}} e^{-E_S/kT_R}$, giving a normalized probability that has Gibbs' canonical form : $e^{-E_i/kT_R} / \sum_j e^{-E_j/kT_R}$.

The normalization of the relative state probabilities $\{\, e^{-E_S/kT}\,\}$ is the canonical partition function $Z = \sum_j e^{-E_j/kT_R}$. Differentiation of that partition function with respect to volume and temperature provides atomistic expressions for the pressure $P(V,T)$ and the energy $E(V,T)$. This working model of thermodynamics can be pursued and extended to include the nonequilibrium states encountered in fluid flows.

6.6 Continuum Mechanics as an Extension of Gibbs' Work

We have pointed out that the atomistic viewpoint provides exactly the same expressions for the local values of the energy, temperature, pressure tensor, and heat-flux vector. The last two follow from the Virial Theorem and the Heat Theorem.

$$kT = \langle\, mv_x^2 \,\rangle \; ; \; (E/N) = \langle\, (mv^2/2) + (1/2)\sum_j \phi_{ij} \,\rangle \; ;$$

$$(PV/N) = \langle\, mvv + \sum_j (rF)_{ij} \,\rangle \; ;$$

$$(QV/N) = \langle\, \sum_i (ve)_i + \sum_{i<j}(1/2)(v_i + v_j)\cdot(Fr)_{ij} \,\rangle \; .$$

Here the volume V needs to be small enough (but not too small) so that the point properties of continuum mechanics can be evaluated as local averages. Together with conservation of mass, momentum, and energy, these microscopic expressions for the constitutive variables provide all of the ingredients of fluid mechanics, subject to the constraints of granularity and reversibility. We still must model the boundary conditions for fluid mechanics with thermostatting and we still must formulate algorithms for the local variables appearing in the continuum equations. The free expansion and shockwave problems are evidence of the utility of this approach.

A casualty of this correspondence between Gibbs' ensemble averages and atomistic dynamics is Gibbs' entropy, $k\ln\Omega$. At equilibrium Gibbs showed that entropy is proportional to the logarithm of the phase volume. And for reversible processes changes in the entropy can be measured directly from thermodynamics, $dS = (dE/T) + (PdV/T)$. Boltzmann's H Theorem suggested that this concept could be extended to dilute gases away from equilibrium. Evidently all that is required is to augment mechanics with sources and sinks for heat and work. Brad Holian, Harald Posch, and I found that steady nonequilibrium boundary conditions (different temperatures or different velocities at the left and right boundaries of a simulation, for instance) cause the phase volume to vanish, exponentially fast. Extending the thermal part of thermodynamics to microscopic simulation is restricted to temperature and energy with the pressure tensor and the heat flux vector. The entropy and the free energies are missing. This complicates the search for nonequilibrium constitutive relations though it simplifies understanding the Second Law of Thermodynamics through the rarity of nonequilibrium states.[8]

6.6.1   *N! and Gibbs' Paradox*

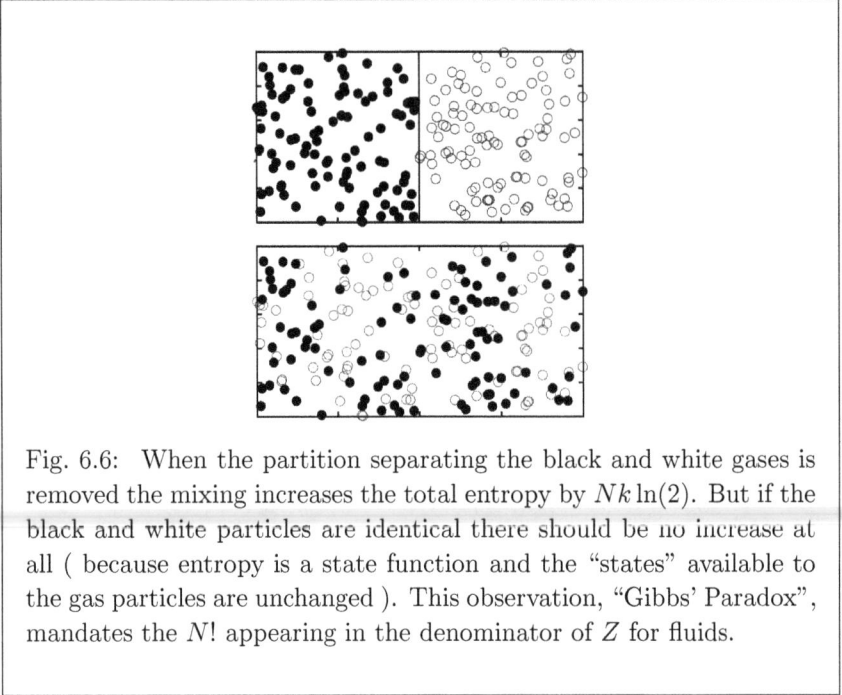

Fig. 6.6:   When the partition separating the black and white gases is removed the mixing increases the total entropy by $Nk\ln(2)$. But if the black and white particles are identical there should be no increase at all (because entropy is a state function and the "states" available to the gas particles are unchanged). This observation, "Gibbs' Paradox", mandates the $N!$ appearing in the denominator of Z for fluids.

A container of gas, partitioned into two similar parts, has a configurational state sum or integral proportional to $(V/2)^N$. Removing the partition changes the configurational part as the volume accessible to each particle increases by a factor of two, $(V/2)^N \to V^N$, suggesting that the Helmholtz Free energy has decreased by $NkT\ln(2)$. Thermodynamics does not distinguish the $N!$ permutations of particle positions included in Z so long as the particles are identical in type, "indistinguishable". Avoiding the overcounting of states by the replacement $Z \to Z/N!$ eliminates this error. For solids it seems obvious to choose only one of the $N!$ arrangements but the reasoning is exactly the same for fluids. If one cares to consider all the permutations of the particles among $N!$ sites then solids too require the division of Z by $N!$. This free expansion "thought experiment" is reminiscent of the dog-flea model. By using two species of fleas one can model the entropy of mixing when the fleas become free to emigrate.

6.6.2   *Jaynes' Maximum Entropy Principle*

"Maximum entropy" was put forward by Ed Jaynes in 1957.[10,11] His idea was that in the absence of definite information the most likely solution to a problem is that with the maximum entropy. This useful idea is typically applied to problems where the "entropy" is Shannon's "information entropy" rather than Gibbs' or Boltzmann's thermodynamic entropy. We see that in certain situations, like the piston problems' details, the maximum-entropy solution can be misleading. Likewise, the Rayleigh-Bénard fluid-flow problem provides a wide variety of gravitationally and thermally-driven convective flows.[12] Even for fixed boundary conditions many different flows can result. That which appears in these cases can be sensitive to the initial conditions. Evidently the Maximum Entropy idea does not describe nonequilibrium flows.

6.6.3   *Distinguishing Gibbs' and Boltzmann's Entropies*

"The experimental entropy cannot decrease in a reproducible adiabatic process that starts from a state of complete thermal equilibrium" appears on page 397 of Jaynes' Reference 11. Gibbs' entropy is constant in Hamiltonian flows according to Liouville's Theorem. The phase-space probability density flows through the space unchanged by an adiabatic process. Boltzmann's single-particle entropy, $\langle -k \ln f_1(q,p) \rangle$, can behave differently. It responds to collisional changes in the velocity distribution of nearly-ideal gases. According to the Boltzmann Equation this single-particle entropy can only increase in adiabatic processes. This was an outstanding accomplishment, Boltzmann's H Theorem.[2] The Theorem is Boltzmann's version of the Second Law of Thermodynamics, valid for dilute gases.

It is clear that Boltzmann's entropy is only an approximation. It includes only the kinetic part of the true entropy. For hard spheres it it easy to compute the dependence of Gibbs' entropy on the density :

$$(S - S_{\text{ideal}})/Nk = -B_2\rho - (B_3/2)\rho^2 - (B_4/3)\rho^3 - (B_5/4)\rho^4 - \ldots \ .$$

Differentiation gives the usual virial expansion of the pressure as a power series in the density :

$$P = T(\partial S/\partial V)_E \rightarrow (PV/NkT) = 1 + B_2\rho + B_3\rho^2 + B_4\rho^3 + B_5\rho^4 + \ldots \ .$$

For hard spheres the nonideal part of the entropy varies from 0 at low density to just less than $-4.85Nk$ at the freezing density.[13] This means that at the freezing point the Boltzmann entropy overestimates the phase volume by a factor of $e^{4.85N} \simeq 128^N$. The freezing point is very close to two-thirds the maximum number density for hard spheres, where spheres of diameter σ have a maximum packing density of $(N/V) = \sqrt{2}/\sigma^3$.

For numerical work[13, 14] it is more convenient to work with closed-form expressions for thermodynamic properties. For instance the hard-sphere pressure and entropy (relative to that of an ideal gas at the same temperature and density) can be written as "Padé approximants" :

$$\frac{PV}{NkT} - 1 = \frac{x(1 + 0.063507x + 0.017329x^2)}{1 - 0.561493x + 0.081313x^2} \; ;$$

$$\frac{[\, S(V,T) - S_{\text{ideal}}(V,T) \,]}{Nk} = \frac{-x(1 - 0.11075186x + 0.00469232x^2)}{1 - 0.42325186x + 0.04130870x^2} \; .$$

Here the dimensionless density x is the product of the number density (N/V) ratio and the second virial coefficient $B_2 = (2/3)\pi\sigma^3$. The Taylor series expansions are consistent with the coefficients $B_2 \ldots B_6$.

Thought Experiment : Consider the adiabatic expansion of a hard-sphere fluid at the freezing density to a density so low that the fluid can be considered ideal. The change of entropy in the ideal gas is zero so that $VT^{3/2}$ is constant. A thousand-fold expansion (increasing the sidelength by a factor of 10) would decrease the temperature of an ideal gas by a factor of 100, leaving the entropy unchanged. On the other hand a thousand-fold expansion of a freezing hard-sphere gas gives an additional increase in the entropy [relative to ideal] of about $4.85Nk$. Accordingly the expanded temperature for the hard-sphere gas should be $(T/100)e^{(2/3)4.85} \simeq (T/4)$ rather than $(T/100)$. A confirming simulation would make a nice demonstration project. A simpler variant would use Lucy's potential (modelling an ideal gas) confining the gas with an external force decreasing with time. As the rate decreases from sudden (free expansion) to slow (adiabatic) the states covered should range from constant energy to constant entropy (adiabatic).

6.7 The Mayers' Density Expansion of the Pressure

The canonical partition function,[15] with its exponential weighting of the energy, $e^{-K/kT}e^{-\Phi/kT}$, can be expanded as a power series in the density, giving useful expressions for the pressure and other thermodynamic properties of fluids. The Mayers' developed this idea by treating the difference between $e^{-\phi/kT}$ and 1 as a small perturbation. Here ϕ is a pair potential. It is assumed that the total potential energy of the fluid is a sum of $N(N-1)/2$ pair terms, $\Phi = \sum \phi_{i<j}(r_{ij})$. In the gas phase, where typical densities are a thousand times less than the liquid density, this idea looks reasonable. The "virial series" is a series expansion in the density. Truncating the series after a few terms looks promising at densities near the critical density (expanded to about one third the liquid density).

If we take just three terms,

$$(PV/NkT) = 1 + B_2\rho + B_3\rho^2 \; ; \; \rho \equiv (N/V) \;,$$

and determine the conditions at which the first and second derivatives of pressure with respect to density vanish (these are the conditions defining the "critical point") we find

$$1 + 2B_2\rho + 3B_3\rho^2 = 0 \; ; \; 2B_2 + 6B_3\rho = 0 \rightarrow$$

$$B_2\rho = -1 \; ; \; B_3\rho^2 = (1/3) \; ; \; (PV/NkT)_{\text{critical}} = (1/3) \;.$$

The dimensionless compressibility factor, here $(1/3)$, is about ten percent higher than the values found experimentally for simple fluids like the rare gases. As this approach looks promising let us look at the details for the simplest model fluids, hard rods (one dimension), squares and disks (two dimensions), cubes and spheres (three dimensions).

If we expand the Boltzmann factor in the partition function,

$$-(A_\Phi/kT) = (1/N) \ln \sum e^{-\Phi/kT} = (1/N) \ln \langle \; \prod [\, 1 + f \,] \; \rangle$$

the result is quite complicated. The leading term is a sum of integrals, $N(N-1)/2 = \binom{N}{2}$ of them, with just a single f function. In expanding the logarithm the next term includes half the square of the first, $(1/2)\binom{N}{2}^2$. Also included are integrals of two f functions $f_{ij}f_{kl}$ as well as $\binom{N}{3}$ terms involving $f_{ij}f_{jk}f_{ki}$ and three times as many with only two linked f functions, like $f_{ij}f_{jk}$. Already the situation looks complicated. But the Mayers persevered. Considering the large-N limit and taking a tremendous amount of cancellation into account, the final series for the excess (measured relative to the ideal gas) Helmholtz free energy is relatively simple :

$$(-\Delta A/NkT) = (\rho/2) \int f_{12}dr_2 + (\rho^2/6) \int \int f_{12}f_{23}f_{31}dr_2dr_3 + \dots \;.$$

In the end each virial coefficient in the density series contains all of the n-particle "star-graphs" multiplied by $(1-n)/n!$. Star graphs are graphs in which every pair of particles is linked by at least two independent paths. Their number is a rapidly increasing function of n. Up until Francis Ree and I took up the problem only the five-particle graphs had been integrated for hard spheres. We eventually calculated the virial series through the seven-particle integrals for disks and spheres.[13-16] We simplified the disk and sphere calculations considerably by introducing graphs with two kinds of bonds, the Mayer f-functions and the Boltzmann factor $e^{-\phi/kT}$.

Forty years later Clisby and McCoy used this idea to extend the disk and sphere calculations through the ten-particle graphs.[17] There are 45 different f functions so that the possible number of different graphs is 2^{45} = 35 184 372 088 832. Permutations of the particles could mean that as many as 10! = 3 628 800 topologically identical graphs exist, giving a minimum number of topologically different graphs, mostly star graphs : $2^{45}/10! \simeq 9\ 695\ 870$. Clisby and McCoy found 9 743 542 star graphs, so that this approximation is a good one. The number of graphs required is roughly cut in half by using the two types of bonds, $\{f\}$ and $\{e^{-\phi/kT}\}$.

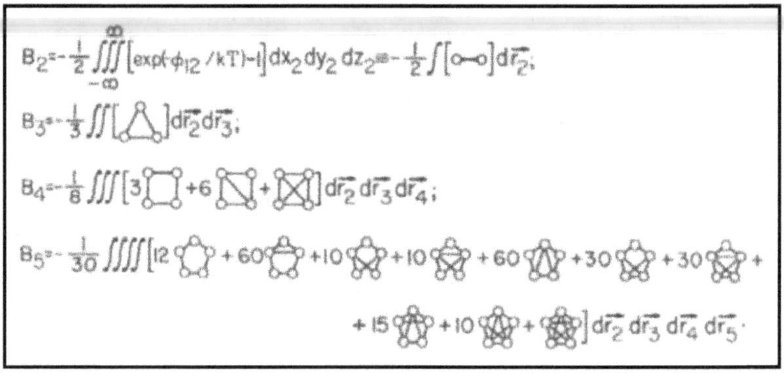

Fig. 6.7: Star-graph integrals for virial coefficients $B_2 \ldots B_5$.

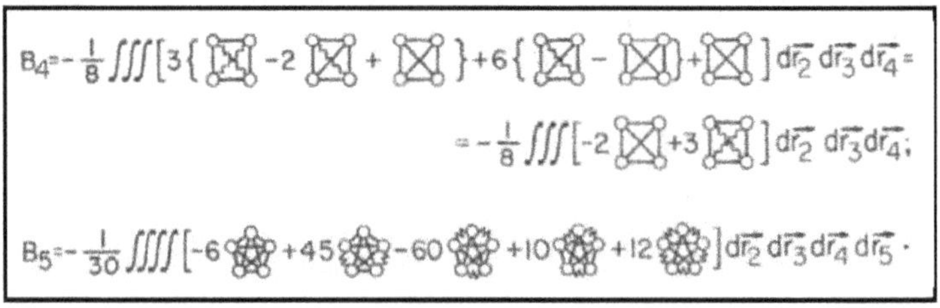

Fig. 6.8: The wiggly-line $[e^{-\phi/kT}]$ integrals for B_4 and B_5.

There are 4 980 756 Ree-Hoover graphs contributing to B_{10} all of which are much smaller than the corresponding Mayer graphs with some of them vanishing altogether. In one dimension any diagram containing $e^{-\phi/kT}$ requires that two of the particles are separated by a distance greater than 1. For hard rods only one of the Ree-Hoover graphs, the one with $\binom{N}{2}$ f bonds, ends up contributing to the Nth virial coefficient, providing a useful check. Lewi Tonks[18] had considered that model and shown that its equation of state is the same as the repulsive part of van der Waals' equation.

6.7.1 Lewi Tonks' Equation of State for Hard Rods

Following Tonks we consider N hard rods of unit length in a one-dimensional volume V. The configurational integral, provided that the rods are kept in order and that the coordinates are measured from their leftmost position, is $(V-N)^N/N!$. This follows easily by extending the three-rod example, from which the general case is obvious by induction :

$$Z_3 \propto \int_0^{V-3} dx_1 \int_{x_1}^{V-3} dx_2 \int_{x_2}^{V-3} dx_3 = \int_0^{V-3} dx_1 \int_{x_1}^{V-3} dx_2 (V-3-x_2)$$

$$= \int_0^{V-3} dx_1 (V-3-x_1)^2/2! = (V-3)^3/3! \longleftrightarrow (V-N)^N/N! \ .$$

The equation of state *is* the repulsive part of van der Waals' equation :

$$(PV/NkT) = (1/N)(\partial \ln Z/\partial \ln V)_T = V/(V-N) = 1+\rho+\rho^2+\rho^3+\dots \ ,$$

so that each of the virial coefficients is equal to unity and the pressure diverges at the close-packed density $\rho = 1$.

The hard-rod model is an excellent example problem for getting acquainted with virial series. Hard parallel squares and cubes are more challenging from the standpoint of theory although the independence of the velocity distribution and the configurational integral allow for a simpler dynamics[19] in both these cases, with all particles travelling with velocity components ± 1. The virial series for squares and cubes as well as for disks and spheres are given below :

	B_2	B_3	B_4	B_5	B_6	B_7
Squares	2.000	3.000	3.667	3.722	3.025	1.651
Cubes	4.000	9.000	11.333	3.160	-18.880	-43.505
Disks	1.814	2.573	3.176	3.610	3.904	4.090
Spheres	2.962	5.483	7.456	8.486	8.864	8.794

The units for all four models have been chosen such that the compressibility factor is the sum of the B_n multiplied by powers of the ratio of density to close-packed density. It is interesting that the virial coefficients for the hard cubes become negative. The hard squares and hard cubes appear to have continuous phase transitions though the complex number-dependence of their pressures makes definite conclusions difficult.

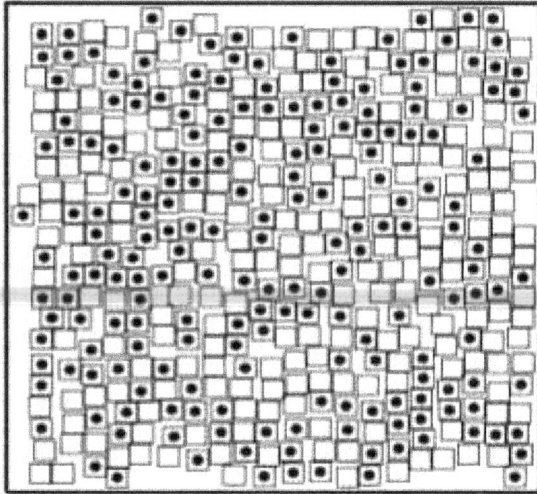

Fig. 6.9:   A periodic configuration of 400 hard parallel squares at two-thirds the close-packed density. The decorated particles originally occupied even-numbered rows in the initial condition, a perfect square lattice. The melting transition for squares appears to be second order.[19]

6.8   Equations of State for Hard Disks and Spheres

Ever since Metropolis, the Rosenbluths, and the Tellers carried out Monte Carlo measurements of the equation of state of hard disks,[20] the details of the equation of state and of the virial coefficients have gradually become better known. Lyberg presents good evidence that it is possible to compute analytic values for the virial coefficients for hard spheres.[21] Clisby and McCoy[17] tabulated virial coefficients for hard spheres of dimensionality 2 through 8 including the virial coefficients $B_2 \ldots B_{10}$.

Despite this considerable body of theoretical and computational work the precise natures of the virial equations of state in two or more dimensions are unknown. It appears that hard spheres with dimensionality greater than two have a sharp melting-freezing phase transformation that is first order (with a coexistence range within which the two phases can coexist). The two-dimensional hard-disk fluid freezes to form a hexatic phase which then continuously compresses into a triangular-lattice solid phase. Recent definitive work has shown convincing van der Waals' loops for 2^{16}, 2^{18}, and 2^{20} hard disks.[22] It is interesting that the correct transition pressure differs from the estimates made in the early 1960s by only a few percent. The nature of the transition had been a long-standing research goal since the earliest work by Alder, Wainwright, and Wood in the 1950s.

6.9   Entropy of Harmonic Crystals and Their Defects

The Gaussian integral $\int_{-\infty}^{+\infty} e^{-\alpha x^2/2} dx = \sqrt{2\pi/\alpha}$ can be generalized to the situations where α is replaced by a symmetric $n \times n$ matrix, in which case the n-dimensional integral becomes $(2\pi)^{n/2}/\sqrt{\det A}$. This makes it possible to compute the partition function for one- or two- or three-dimensional harmonic crystals, with or without lattice defects such as vacancies, dislocations, stacking faults, and surfaces. These calculations show that the simple Einstein model for the crystal partition function is too small. The Einstein model sets the Nth root of the configurational partition function equal to $(kT/\nu_{\text{Einstein}})$ where the Einstein frequency is computed with all the particles but one held fixed at their lattice sites. In one dimension the exact partition function is $(kT/\nu)^N$ where $\nu = \sqrt{\kappa/m}$. According to the Einstein model the restoring force on the single moving one-dimensional particle is $-2\kappa\delta x$ so that the frequency is too high by $\sqrt{2}$ and the entropy too low, by $0.34657Nk$. Dale Huckaby[23] solved the two-dimensional lattice dynamics problem for the triangular structure analytically and found $0.27326Nk$ for $(S - S_{\text{Einstein}})/Nk$ while I found that the three-dimensional face-centered cubic crystal entropy exceeds the Einstein prediction by $0.24689Nk$; with the three-dimensional hexagonal lattice slightly less stable at $0.24541Nk$.[24]

6.10   Validation of the Macroscopic Viewpoint

Given the straightforward nature of Gibbs' work it is surprising to find any of its consequences denied and denigrated as false. But a century after

Gibbs' book and long past Irving and Kirkwood's 1950 paper[25] formulating
the local stress as a function of the (q, p) variables there are still papers in
respectable journals claiming that the virial theorem stress is incorrect[26]!
The best way to counter such a claim is to formulate and solve good example problems. With an understanding of Gibbs' ideas relating microscopic
states to macroscopic behavior through the canonical distribution let us
review two mesoscopic (intermediate scale) problems in order to see how
well the two points of view can coincide.[27] Both problems are relatively
simple, requiring just a hundred lines of FORTRAN. The first of the two
problems is the equilibration of a fluid in a constant vertical gravitational
field g as shown in **Figure 6.10**.

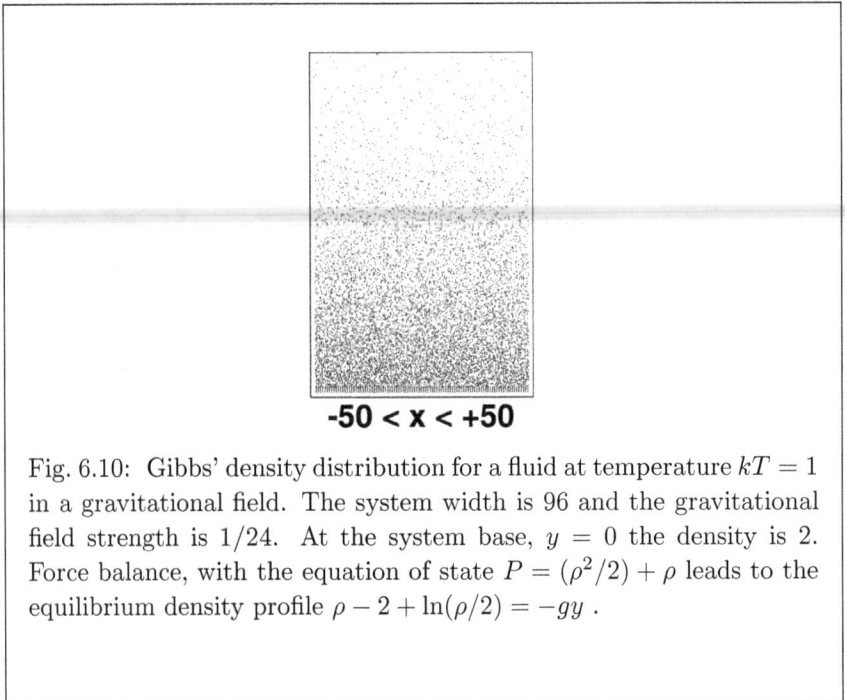

-50 < x < +50

Fig. 6.10: Gibbs' density distribution for a fluid at temperature $kT = 1$
in a gravitational field. The system width is 96 and the gravitational
field strength is 1/24. At the system base, $y = 0$ the density is 2.
Force balance, with the equation of state $P = (\rho^2/2) + \rho$ leads to the
equilibrium density profile $\rho - 2 + \ln(\rho/2) = -gy$.

For simplicity we choose a short-ranged repulsive potential
$$\phi(r < h = 3) = (10/\pi h^2)[\ 1 - (r/h)\ ]^3\ ,$$
with $kT = 1$. This pair potential is normalized for any choice of h :
$$\int_0^h 2\pi r\phi(r)dr = (10/\pi) \int_0^1 2\pi x(1 - 3x + 3x^2 - x^3)dx =$$
$$20[\ (1/2) - (3/3) + (3/4) - (1/5)\ ] \equiv 1\ ,\text{ where } x \equiv (r/h)\ .$$

This choice guarantees that the potential energy of any particle is equal to $(\rho/2)$ where ρ is the local density at that particle. We choose $96 \times 96 = 9216$ moving particles with an additional $6 \times 96 = 576$ motionless particles in a horizontal strip of area 3×96 corresponding to a boundary density of 2. An additional boundary force, $-100y^3$ is applied to any of the moving particles which is able to penetrate into the boundary strip at the bottom of the system.

The continuum version of this molecular dynamics problem is relatively easy to solve. The fluid's equation of state is $P = (\rho^2/2) + \rho$ (remember that $T = 1$) and the gravitational field is chosen to be $g = 4/96$ so that the downward force per unit length is 4 (the "weight" of a column of 96 particles) and is equal to the pressure from the equation of state at $y = 0$ where $\rho = 2$. Force balance at an elevation y provides a differential equation for the density profile :

$$(dP/dy) = -\rho g = (dP/d\rho)(d\rho/dy) = (\rho + 1)(d\rho/dy) = -\rho g \longrightarrow$$

$$\rho - 2 + \ln(\rho/2) = -gy \; ; \; P = g \int_y^\infty \rho(y')dy'.$$

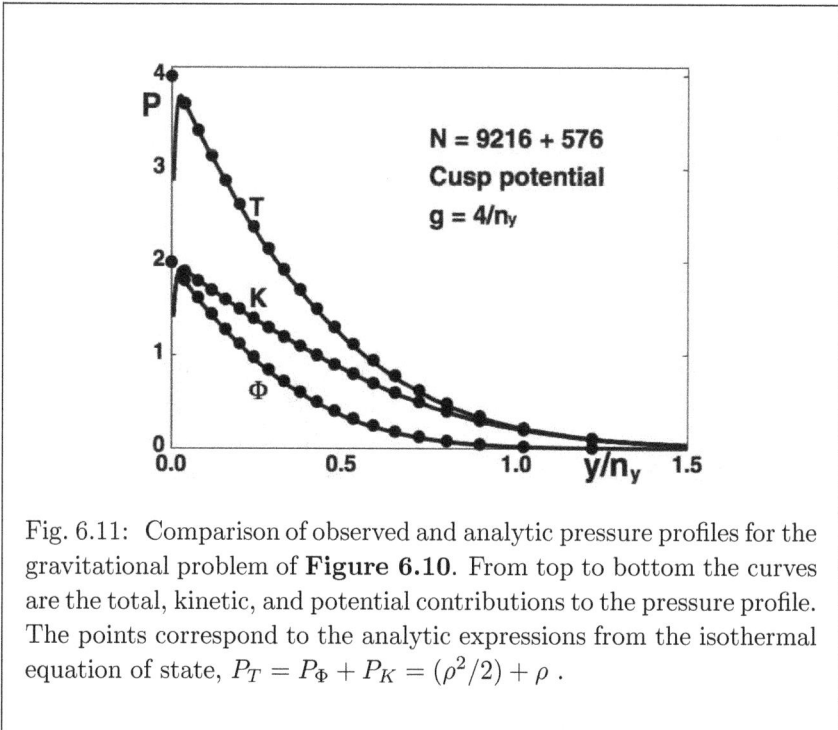

Fig. 6.11: Comparison of observed and analytic pressure profiles for the gravitational problem of **Figure 6.10**. From top to bottom the curves are the total, kinetic, and potential contributions to the pressure profile. The points correspond to the analytic expressions from the isothermal equation of state, $P_T = P_\Phi + P_K = (\rho^2/2) + \rho$.

Figure 6.11 shows the result of a computer simulation with conventional molecular dynamics. The kinetic and potential contributions to the total pressure are functions of the vertical coordinate y and plotted as continuous curves. The points follow from the continuum equations. The analytic relation is $\rho - 2 + \ln(\rho/2) = -gy$. We see that the claim that the kinetic stress should be ignored in continuum mechanics is false.

6.11   Stress in Steadily-Rotating Solids

A second demonstration problem where the kinetic contribution to stress is important is the steady rotation of a solid structure. We construct a model solid by choosing those particles from a perfect triangular lattice which lie within a maximum distance of $\sqrt{637} = 25.239$. With the nearest-neighbor Hooke's-Law force constant and the equilibrium spacing between neighboring particles both equal to unity we rotate the solid at an angular velocity $\omega = 0.01$ and measure the mean stress from the Virial Theorem.

It is "natural", because it is simplest, to solve the continuum equations for rotating disks in polar (cylindrical) coordinates as we do here.[27] The triangular lattice is elastically isotropic so that there is a single shear modulus, $\eta = \sqrt{(3/16)}$. The bulk modulus, $\lambda + \eta$, is twice as large so that the two Lamé constants are equal, $\eta = \lambda$. The force balance between the centrifugal force and the restoring elastic tensile forces is best expressed in polar coordinates :

$$0 = +\rho\omega^2 r + (\partial\sigma_{rr}/\partial r) + (\sigma_{rr} - \sigma_{\theta\theta})/r .$$

The radial stress vanishes at the outer edge of the rotating disk, $r = R$. The displacement u_r provides a radial and a circumferential strain, $\epsilon_{rr} = (du_r/dr)$, and $\epsilon_{\theta\theta}$. The stresses convert the force-balance just described into an ordinary differential equation to be solved for $u(r)$:

$$r^2(d^2u/dr^2) + r(du/dr) - u = -\omega^2 r^3/3 \longrightarrow u(r) = (\omega^2 r/18)(5R^2 - 3r^2) .$$

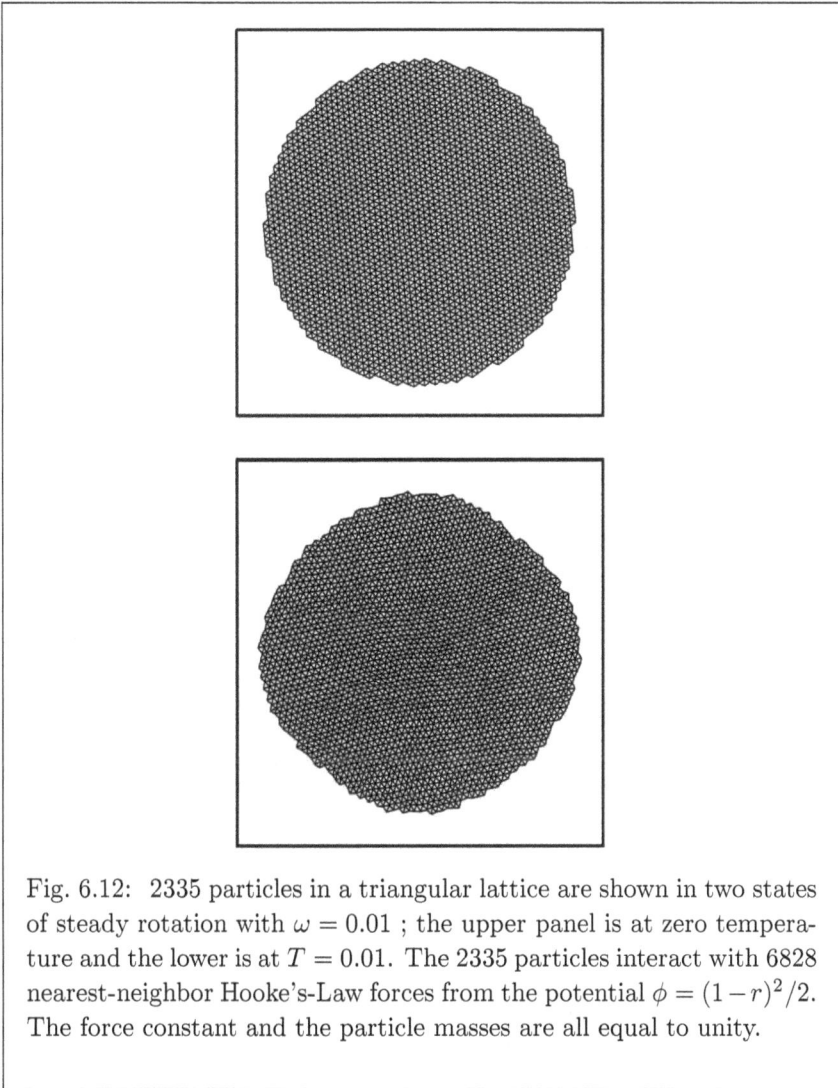

Fig. 6.12: 2335 particles in a triangular lattice are shown in two states of steady rotation with $\omega = 0.01$; the upper panel is at zero temperature and the lower is at $T = 0.01$. The 2335 particles interact with 6828 nearest-neighbor Hooke's-Law forces from the potential $\phi = (1 - r)^2/2$. The force constant and the particle masses are all equal to unity.

Exercise : Show numerically that the bulk modulus of the triangular lattice is equal to $\sqrt{(3/4)}$ when the mass, force constant, and Hooke's-Law restlength are all equal to unity.

The radial displacement gives the stress tensor :

$$\sigma_{rr} = (\rho\omega^2/12)(5R^2 - 5r^2) \; ; \; \sigma_{\theta\theta} = (\rho\omega^2/12)(5R^2 - 3r^2) \, .$$

The stresses in polar coordinates from molecular dynamics are most easily calculated by first computing $P_{xx}V$ and $P_{yy}V$ from the virial theorem at each particle. Then a rotation matrix R and its transpose R^t provide the stresses in polar coordinates :

$$(PV)_{\text{polar}} = R \cdot (PV)_{\text{Cartesian}} \cdot R^t$$

where

$$R_i = \begin{bmatrix} +\cos(\theta_i) & +\sin(\theta_i) \\ -\sin(\theta_i) & +\cos(\theta_i) \end{bmatrix} \; ; \; \theta_i = \arctan(y/x)_i \, .$$

Figure 6.13 shows the variation of stress at the two temperatures, 0.00 and 0.01. Except near the outer boundary the points (from molecular dynamics) agree very nicely with the lines (from linear elasticity, as just outlined). Once again this good agreement shows that the contribution to the stress from the momenta (the upper line in the righthand panel) must be included if the microscopic and macroscopic stress tensors are to agree.

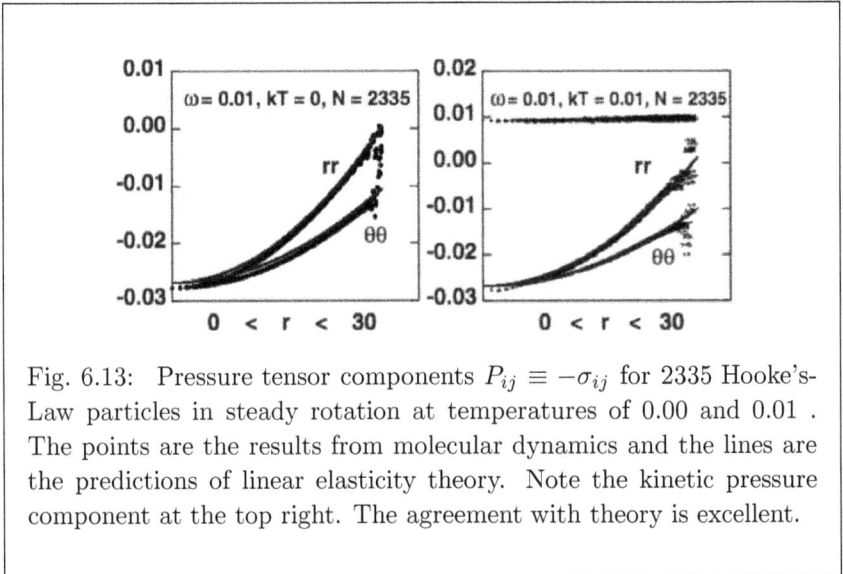

Fig. 6.13: Pressure tensor components $P_{ij} \equiv -\sigma_{ij}$ for 2335 Hooke's-Law particles in steady rotation at temperatures of 0.00 and 0.01 . The points are the results from molecular dynamics and the lines are the predictions of linear elasticity theory. Note the kinetic pressure component at the top right. The agreement with theory is excellent.

6.12 Entropy Away From Equilibrium ?

At equilibrium Gibbs' entropy has the properties of the thermodynamic entropy and is able to distinguish possible from impossible outcomes in some cases. Liouville's Theorem shows very directly that Gibbs' entropy, viewed as a comoving property, is not a practical computational tool. The "states" which matter comes to occupy as it moves through phase space are all of them accessible. As the past disappears and the coverage of phase points becomes uniform the Gibbs entropy should consider the time-averaged phase-space occupancy rather than an instantaneous one. The deviations from local thermodynamic equilibrium are relatively brief, depending as they do on local collisions rather than diffusive processes.

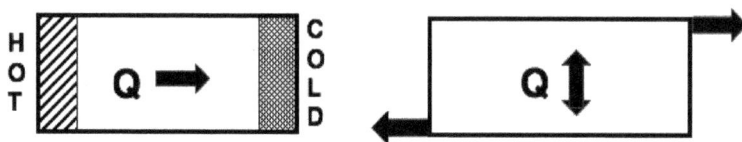

Fig. 6.14: On the left steady heat flow causes a loss in entropy proportional to the square of the heat flux. On the right steady isoenergetic shear flow causes a loss in entropy that varies as the square of the strain rate. These unavoidable consequences of energy conservation in continuum mechanics gave rise to the concept of "entropy production", something absent from molecular dynamics, and symptomatic of irreversibility.

Away from equilibrium, especially in steady states, more can be said about entropy. Consider a conducting rod linking two Nosé-Hoover thermostats, one cold and the other hot, giving a steady left-to-right flow of heat in response to a negative temperature gradient. See the left panel of **Figure 6.14**. There, at the lefthand boundary, the incoming heat, QA where Q is the heat flux and A is the cross-sectional area normal to the x axis, causes a gain of entropy in the rod, at the rate $(+QA/T_H)$. Meanwhile at the cold end of the rod heat exits at the same rate but with a different consequence for entropy, $(-QA/T_C)$. Evidently the entropy of the rod is diminishing :

$$Q = -\kappa(dT/dx) \longrightarrow \dot{S} \simeq \kappa A[\ (T_H - T_C)/L\ ][\ (1/T_H) - (1/T_C)\ ] < 0\ .$$

The loss rate varies as the square of the temperature gradient or alternatively as the square of the heat flux. But surely there can be no change in a steady state. What is wrong ? Another example may prove helpful. See the right panel of **Figure 6.14** for a fluid under steady shear, with $P_{yx} = -\eta(du_x/dy)$. Because work is done *on* the fluid it must lose an equivalent quantity of heat to its surroundings :

$$\dot{S}_{\text{fluid}} = P_{yx}\Delta x(du_x/dy)\Delta y/T = -P_{yx}^2 V/(\eta T) < 0 \ .$$

Again the system's interaction with the outside world is an entropy loss proportional to the square of the gradient driving the heat flow.

Standard procedure in both cases (so as to conform to the Second Law of Thermodynamics) is to avoid a violation by ascribing "entropy production" to the conducting rod and to the sheared sample of viscous fluid. In both cases the entropy production is simply set equal to the rate at which entropy is lost at the boundaries.

Problem : Suppose that the righthand flow in **Figure 6.14** represents a steady shear. How can one tell whether the element is doing work on its neighbors or is having work done on it by the neighbors? What is/are the direction(s) of heat flow in these cases? Are both possible?

One can see that this is "wrong" to the extent that Gibbs' entropy is "right". In simulations of heat or viscous flows temperature is controlled by frictional thermostat forces and entropy is extracted in the form of heat, giving exactly the entropy change that is conventionally offset by "entropy production".

My own conclusion (which dates back to 1987[8]) is that the Gibbs entropy (which arguably *is the* entropy) progresses toward $-\infty$ in a steady state. As this transformation occurs the phase-space description of the flow becomes more and more complicated. The distribution is becoming more and more "fractal" rather than smooth, and showing structure on all scales. This insight into the Second Law is best illustrated with a simple model. We will devote the next Section to introducing and analyzing the Baker Map, a bare-bones model of deterministic time-reversible fractal dynamics.

6.13   The Equilibrium and Nonequilibrium Baker Maps

The equilibrium Baker Map[28, 29] maps a 2×2 diamond-shaped region into itself. The stationary distribution that results is shown at the left in **Figure**

6.15. Here is the corresponding map :

$$(q < p) : q = (5q/4) - (3p/4) + \sqrt{9/8} \; ; \; p = -(3q/4) + (5p/4) - \sqrt{1/8} \; .$$

$$(q > p) : q = (5q/4) - (3p/4) - \sqrt{9/8} \; ; \; p = -(3q/4) + (5p/4) + \sqrt{1/8} \; .$$

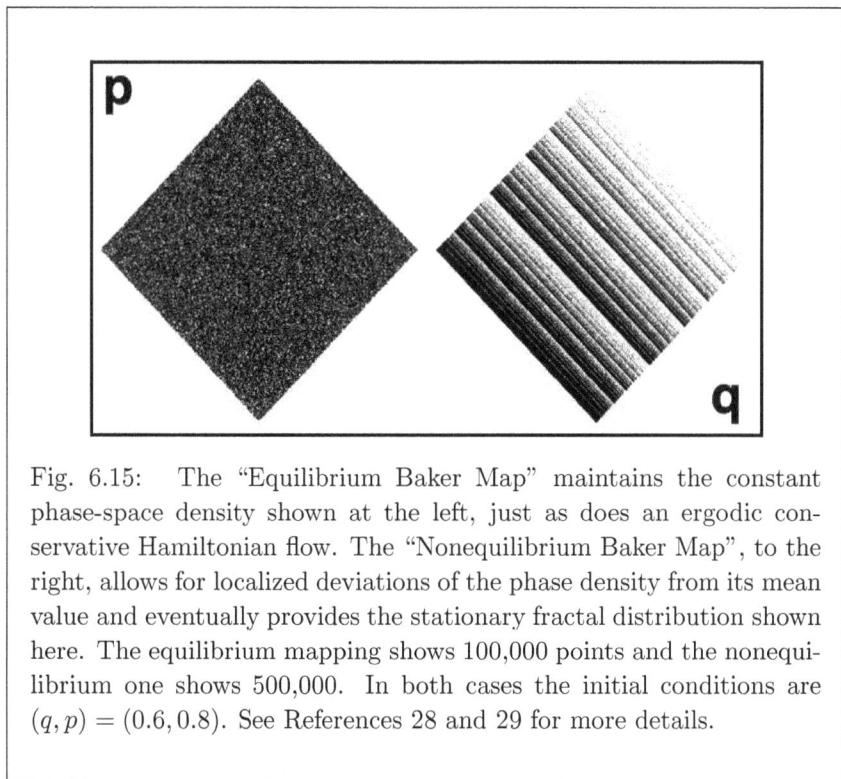

Fig. 6.15:   The "Equilibrium Baker Map" maintains the constant phase-space density shown at the left, just as does an ergodic conservative Hamiltonian flow. The "Nonequilibrium Baker Map", to the right, allows for localized deviations of the phase density from its mean value and eventually provides the stationary fractal distribution shown here. The equilibrium mapping shows 100,000 points and the nonequilibrium one shows 500,000. In both cases the initial conditions are $(q, p) = (0.6, 0.8)$. See References 28 and 29 for more details.

The stretching direction is parallel to the line $q + p = 0$ while the shrinking direction is parallel to $q - p = 0$. The map is a useful caricature of ergodic, chaotic, time-reversible Hamiltonian mechanics. Two nearby points tend to separate from one another exponentially fast, and all the states in the diamond-shaped (q, p) phase space are approached as time goes on. There are no "holes" in the distribution function and all $dqdp$ states are equally likely. Both the fixed points $(q, p) = (0, \pm\sqrt{2})$ are unstable.

The paradoxical behavior found in nonequilibrium steady states can be seen in simple models of phase-space distributions. Because specifying gradients in the motion equations requires changes in the comoving phase volume caused by thermostat forces a useful model for nonequilibrium steady flows is the Baker Map. The distribution generated by the nonequilibrium map is at the right of **Figure 6.15**. Iterations of the map turn out to be ergodic, with no holes. The "Nonequilibrium Baker Map" is as follows :

$$\text{if } (q < p - \sqrt{(2/9)}) \text{ then}$$

$$q = (11q/6) - (7p/6) + \sqrt{49/18} \; ; \; p = (11p/6) - (7q/6) - \sqrt{25/18} \; .$$

$$\text{or if } (q > p - \sqrt{(2/9)}) \text{ then}$$

$$q = (11q/12) - (7p/12) - \sqrt{49/72} \; ; \; p = (11p/12) - (7q/12) - \sqrt{1/72} \; .$$

The nonequilibrium map also incorporates stretching and shrinking in the same two orthogonal directions as before, but allows for local changes in area providing a steady state that has the nonuniform singular density shown at the right in **Figure 6.15**. Both mappings are time-reversible in the sense that either Baker mapping B, followed by time-reversal T, another mapping B, and a final reversal T returns to the original (q, p) point :

$$TBTB(q,p) = (q,p) \longleftarrow T(q,p) = (q,-p) \; .$$

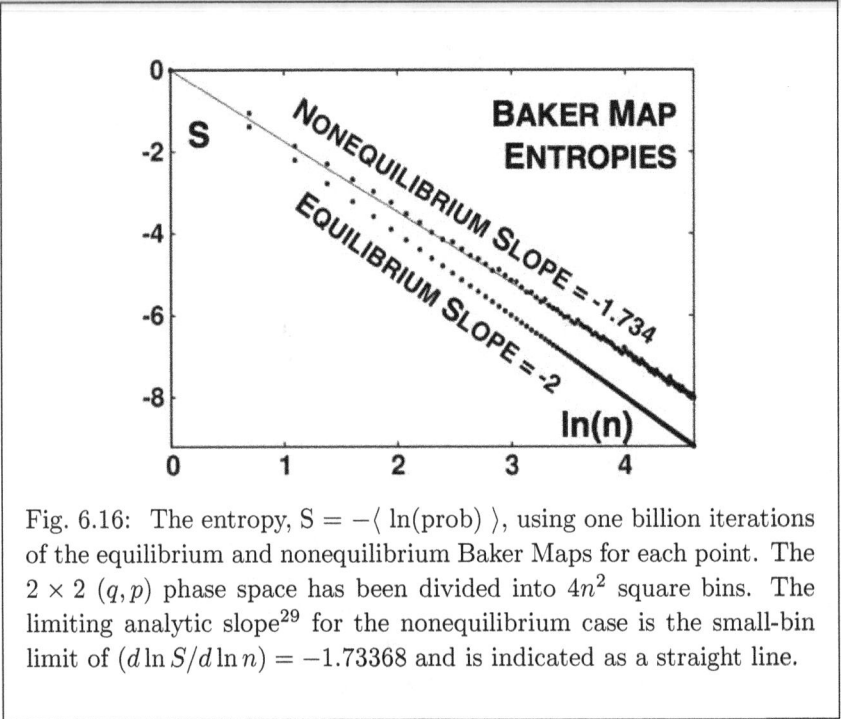

Fig. 6.16: The entropy, $S = -\langle \ln(\text{prob}) \rangle$, using one billion iterations of the equilibrium and nonequilibrium Baker Maps for each point. The 2×2 (q,p) phase space has been divided into $4n^2$ square bins. The limiting analytic slope[29] for the nonequilibrium case is the small-bin limit of $(d \ln S/d \ln n) = -1.73368$ and is indicated as a straight line.

The nonequilibrium map provides a stationary nonuniform density everywhere in the diamond-shaped phase space. The mapping is still ergodic, in the sense that it goes everywhere, but now the entropy steadily decreases, just as in the case of steady heat flow or steady shear. **Figure 6.16** shows the dependence of $-\langle \ln \text{prob} \rangle$ averaged over the 2×2 phase space using a sampling length varying from 1 to 0.01 so that the number of "bins" in which the probability $\text{prob}(q, p)$ is measured varies from 4 to 4×10^4. Notice that the Gibbs' entropy depends upon the scale of observation and the granularity of the distribution. The nonequilibrium entropy varies as $n^{-1.734}$ rather than n^{-2}. This means that the distribution is fractal, with most of the probability concentrated on a "strange attractor" of information dimension 1.734. This result comes from the preponderance of shrinking phase area over expansion. In each iteration of the map (in the steady state) two-thirds of the "natural measure" or probability is mapped to a twofold smaller area while one-third is mapped to a twofold larger area. Because the mapping expands in the direction parallel to $q + p = 0$ and contracts in the perpendicular direction parallel to $q - p = 0$ there is a uniform stretching, which makes the shrinking onto an attractor "strange". We will come back to this model in more detail in connection with Lyapunov instability, which will allow us to find the information dimension analytically.

6.14   Summary of Lecture 6

Thermodynamics and Hydrodynamics are irreversible and describe all the field variables as point functions, usually continuous and differentiable in space and time. This contrasts with the time-reversible nature of atomistic simulations in which particles interact over a distance of the order of the interparticle spacing. A closer examination of these differences reveals that the instantaneous response of the continuum equations is an approximation. Linear response theory is a useful tool for understanding the differences in time symmetry. Gibbs' entropy is constant for Hamiltonian evolutions, suggesting that boundary conditions must involve nonequilibrium forms of dynamics. The time-reversible thermostats are enough to support nonequilibrium steady states with atomistic mechanics.

The Mayers' systematic density expansion of the Helmholtz free energy provides a useful understanding of fluid behavior. As many as ten terms are calculable. On the other hand equilibrium simulations are so rapidly convergent and relatively inexpensive as to have replaced analytic approaches. The agreement of particle and continuum mechanics for steady problems with linear departures from equilibrium is convincing. Beyond linear response the fractal phase-space structure revealed by the nonequilibrium Baker Map signals the need for understanding complex structures in phase space. As a fringe benefit these zero-volume Lyapunov unstable structures provide a clear understanding of the Second Law of Thermodynamics.

References

1. J. W. Gibbs, *Elementary Principles in Statistical Mechanics* (Chas. Scribner's, 1902).
2. L. Boltzmann, *Lectures in Gas Theory*, Steve Brush's translation of the 1896 book (University of California Press, Berkeley, 1964) is also available as a Dover 1995 book.
3. Wm. G. Hoover and B. Moran, "Pressure-Volume Work Exercises Illustrating the First and Second Laws [of Thermodynamics]", American Journal of Physics **47**, 851-856 (1979) .
4. Wm. G. Hoover, B. Moran, R. M. More, and A. J. Ladd, "Heat Conduction in a Rotating Disk *via* Nonequilibrium Molecular Dynamics". Physical Review A **24**, 2109-2115 (1981).
5. S. Nosé, "A Molecular Dynamics Method for Simulations in the Canonical Ensemble", Molecular Physics **52**, 255-268 (1984).
6. S. Nosé, "A Unified Formulation of the Constant Temperature Molecular Dynamics Methods", The Journal of Chemical Physics **81**, 511-519 (1984).
7. Wm. G. Hoover, "Canonical Dynamics: Equilibrium Phase-Space Distributions", Physical Review A **31**, 1695-1697 (1985).
8. B. L. Holian, Wm. G. Hoover, and H. A. Posch, "Resolution of Loschmidt's Paradox : The Origin of Irreversible Behavior in Reversible Atomistic Dynamics", Physical Review Letters **59**, 10-13 (1987).
9. Wm. G. Hoover, *Smooth Particle Applied Mechanics – The State of the Art* Advanced Series in Nonlinear Dynamics Volume **25** (World Scientific, Singapore, 2006).
10. E. T. Jaynes, "Information Theory and Statistical Mechanics", Physical Review **106**, 620-630 and **108**, 171-190 (1957).
11. E. T. Jaynes, "Gibbs *versus* Boltzmann Entropies", American Journal of Physics **33**, 391-398 (1965).
12. V. M. Castillo, Wm. G. Hoover, and C. G. Hoover, "Coexisting Attractors in Compressible Rayleigh-Bénard Flow", Physical Review E **55**, 5546-5550 (1997).
13. F. H. Ree and Wm. G. Hoover, "Fifth and Sixth Virial Coefficients for Hard Spheres and Hard Disks", The Journal of Chemical Physics **40**, 939-950 (1964).
14. Wm. G. Hoover and F. H. Ree, "Melting Transition and Communal Entropy for Hard Spheres", The Journal of Chemical Physics **49**, 3609-3617 (1968).
15. J. E. Mayer and M. G. Mayer, *Statistical Mechanics* (John Wiley & Sons,

New York, 1940).

16. F. H. Ree and Wm. G. Hoover, "Seventh Virial Coefficients for Hard Spheres and Hard Disks", The Journal of Chemical Physics **46**, 4181-4197 (1967).

17. N. Clisby and B. M. McCoy, "New Results for Virial Coefficients of Hard Spheres in D Dimensions", Pramana - Journal of Physics **64**, 775-783 (2005) = arχiv 0410511.

18. L. Tonks, "The Complete Equation of State of One, Two, and Three-Dimensional Gases of Hard Elastic Spheres", Physical Review **50**, 955-963 (1936).

19. Wm. G. Hoover, C. G. Hoover, and M. N. Bannerman, "Single-Speed Molecular Dynamics of Hard Parallel Squares and Cubes", Journal of Statistical Physics **136**, 715-732 (2009) = arχiv 0905.0293.

20. N. Metropolis, A. W. Rosenbluth, M. N. Rosenbluth, A. H. Teller, and E. Teller, "Equation of State Calculations by Fast Computing Machines", The Journal of Chemical Physics **21**, 1087-1092 (1953).

21. I. Lyberg, "The Fourth Virial Coefficient of a Fluid of Hard Spheres in Odd Dimensions", Journal of Statistical Physics **119**, 747-764 (2005) = arχiv 0410080.

22. M. Engel, J. A. Anderson, S. C. Glotzer, M. Isobe, E. P. Bernard, and W. Krauth, "Hard-Disk Equation of State : First-Order Liquid-Hexatic Transition in Two Dimensions with Three Simulation Methods", Physical Review E **87**, 042134 (2013).

23. D. A. Huckaby, "Exact High-Temperature Harmonic Free Energy for the Triangular Lattice", The Journal of Chemical Physics **53**, 3753-3754 (1970).

24. Wm. G. Hoover, "Entropy for Small Classical Crystals", The Journal of Chemical Physics **49**, 1981-1982 (1968).

25. J. H. Irving and J. G. Kirkwood, "The Statistical Mechanical Theory of Transport Processes. IV. The Equations of Hydrodynamics", The Journal of Chemical Physics **18**, 817-829 (1950).

26. M. Zhou, "A New Look at the Atomic Level Virial Stress : on Continuum-Molecular System Equivalence", Proceedings of the Royal Society of London **459A**, 2347-2392 (2003).

27. Wm. G. Hoover, C. G. Hoover, and J. F. Lutsko, "Microscopic and Macroscopic Stress with Gravitational and Rotational Forces", Physical Review E **79**, 036709 (2009) = arχiv 0901.2071.

28. Wm. G. Hoover, *Time Reversibility, Computer Simulation, and Chaos* (World Scientific, Singapore, 1999).

29. J. Kumičák, "Irreversibility in a Simple Reversible Model", Physical Review E **71**, 016115 (2005).

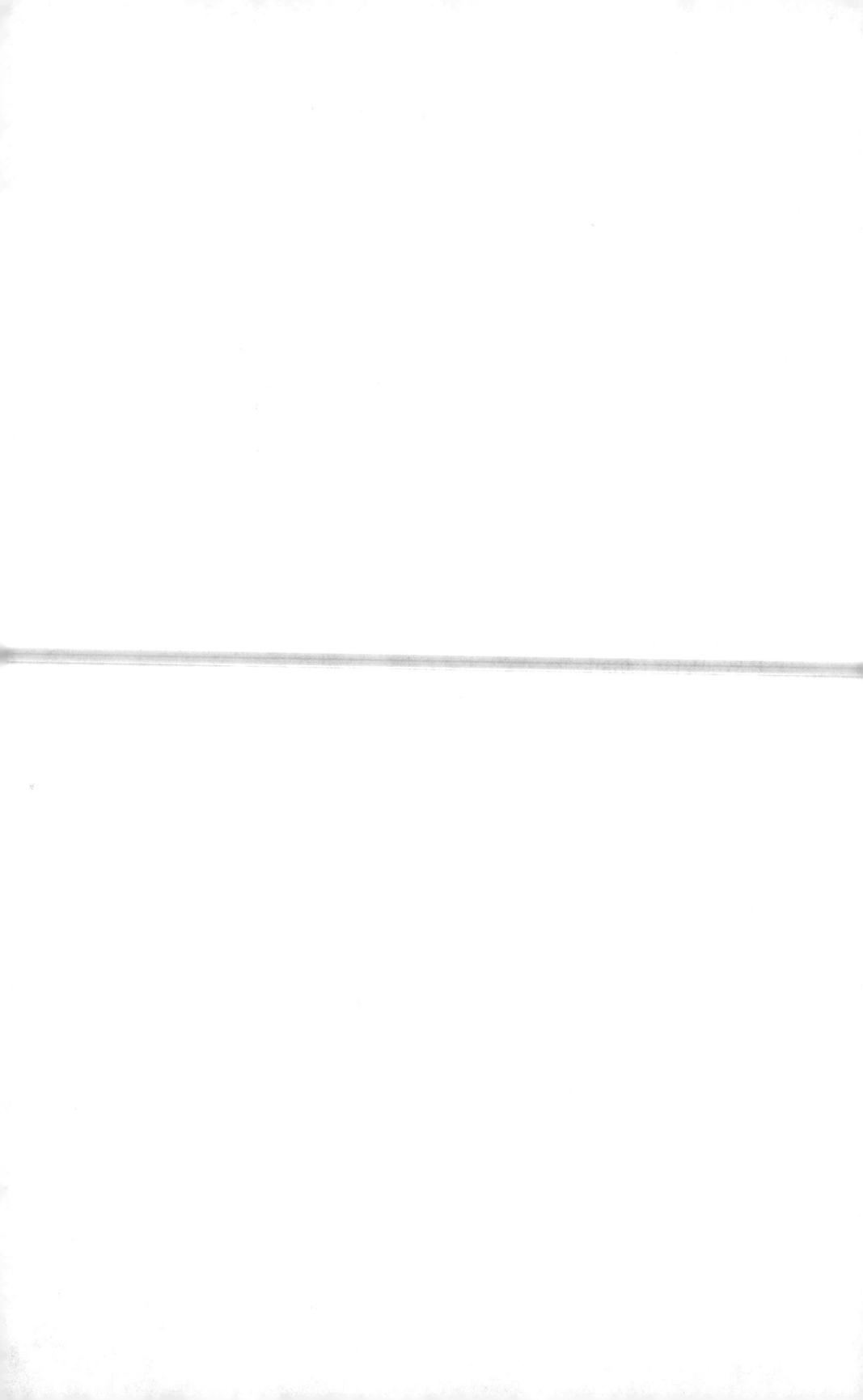

Chapter 7

Statistical Mechanics of Small Systems

/ Gibbs' Statistical Mechanics Cell Models / Free Volume / Percolation Transition / Correlated Cell Model / Hard Disk Dynamics / Galton Board / Galton Staircase / Conducting Oscillator / Tori and Limit Cycles / Mexican Hat Thermostat / Baker Maps / Summary of Lecture 7 /

7.1 Gibbs' Statistical Mechanics for Small Systems

In principle Gibbs' statistical mechanics can be applied to any Hamiltonian system that has a known thermodynamic state. For fluids the state variables, in addition to the composition of the system, include volume, energy, pressure, temperature, and entropy . Gibbs' ensemble idea showed that averaging over "all" the (q, p) microstates consistent with the given state variables would provide a complete thermodynamics. In practice, the catch is that the microstates are too numerous and too complicated. Prior to fast computers Gibbs' formal averaging could only be carried out for simple systems, like ideal gases or harmonic crystals. The main applications of his ideas were to dilute-gas spectroscopy and to phase diagrams.

About fifty years later, as computers were replacing hand-cranked or electric calculators, his idea began to be applied to solving the "manybody" problem. At Los Alamos Fermi's anharmonic chain simulations with 16 and 32 particles and the Metropolis-Rosenbluths-Tellers' Monte Carlo simulations of 56 and 224 hard disks[1] gave a series of small-scale previews of the two independent routes suggested by Gibbs' ensemble ideas. In principle the Monte Carlo approach could trudge through *many* states. Averages had uncertainties of a few percent. An exaflop machine, now the "next phase" of the state of the art will perform more than 10^{25} operations per year.

But because the number of microstates varies *exponentially* in the number of particles it is clear that our interest must be understanding "typical" states rather than generating exhaustive catalogs including all of them. Very large systems are not really necessary, provided that periodic boundaries are used. Metropolis *et alii* introduced periodic boundaries in the second paragraph of their paper. They found that the resulting pressures for 56 and 224 hard disks were within statistical errors of each other.

The exponential nature of microstates suggests the possibility of a product approximation, $\Omega(N) \simeq \prod \Omega(1)$. This works for the ideal gas and the harmonic chain. What about something more challenging? What about hard disks? The Monte Carlo method produced crude few-percent pressures for 56 and 224 hard disks. Within a decade Alder and Wainwright's molecular dynamics simulations revealed almost all there was to know about hard disks with just 870 particles.[2] The most recent simulations[3] have used as many as $2^{20} = 1,048,576$. It is useful to see how well few-particle models can reproduce the many-particle manybody results. We will look in detail at the hard-disk work. That simplest of systems suggests what can be done for the more "realistic" models that are at the state of the art today.

In this Lecture we will take a relatively thorough look at hard-disk systems. The smallest one-body cell models emphasize the importance of the "free volume" and its relationship to the thermodynamic equation of state. The free volume itself undergoes an extensive to intensive phase transition which has no obvious relationship to thermodynamics. This is the "percolation transition" in which the low-density free volume, which is on the order of the total volume V, is transformed by compression to an "intensive" volume, on the order of (V/N).[4] See **Figure 5.2** on page 168.

Another dynamical model involving hard disks and the motion of a single particle is the "Galton Board", in which a single mass point driven by an external field moves through an array of scatterers.[5] This model provides a comprehensive introduction to time-reversible nonequilibrium systems with only a few degrees of freedom. We will see how fractals form in phase space, arbitrarily close to Gibbs' continuous distributions and how their fractal fractional-dimensional nature can be described numerically.

Generalizing these ideas to systems with smooth forces leads us to consider [1] the Galton staircase problem,[6] field-driven mass motion in a sinusoidal potential ; [2] the transport of heat through the vibrations of a harmonic oscillator exposed to a temperature gradient.[7] We revisit also the nonequilibrium Baker Map,[8] a simple but fractional-dimensional caricature of time-reversible motion far from equilibrium. This, the simplest of our idealized models shares all of the qualitative features exhibited by many-body nonequilibrium steady states. Let us begin our small-system studies with hard-disk models, chosen for investigation in the 1950s and 1960s by the pioneers of Monte Carlo at Los Alamos and molecular dynamics at Livermore.[1,2]

7.2   One Particle at a Time – The Cell Model for Hard Disks

To set the stage for "cell models" where a single particle moves with all the rest fixed consider Gibbs' canonical partition function $Z(m)$, for a two-dimensional system in which $N-1$ particles have unit mass while only the "first" particle has a mass m :

$$(N-1)! \times Z(m) = \prod [\int dq_i \int dp_i \ ] e^{-K(m)/kT} e^{-\Phi/kT} .$$

$$K(m) \equiv (p_1^2/2m) + \sum_2^N (p^2/2) .$$

If we evaluate $\frac{\partial \ln Z}{\partial \ln m}$ all of the integrals over coordinates and all but one of those over the momenta (p_x, p_y) cancel (they occur in both the numerator and the denominator) leaving just the two-dimensional integral over dp_1 :

$$\frac{K(m)}{kT} = \frac{\int (p_1^2/2mkT) e^{-p_1^2/mkT} dp_1}{\int e^{-p_1^2/mkT} dp_1} \equiv \langle \ p_1^2/2mkT \ \rangle \equiv 1$$

$$\longrightarrow v_{\text{rms}} = \sqrt{(2kT/m)} .$$

Evidently, as m approaches zero the rms velocity $\sqrt{\langle v^2 \rangle} \propto \sqrt{(1/m)}$ becomes arbitrarily large. The resulting dynamics in the vicinity of Particle 1 "looks like" a cell model with a single particle moving while the rest stay fixed. Because the potential energy gives the same weighting, $\propto e^{-\Phi/kT}$, no matter what the masses, the contribution of the single moving particle to the energy and pressure is, on average, exactly the same as in Gibbs' many-body ensemble average. The exact relationship between the "free volume" (or "free length" or "free area") and the pressure has been confirmed in detail for hard rods, disks, and spheres. Here we consider the ordinary cell model for hard disks and also include a variation that captures the physics of melting in a novel way, the "correlated" cell model.

Lewi Tonks worked out the details for one-dimensional hard particles analytically in 1936, well before computers.[9] His hard-rod model shows $(1/N)$ consequences of the boundary conditions (rigid or periodic) on the pressure, and $(1/N)\ln(N)$ effects of fluctuations on the entropy and free energies. Tonks' finding that the configurational part of the partition function, $(V - N)^N/N!$ looks like the Nth power of a one-dimensional particle-in-a-cell model suggests trials in two and three dimensions. Two-dimensional models are more interesting. They introduce the possibility of diffusion and of phase transitions and are advantageous when it comes to visualization.

In a system with two Hamiltonian degrees of freedom (x, y, p_x, p_y) energy conservation restricts the motion to a three-dimensional energy "surface" which can be covered in its entirety if the system is sufficiently ergodic. Let us look at the details of the cell model for a single hard disk confined to a cell by either four or by six fixed neighbors arranged in the triangular lattice structure.[10] See **Figure 7.1** for both these cases.

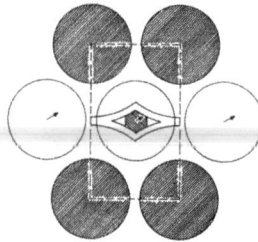

Fig. 7.1:   The 1963 Correlated Cell Model incorporates the hexatic sliding motion which leads to the melting of hard disks. The Ordinary Cell Model restricts the motion of the central "wanderer" particle to the dark hexagon at the cell center. When the two white neighbors move cooperatively with the central disk the diamond-shaped "free area" is increased by a factor of about $(4/3)$ relative to the uncorrelated hexagon. At $(4/3)$ the close-packed volume the wanderer escapes. Pressures for the two cell models[10] are shown in **Figure 7.5** on page 237 and compared to the results of several different model calculations.

It is convenient to use the nearest-neighbor spacing d in the triangular lattice to describe the area of the hard-disk system. The area per particle is $\sqrt{(3/4)}d^2$ with free area a_{cell}. The ordinary cell-model approach chooses the integral of one "wanderer" particle (with all the rest fixed at their cell centers) as the Nth root of Gibbs' configurational partition function[10] :

$$Z_{\text{Cell}}^{(1/N)} \propto a_{\text{cell}} = (3/2)\sqrt{3d^4} - (3/2)\sqrt{4d^2 - d^4} + 6\cos^{-1}(\sqrt{(d^2/4)}) - 2\pi \ .$$

Gibbs' relation for the compressibility factor $(PV/kT) = \frac{\partial \ln(Z)}{\partial \ln(V)}$ is most easily evaluated numerically directly from this formula, using a centered difference for both a_{cell} and d^2 where $a_\pm = a \pm 0.000001$ and $d^2_\pm = d^2 \pm 0.000001$.

"Computer-generated oscilloscope traces of disk trajectories" showed "sliding of rows of particles past each other".[10] This same property characterizes what some call the "hexatic" phase[3] and is the mechanism for the melting of the hard-disk solid phase. Berni and Tom and I had the idea of modelling this mechanism with a Correlated-Cell Model, as sketched in **Figure 7.1**. In this correlated variation the white particles move cooperatively over a larger area than that of the ordinary cell model, larger by a factor of $(4/3)$ at high density. The corresponding entropy increase $Nk \ln(4/3)$ greatly exceeds the Hoover-Ree estimate of the manybody excess entropy for disks (relative to that of the ordinary cell model) of just $0.05Nk$.

In the solid phase the correlated-cell free area is the sum of five terms :
$$Z_{CC}^{(1/N)} \propto a_{CC} = \sqrt{3d^4} - (1/2)\sqrt{4d^2 - d^4} - 2\sin^{-1}(\sqrt{(d^2/4)})$$
$$-(1/2)\sqrt{12d^2 - 9d^4} + 2\cos^{-1}(\sqrt{(3d^2)/4})) \text{ for } (1 < d^2 < (4/3)) .$$
In the fluid phase the free area is just the first three terms (the top row) of the solid-phase expression.[10]

The motion of a hard disk confined to the hexagonal or diamond-shaped regions of **Figure 7.1** provides two examples of ergodic motion where the direction of the velocity coupled with two spatial coordinates defines a three-dimensional phase volume that can be covered by the disk's chaotic dynamics. The Galton Board problem[5] provides two such chaotic examples, both of them with periodic boundary conditions. The first is ordinary Newtonian mechanics with energy constant. The second is novel, Gaussian mechanics with an accelerating field which keeps the kinetic energy constant by incorporating a time-reversible constraint force.

In many situations, such as the harmonic crystal, the system's dynamics is not sufficiently chaotic to reach all of its states, even in an *infinite* time, not just the age of the Universe. In other cases, where the motion *is* ergodic in an infinite length of time, time can be too short ever to forget the influence of the initial conditions, let alone to access all of the states counted up by Gibbs. Lack of ergodicity and finite sampling times may not matter much. Typical equilibrium molecular dynamics simulations can shortly produce averages with two-, three-, or four-figure accuracy because fluctuations (of order $1/\sqrt{t}$) can easily be made "small" in simulations of practical length.

State-counting is problematic when irreversible processes rather than equilibria are considered. A low-density hard-disk gas, slowly but inexorably compressed, will eventually reach a "jammed" configuration with the actual state reached being only one of $N!$ equivalent configurational states obtained by permuting the particle indices.

Let us examine how well Gibbs' all-states averaging applies as the phase-space dimensionality is increased from two to a few, recognizing that all states can never be realized once the number of degrees of freedom is of order 10. The simplest one-dimensional systems with a two-dimensional (q, p) phase space, like the oscillator or the pendulum, are ergodic because a one-dimensional trajectory *defines* the one-dimensional "energy surface".

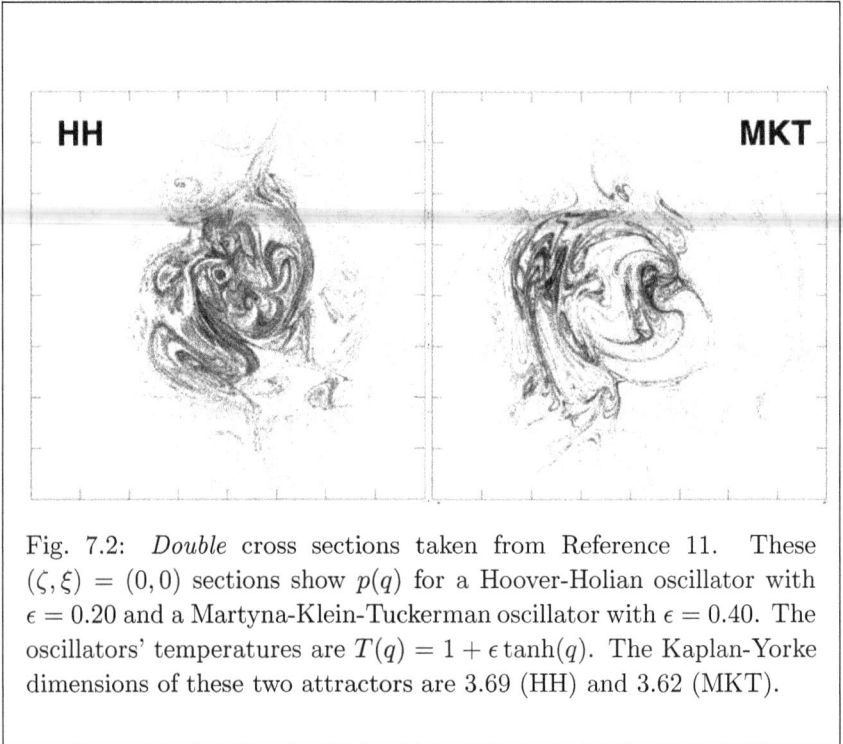

Fig. 7.2: *Double* cross sections taken from Reference 11. These $(\zeta, \xi) = (0, 0)$ sections show $p(q)$ for a Hoover-Holian oscillator with $\epsilon = 0.20$ and a Martyna-Klein-Tuckerman oscillator with $\epsilon = 0.40$. The oscillators' temperatures are $T(q) = 1 + \epsilon \tanh(q)$. The Kaplan-Yorke dimensions of these two attractors are 3.69 (HH) and 3.62 (MKT).

Three-dimensional motion can be examined for ergodicity by constructing cross sections of the flow. In the absence of fixed points a cross section of the motion in a three-dimensional space can generate closed curves in the case of a torus or an irregular fractal distribution of points in the case of a chaotic orbit. Tori generate closed curves on such a section while a chaotic sea can generate a two-dimensional area or a fractional-dimensional cross section. A fractional-dimensional section can be ergodic, as in the Galton-Board case where following the time dependence of a grid of cells covering the equilibrium section shows, from the statistical standpoint, that every cell will eventually include penetrations provided that the trajectory is sufficiently long. At present it is quite possible to generate "*double* cross sections" where the values of *two* out of four variables are specified rather than just one out of three. Two examples are shown in **Figure 7.2**.

In the early days of computer simulation two ways of studying small systems were widely accepted : [1] Hamiltonian mechanics, with the results, chaotic or not, dependent upon the initial conditions; [2] ergodic Monte Carlo simulations, where the results must in principle agree with Gibbs' ensemble theory. The "in principle" caveat conceals the fact that an exhaustive Monte Carlo analysis can only be carried out for systems with just a few degrees of freedom. Nosé's 1984 development of deterministic thermostats opened up a third way, which is now widely accepted too. The use of "thermostat variables" makes it possible to study equilibrium Gibbs'-ensemble properties with molecular dynamics. This idea has also been extended to nonequilibrium systems. By using space-dependent thermostats simulations with gradients can be included. To illustrate these ideas we next consider all three approaches for simple potential models ranging from a single degree of freedom up to relatively small "manybody" problems.

7.3 Cell Model, Free Volume, and Percolation Transition

When molecular dynamics was in its infancy there were many approximate paths leading to thermodynamic equations of state. Integral equations for the pair distribution function and truncated or extrapolated virial series are examples. The "cell model" and its variants are interesting examples because this model has physical connections to molecular dynamics and Monte Carlo. Gibbs' statistical mechanics for "simple fluids" with pairwise-additive forces provides a natural separation of the energy and the pressure tensor into potential and kinetic parts :

$$\mathcal{H} = \Phi + K = \sum_{\text{pairs}} \phi_{i<j} + \sum_{i}(p_i^2/2m) \; ; \; PV = \sum_{\text{pairs}}(rF)_{i<j} + \sum_{i}(pp/m) \; .$$

The canonical partition function factors into two parts, the mass-independent configurational integral and the momentum integral :

$$Z(N,V,T) = \prod^{\#} \int dq e^{-\Phi/kT} \times \prod^{\#} \int dp e^{-K/kT} \; .$$

All of the mass dependence is included in the latter integral, $(2\pi mkT)^{\#/2}$ for $\#$ degrees of freedom. In approximate models of the solid phase the partition function becomes a product of normal-mode partition functions :

$$Z(N, V, T)_{\text{harmonic}} = e^{-\Phi_0/kT} \prod^{\#} (kT/h\nu) .$$

In harmonic models the distribution of frequencies $\{\, \nu \,\}$ gives the Helmholtz free energy while the pressure depends upon the "cold curve" $\Phi_0(V)$ and the density dependence of the frequencies. Today these ideas have been rendered obsolete by simulation. Given a recipe for the forces it is far simpler to measure the pressure tensor using the virial theorem than it is to expand the energy into a quadratic form.

7.3.1   *Free Volume*

The notion of "free volume", the volume (area in two dimensions) accessible to a particle if all the others are held fixed, gives an approximation to the configurational integral similar to the Einstein model for a harmonic crystal with DN vibrational modes :

$$Z \propto v_f^N \propto (kT/h\nu_{\text{Einstein}})^{DN} .$$

In **Lecture 5** I pointed out that this notion becomes exact for hard rods, disks, and spheres where one particle is much lighter than the rest.[4] This light "wanderer" particle could trace out its free volume while its heavier neighbors are essentially motionless. Because *any* particle could be the light one it is clear that the equilibrium pressure for hard particles can be obtained by averaging the collision rates of all the particles in the system,

$$PV/NkT = 1 + (\sigma/2D)\langle\, s_f/v_f \,\rangle .$$

For hard particles there is an alternative formulation[1,12] of the equation of state in terms of the pair distribution function evaluated at the contact distance σ :

$$PV/NkT = 1 + B_2(N/V)g(\sigma) ,$$

where $g(r)$ is the ratio of the average density of particles at a distance r from a central particle to the corresponding density there for an ideal gas.

7.3.2   *Microscopic Free Volumes – Percolation Transition*

The distributions of free volume and surface are relatively easily generated in two dimensions and not so easily in three. At low density the excluded area around each disk is $\pi\sigma^2$ so that $a_f = V - N\pi\sigma^2 = V - 2NB_2$. As the density becomes higher there are two alternative definitions of free volume. One of them can be based on examining an equilibrium snapshot, looking for places large enough to contain an additional disk. Another is the simple measurement of the free area around each disk computed as a contour integral $\oint y\,dx$ or $\oint x\,dy$.

At high density the free area varies as $(V/N)[\,(V/V_{\rm cp})-1\,]^2$. Evidently the free volume changes from extensive, of order V, to intensive, of order (V/N), either gradually or suddenly. A numerical investigation[4] revealed a sudden "percolation transition" at a density one fourth of the close-packed. Two views of hard disks at the transition density appear in **Figure 7.3**.

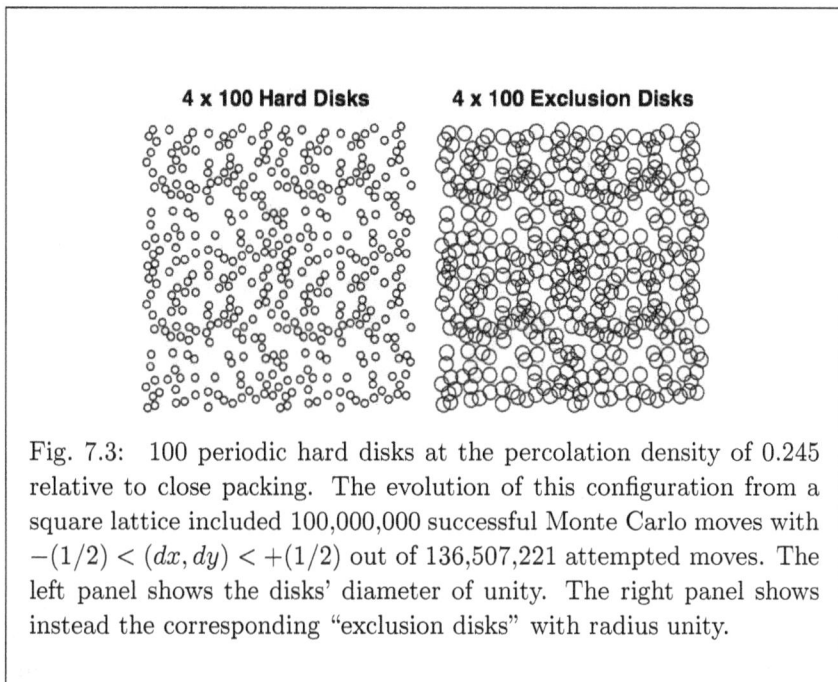

4 x 100 Hard Disks   **4 x 100 Exclusion Disks**

Fig. 7.3: 100 periodic hard disks at the percolation density of 0.245 relative to close packing. The evolution of this configuration from a square lattice included 100,000,000 successful Monte Carlo moves with $-(1/2) < (dx, dy) < +(1/2)$ out of 136,507,221 attempted moves. The left panel shows the disks' diameter of unity. The right panel shows instead the corresponding "exclusion disks" with radius unity.

7.4 Molecular Dynamics for Hard Disks from 1962 to 2013

Metropolis, along with the Rosenbluths and Tellers, chose hard disks to demonstrate their Monte Carlo simulation ideas for a system both simple

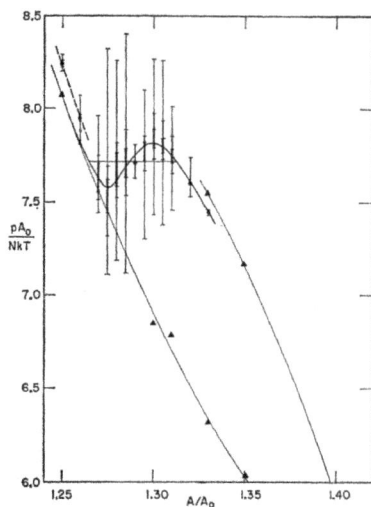

Fig. 7.4: Alder and Wainwright's 1962 hard-disk pressures for 72 disks (triangles) and 870 hard disks. The estimated transition pressure of 7.72 lies below the accurate value of 7.95 from Reference 3. An entropy-based calculation of 8.27 ± 0.13 is described in Reference 13.

and challenging.[1] Berni Alder and Tom Wainwright chose to develop molecular dynamics for this same system and it turned out to be an excellent choice. The dynamics programming is more elaborate than the Monte Carlo. The molecular dynamics program repeats two steps over and over, a "long cycle" followed by many "short cycles". In the long cycle an ordered list of times-to-collision is computed containing all such times for pairs of disk close enough that their collision is possible during a long-cycle time. This compilation is followed by a series of short cycles calculating one collision each. In the short cycle the pair of particles with the shortest time to collision is brought to that collision (subtracting that time from all the stored times for other pairs) and in which the collision times for pairs involving the two colliders are recalculated. In the course of the computation the collision rate Γ is measured and is directly proportional to the nonideal part of the pressure :

$$(PV/NkT) = 1 + (\Gamma/\Gamma_o)B_2\rho \; ; \; B_2 = (\pi\sigma^2/2) \; .$$

The low-density kinetic-theory collision rate Γ_o and its number dependence are explained in Reference 12. An alternative route to the pressure comes from the "virial", $\sum r_i \cdot F_i$, which simplifies in D dimensions to a time-averaged transfer of radial momentum over a long time τ :

$$PV = NkT + (D/2) \sum_{i<j} (r_i - r_j) \cdot (v_i - v_j)'/\tau \ .$$

The prime indicates the post-collision velocities. The two ways of measuring the pressure differ by terms of order $(1/N)$ due to the conservation of overall momentum and to the finite number of possible pairs, $\binom{N}{2}$.[12] This difference is completely negligible relative to the ten percent pressure fluctuations found in 50-hour runs of 100,000,000 collisions each shown in **Figure 7.4**.

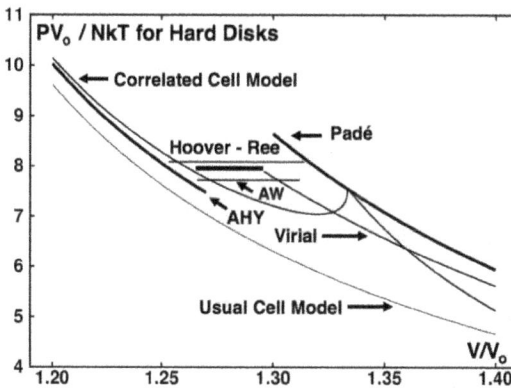

Fig. 7.5: The pressures from numerical differentiation of the cell and correlated cell models[10] are compared to a "Padé" Approximant reproducing the first six virial coefficients,[14] to Clisby and McCoy's ten-term "Virial" series,[15] and to the solid-phase pressures "AHY" from Reference 16. The heavy horizontal tie-line between the Ree-Hoover and AW = Alder-Wainwright estimates is the accurate melting-freezing pressure found in Reference 3. The pressure from the correlated cell model has a discontinuous slope at three-fourths the close-packed density.

In the fifty-year period separating Alder and Wainwright's disk simulations from the modern complete and efficient ones the practical run length
has changed from 10^8 to 10^{12} collisions, consistent with a Moore's-Law doubling time of about four years. A bit less than half the improvement is due
to the development of more efficient algorithms. Reference 3 describes the
several algorithms well. The hard-disk melting transition can at last be labelled "case closed". Looking at **Figure 7.5** we see that the Clisby-McCoy
virial series and the correlated-cell solid models nicely fit the accurate coexistence pressure determined by Engel, Anderson, Glotzer, Isobe, Bernard,
and Krauth.

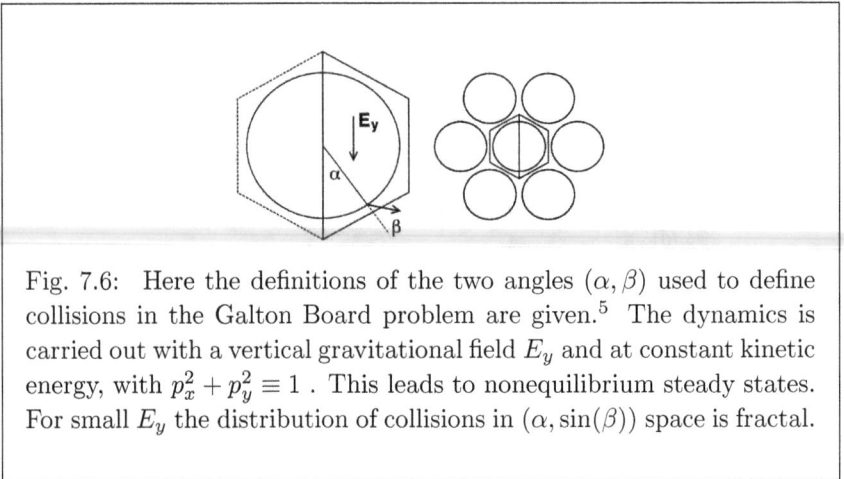

Fig. 7.6: Here the definitions of the two angles (α, β) used to define
collisions in the Galton Board problem are given.[5] The dynamics is
carried out with a vertical gravitational field E_y and at constant kinetic
energy, with $p_x^2 + p_y^2 \equiv 1$. This leads to nonequilibrium steady states.
For small E_y the distribution of collisions in $(\alpha, \sin(\beta))$ space is fractal.

7.5   The Galton Board, a Simple Nonequilibrium System

A simple disk-based dynamical model describes the chaotic motion of a single moving mass point in a periodic lattice of hard-disk scatterers.[5, 17] We
set the hard-disk diameter equal to unity for convenience. Then the maximum density for a triangular lattice with each of the scatterers touching
six neighbors is $\sqrt{(4/3)}$. At that density a mass point wanderer is trapped
in a roughly triangular region bounded by three scatterers. At densities
less than $\sqrt{(3/4)}$ [three quarters of the closest packed density] trajectories parallel to the x axis can travel ballistically, without any scattering.
Figure 7.6 shows a unit cell with the scatterers at a density of four-fifths
of close packing, high enough that scattering is inevitable.

> **Exercise :** Show that the maximum density for a triangular lattice of hard disks of diameter 1 is $\sqrt{(4/3)}$ and that correlated motion is possible at densities of $\sqrt{(3/4)}$ and below.

Machta and Zwanzig[17] calculated the diffusion coefficient of the wanderer point mass near close-packing, when the mass is nearly trapped in one of the roughly triangular free volume regions and makes occasional jumps to an adjacent region. Their zero-field kinetic-theory calculation agrees nicely with the conductivity extrapolation from dynamical simulations. I have chosen to refer to this model of a scattering mass point as the "Galton Board" rather than the "Lorentz Gas" as Lorentz' theoretical work occurred about 30 years later than Galton's use of the Board as a classroom probability demonstrator.

Working with Bill Moran[5, 18] I had the idea that this Model could stabilize an interesting nonequilibrium steady state by including a thermostat variable ζ from Gauss' Principle in the set of thermostatted isokinetic equations of motion :

$$\{ \dot{x} = p_x \; ; \; \dot{y} = p_y \; ; \; \dot{p}_x = -\zeta p_x \; ; \; \dot{p}_y = -E_y - \zeta p_y \} \; .$$

$$\text{where } p_x^2 + p_y^2 \equiv 1 \; .$$

The collisions with the hard-disk scatterers are elastic and impulsive, reflecting the radial momentum and leaving the tangential momentum unchanged. In the streaming motion between collisions we keep the kinetic energy fixed using Gauss' friction coefficient ζ as follows :

$$\dot{K} = \dot{p}_x p_x + \dot{p}_y p_y = 0 = -\zeta p_x^2 + (-E_y - \zeta p_y)p_y \longleftrightarrow$$

$$\zeta = -E_y p_y \longrightarrow \{ \dot{p}_x = +E_y p_x p_y \; ; \; \dot{p}_y = -E_y + E_y p_y^2 \equiv -E_y p_x^2 \} \; .$$

It is worth noting that the comoving phase-volume change using Liouville's continuity equation is *not* $(\partial \dot{p}_x / \partial p_x) + (\partial \dot{p}_y / \partial p_y) = 2\zeta$. The velocity constraint makes the Galton Board phase space three-dimensional rather than four. A correct three-dimensional approach formulates the momentum space with a single angle θ with $(p_x, p_y) = (\cos(\theta), \sin(\theta))$. The two motion equations for $\dot{p}_x$ and $\dot{p}_y$ then give exactly the same motion equation for $\dot{\theta}$:

$$\dot{p}_x = -\dot{\theta} \sin(\theta) = +E_y \sin(\theta) \cos(\theta) \to \dot{\theta} = -E_y \cos(\theta) \; ;$$

$$\dot{p}_y = \dot{\theta} \cos(\theta) = -E_y \cos^2(\theta) \longrightarrow \dot{\theta} = -E_y \cos(\theta) \; .$$

The change of phase volume in (x, y, θ) space is now correctly given as :

$$(\partial \dot{\theta} / \partial \theta) = E_y \sin(\theta) = E_y p_y = -\zeta \longrightarrow \otimes(t) = \otimes(0) e^{-\int_0^t \zeta' dt'} \; .$$

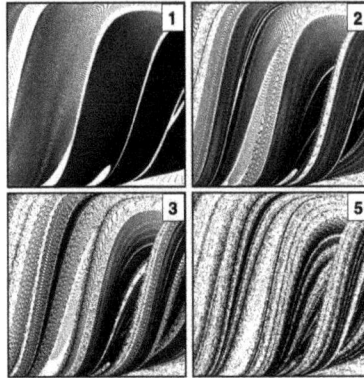

Fig. 7.7:   Poincaré sections after 1, 2, 3, and 5 collisions with a field
strength of 3. After ten collisions little further change is visible. Here
the progress of a square array of 320×320 equally-spaced points has
been followed. The ordinate is $-1 < \sin(\beta) < +1$ and the abscissa is
$0 < \alpha < \pi$ following the definitions of **Figure 7.6**.

As time goes on the reversible friction coefficient ζ is predominantly
positive. Thus the comoving phase volume $\otimes$ shrinks to zero. The probability density f obeys the conservation relation $d(f\otimes)/dt \equiv 0$ so that
the probability density diverges as $\otimes$ vanishes. This singular behavior is
characteristic of nonequilibrium fractals. Gauss' isokinetic dynamics is precisely time-reversible. The frictional force $-\zeta p$ is unchanged in the reversed
motion as both the friction coefficient and the momenta change signs.

Bill Moran and I had expected to see an interesting distribution of
collision types as a function of the driving E_y field, amenable to Fourier-
transform analysis. Instead we found that an initially uniform distribution
in $(\alpha, \sin(\beta))$ space develops rapidly into an elaborately striped distribution,
evidently discontinuous not just here and there, but almost everywhere !

Figure 7.7 illustrates what we found. The figure shows four snapshots
in the history of a 320×320 square grid of initial conditions after a single
collision, after two and three collisions, and finally after five collisions. The
process soon creates features on a scale too small to be seen. The apparent
steady-state distribution is not very different to what we see after just five
or ten collisions. Furthermore the distribution is evidently ergodic for field

strengths below 3. The statistics gathered in what were long runs in 1987 were fully consistent with distributions that would eventually fill every box.

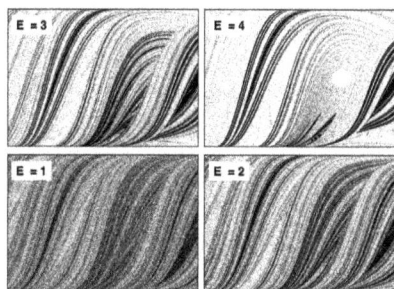

Fig. 7.8: A comparison of fractal distributions found with field strengths of 1, 2, 3, and 4. In the last case a new feature can be seen, a nearly circular hole. Simulations with initial conditions chosen "in" the hole turn out to be quasiperiodic tori. The details of three of them appear in **Figure 7.9**.

Figure 7.8 shows the effect of changes in the field strength driving the current. The density becomes increasingly striated but remains ergodic until a field strength of 3.69. At that field strength the motion seeks out a 20-bounce limit cycle which is attractive and stable. As the field is increased further, another surprise occurs at a field strength of 4. There is an isolated hole in the phase-space cross section, visible in the Figure. An initial point chosen within the hole corresponds to a conservative torus describing the wanderer's stable bouncing back and forth as shown in **Figure 7.9**. The remaining part of the section is chaotic and dissipative with overall motion in the vertical direction. There are no doubt an infinite number of stable multiple-bounce limit cycles obtained at higher fields. Many are tabulated in Reference 5. A five-bounce repeating sequence is shown in **Figure 7.9**.

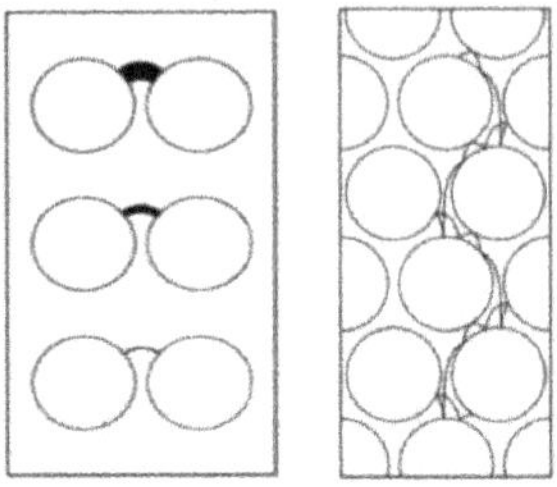

TORUS and LIMIT CYCLE
E = 4   E = 6

Fig. 7.9: In addition to its chaotic sea the Galton Board exhibits conservative tori adjacent to the sea as well as an unbounded number of attractive limit cycles. Examples of these two solution types are shown here. See Reference 5 for more.

How does one best describe the fractal objects found in the Galton Board problem? One set of descriptors is the Lyapunov spectrum – there are three exponents in the spectrum as the phase space is three-dimensional. The largest Lyapunov exponent can be measured by studying the tendency of two nearby phase points to separate as time goes on. The calculation is complicated by the jumps in momentum which occur at the collisions. Harald Posch and Christoph Dellago developed a mapping technique taking the jumps into account. This is described in Christoph's Ph D thesis at the University of Vienna. There are three Lyapunov exponents. With chaos one is positive, one zero, and the last sufficiently negative that the sum equals the mean value of $-\zeta$. ζ is again the friction coefficient that describes the change of phase volume with time.

The Galton Board problem is similar in complexity to the Baker Map discussed in Section 6.13. In the Baker Map the stretching and shrinking directions were perpendicular. In the Galton Board sections of **Figures 7.7 and 7.8** it is evident that the characteristic directions vary and are also discontinuous. In view of the time-reversibility of the Galton Board motion equations it is paradoxical that the motion invariably seeks out the attractor in the three-dimensional phase space rather than the repellor, which is the mirror image of the attractor, with the angle β changed in sign. In a strictly reversed motion the sum of the exponents would reverse to positive, indicating both a physical and a computational instability. In fact this never happens. The motion seeks out the attractor and avoids the repellor. It is paradoxical that both the attractor and the repellor are typically ergodic, with nonvanishing density throughout the phase space. **Figure 7.15** indicates that the computational irreversibility can furnish an explanation.

7.6 Fractal Dimensions of the Galton Board Distributions

Evidently the distributions are typically *fractal* with fractional dimensionality. Dimensionality can be quantified by determining how the cell probabilities scale with cell size – in the usual nonfractal case the probability within a two-dimensional cell $dxdy$ should vary as the square of the cell size. The probability for a three-dimensional cell $dxdydz$ should vary as the cube, and so on. In fractals this simplicity, which follows from differentiability, is lost. The fractal dimensions provide a quantitative description of this lack of differentiability. To demonstrate the basic concept we first calculate the "natural measure" of each cell in a grid spanning the phase space section. Following my work with Bill Moran,[18] I focus on the field strength $E = 3$ and use a grid of 1024×1024 rectangular cells spanning the section. The "natural measure" of each cell is the fraction of the sampled points (six million in all) in the cell. The variation of the measure with a "small, but not too small" cell size Δ gives the information dimension D_I :

$$D_I \equiv \langle\, p\ln(p)\, \rangle / \ln(\Delta) \simeq (dS/d\ln\Delta) \ .$$

This dimension is analogous to the Gibbs entropy in that it is an average of the logarithm of a probability.

Another fractal dimension, one that is particularly easy to calculate, is the *correlation* dimension, which measures the variation of the number of *pairs* of points with distance. Rather than counting pairs one can simply weight cells in a grid according to the *square* of the number of sampling points, defining the pair measure p_2. The correlation dimension D_C is

$$D_C \equiv \langle\, p_2\ln(p_2)\, \rangle / \ln(\Delta) \ ,$$

again for a propitious range of Δ. The left panel of **Figure 7.10** shows the utility of these measures and dimensions for "reasonable" grid sizes ranging from 16×16 to 1024×1024.

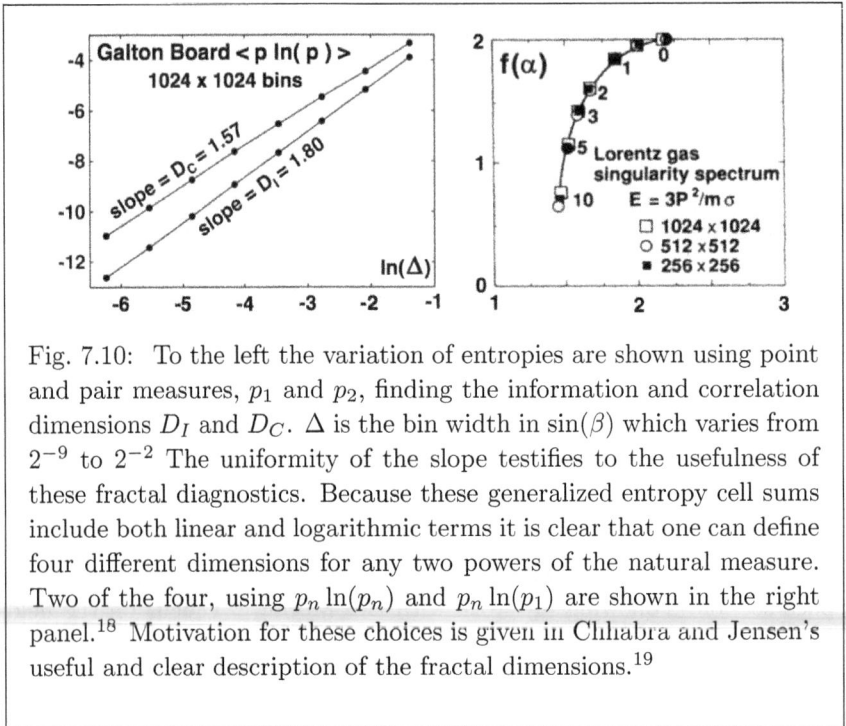

Fig. 7.10: To the left the variation of entropies are shown using point and pair measures, p_1 and p_2, finding the information and correlation dimensions D_I and D_C. Δ is the bin width in $\sin(\beta)$ which varies from 2^{-9} to 2^{-2} The uniformity of the slope testifies to the usefulness of these fractal diagnostics. Because these generalized entropy cell sums include both linear and logarithmic terms it is clear that one can define four different dimensions for any two powers of the natural measure. Two of the four, using $p_n \ln(p_n)$ and $p_n \ln(p_1)$ are shown in the right panel.[18] Motivation for these choices is given in Chhabra and Jensen's useful and clear description of the fractal dimensions.[19]

7.7 The Galton Staircase

On finding the singular zero-volume fractals characterizing the Galton Board I wanted to see how general this shrinking was. Because the symmetry-breaking exhibited by the time-reversible motion equations provides an explanation for irreversible Second-Law behavior these models are of particular interest. It appears that their behavior is outside the usual bailiwick of mathematicians so that there is motivation to develop and analyze more such models. In collaboration with five colleagues at a 1986 CECAM workshop in Orsay (Centre Européen de Calcul Atomique et Moléculaire), just outside Paris, I decided to study a biased cosine potential with the bias strong enough to generate a readily detectable steady-state mass current. That work is described in Reference 6.

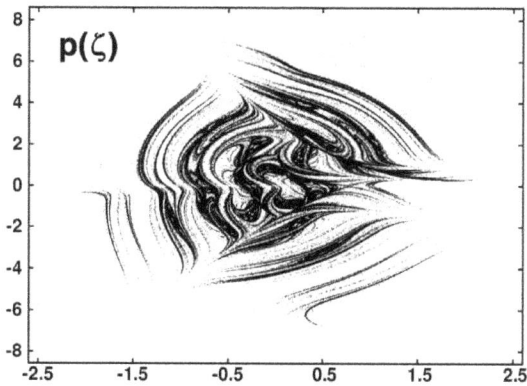

Fig. 7.11: Poincaré section at the plane $q = 0$ for the time-reversible dissipative dynamics of a Galton staircase, introduced in Reference 6 : $\dot{q} = p$; $\dot{p} = 0.3 - \sin(q) - \zeta p$; $\dot{\zeta} = (p^2 - 1)/10$.

It shares many of the features present in the Galton Board simulations, suggesting that the results are of general validity. As we pointed out at the time (1987) there is a long list of the dynamics' features that are worth noting : [1] It is deterministic. [2] It is time-reversible. [3] It is Lyapunov unstable, with small perturbations growing exponentially larger with time. [4] Sufficiently far from equilibrium the motion generates one-dimensional limit cycles in the extended phase space. [5] There is a dissipative conversion of work to heat which satisfies the Second Law of Thermodynamics. [6] There is an overall contraction in the phase volume with a loss of dimensionality. [7] Despite the time-reversibility of the motion equations Lyapunov instability guarantees the relative *instability* of the even less-stable time-reversed solution. Thus a numerical simulation *always* chooses the attractor rather than the mirror-image repellor.

The scarcity of nonequilibrium states – they occupy zero volume in the phase space – was a surprise. Rather than Fourier analysis, *fractal* analysis is required for nonequilibrium steady states. These exciting and unexpected results have motivated much of Carol's and my research over the thirty years since. Let us look at more recent developments based on nonequilibrium generalizations of the one-dimensional harmonic oscillator.[20]

7.8   Oscillator Heat Conduction with a Thermal Gradient

The simplicity of the thermostatted oscillator problem provides a good
starting point for additional studies of nonequilibrium systems. The disadvantage of the Nosé and Nosé-Hoover oscillators, relative to the Galton
Board example is their lack of ergodicity, with most of the phase space occupied by conservative toroidal solutions. Lack of ergodicity means that the
results of linear-response theory based on Gibbs' canonical ensemble will
not apply, a real drawback for any serious study of nonequilibrium systems.

Advances in 2014-2016 led to somewhat more complicated thermostatting than Nosé-Hoover but with compensating advantages : the Gaussian
form of the oscillator's distribution was known, and that Gaussian was
sited in just three dimensions rather than four.[20] Here is the first of the
two examples we will consider, the "0532 Model" :

$$\dot{q} = p \; ; \; \dot{p} = -q - \zeta[\, 0.05p + 0.32(p^3/T)\,] \; ;$$

$$\dot{\zeta} = 0.05[\, (p^2/T) - 1\,] + 0.32[\, (p^4/T^2) - 3(p^2/T)\,] \; [\, 0532 \text{ Model}\,] \, .$$

The coefficients 0.05 and 0.32 were found with a Monte Carlo search once I
thought of applying "weak control" to the oscillator. These 0532 equations
of motion are consistent with the Maxwell-Boltzmann distribution, $e^{-p^2/2T}$
with $\langle\, p^2 - T\,\rangle$ and $\langle\, p^4 - 3p^2T\,\rangle$ both vanishing. A single friction coefficient ζ
cannot enforce both constraints simultaneously. But we are in luck ! Weak
control is enough.

Exercise : Show that the 0532 motion equations are consistent with
the oscillator's canonical distribution at temperature T multiplied by
a Gaussian in ζ .

A computer program choosing the constraining coefficients randomly,
and including also sixth-moment weak control of $\langle\, p^6 - 5p^4T\,\rangle$, found not
only the 0532 Model but also a much stiffer single-thermostat set of motion
equations, likewise ergodic, and controlling both $p^2 - T$ and $p^6 - 5p^4T$:

$$\dot{q} = p \; ; \; \dot{p} = -q - \zeta[\, 1.5p - 0.5(p^5/T^2)\,] \; ;$$

$$\dot{\zeta} = 1.50[\, (p^2/T) - 1\,] - 0.50[\, (p^6/T^3) - 5(p^4/T^2)\,] \; [\, 1.50 \text{ Model}\,] \, .$$

Both the 0532 and 1.50 Model thermostat types were carefully checked for
ergodicity as follows : [1] No "holes", which would indicate tori, were found

in any of the cross sections , (q, p), (p, ζ), or (ζ, q). [2] The deviations from the Gaussian moments,

$$\{ \langle x^{2,4,6} \rangle = 1, \ 3, \ 15 \} \longleftarrow x = (q, p, \zeta) \ ,$$

were statistically insignificant. [3] The largest Lyapunov exponent $\lambda_1 = \langle \lambda_1(t) \rangle$ was independent of the initial conditions. [4] The sections that we obtained were visually independent of the integrator type, RK4, RK5, or adaptive. To look for more detail I separated the (q, p) sections with $\zeta = 0$ into their Lyapunov unstable and Lyapunov stable parts (based on the sign of the local Lyapunov exponent).

The local Lyapunov exponent YAPL during a timestep from TIME to TIME + DT requires only the scaling FACTOR which returns the offset at TIME + DT between satellite and reference back to its desired value. A typical choice for the offset in these oscillator problems is DELTA = 0.000001. Here is the rescaling operation :

```
RR = (QS-QR)*(QS-QR) + (PS-PR)*(PS-PR) + (ZS-ZR)*(ZS-ZR)
FACTOR = DELTA/DSQRT(RR)
QS = QR + FACTOR*(QS-QR)
PS = PR + FACTOR*(PS-PR)
ZS = ZR + FACTOR*(ZS-ZR)
YAPL = - DLOG(FACTOR)/DT ! [ The local Lyapunov Exponent ]
```

The penetrations of the $\zeta = 0$ plane are separated according to the sign of the local exponent and shown in **Figure 7.12**. It is interesting to see that the stable and unstable regions share considerable overlap, particularly for the 1.50 Model. The stiffness of this model results in a penetration rate of the $(q, p, \zeta = 0)$ plane more than seven times that of the 0532 Model. The additional mixing in this stiffer case reduces the correlation between instability and higher energy which is clear near the origin in the smoother slower-paced 0532 sections.

Because, at equilibrium, viewing the oscillator in a mirror changes the signs of both q and p while leaving ζ unchanged, each of the four sections shows inversion symmetry about the origin with the densities at $(+q, +p)$ and $(-q, -p)$ equal for both positive and negative values of the coordinate and momentum.

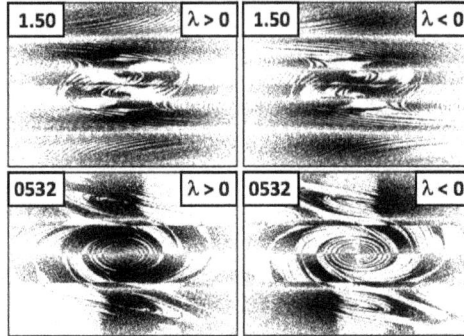

Fig. 7.12: $p(q)$ sections with $\zeta = 0$ for two ergodic oscillator models at equilibrium. The ranges shown are $-3 < q < +3$ for the abscissa and $-4 < p < +4$ for the ordinate. Notice the inversion symmetry corresponding to mirror symmetry for the oscillator. Each section has been divided into two separate panels. 90,000 points are shown with positive and negative values of $\lambda_1(t)$, plotted separately. In fact for a fixed time interval the positive points substantially outnumber the negative ones.

Away from equilibrium, as illustrated in **Figure 7.13** we let the temperature vary smoothly between $1 - \epsilon$ and $1 + \epsilon$: $T(q) = 1 + \epsilon \tanh(q)$. As a result, there is an overall dissipative flow of heat from right to left in response to the positive temperature gradient. This results in entropy production, which can be assessed in either of two ways. In the case of the 0532 model, the comoving time-rate-of-change of the phase volume $\otimes = dqdpd\zeta$ follows from the motion equations :
$$\dot{S}_{Gibbs} = (d\ln\otimes/dt) = -\zeta[\, 0.05 + 0.96(p^2/T)\,]$$
The heat extracted by the thermostat force, divided by the temperature, gives the thermodynamic entropy loss rate :
$$\dot{S}_{thermo} = -\zeta[\, 0.05(p^2/T) + 0.32(p^4/T^2)\,]\,.$$
At the same time the boundedness of the friction coefficient ζ provides the useful (time-averaged) identity :
$$\langle\, \zeta\dot{\zeta}\,\rangle \equiv 0 = \langle\, 0.05\zeta[\,(p^2/T) - 1\,] + 0.32\zeta[\,(p^4/T^2) - 3(p^2/T)\,]\,\rangle \longrightarrow$$
$$\langle\, 0.05\zeta(p^2/T) + 0.32\zeta(p^4/T^2)\,\rangle = \langle\, 0.05\zeta + 0.96\zeta(p^2/T)\,]\,\rangle \longrightarrow$$
$$\langle\, \dot{S}_{Gibbs} = \dot{S}_{thermo}\,\rangle\,!$$

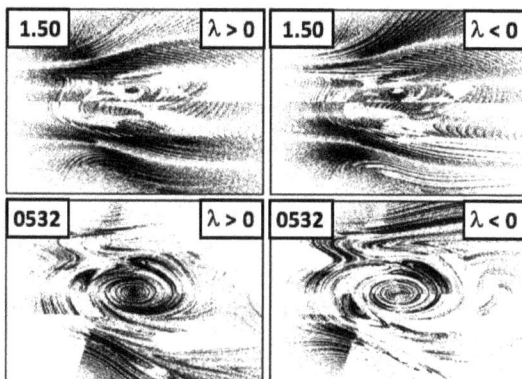

Fig. 7.13: $p(q)$ sections with $\zeta = 0$ for two ergodic oscillator models at $\epsilon = 0.50$. The ranges shown are $-3 < q < +3$ for the abscissa and $-4 < p < +4$ for the ordinate. The temperature gradient has destroyed the symmetry seen in **Figure 7.12**. The sections for the 1.50 model (top) and 0532 model (bottom) are *fractals* divided into two separate panels. 100,000 points are shown in each panel. Points with positive and negative values of $\lambda_1(t)$ are plotted separately. Notice that there is considerable overlap with no sharp boundary between positive and negative points.

Thus, when time-averaged in a stationary nonequilibrium state, the entropy produced by the oscillator and transferred to its surroundings is precisely equal to the change in Gibbs' entropy. $\langle \dot{S}_{Gibbs} = \dot{S}_{thermo} \rangle$. This same identity applies also to the 1.50 Model and to all of the several other time-reversible thermostat models we have studied.[21] John Ramshaw[22] has just recently shown that the equivalence of the two forms of entropy holds for a wide variety of time-reversible deterministic thermostat types. In our work we demonstrated that linear-response theory, due to the ergodicity of the thermostat, relates the nonequilibrium current to the strength of the temperature gradient accurately for $\epsilon = 0.10$. The agreement is spectacular.

7.8.1 *The Mexican Hat Potential as a Small System*

After three decades of failed efforts, we succeeded in generating Gibbs' canonical distribution for the harmonic oscillator potential with a single thermostat variable. That control variable exerted "weak control" over two velocity moments, $\langle\, p^2$ and $p^4\,\rangle$, or $\langle\, p^2$ and $p^6\,\rangle$. Rather than using two thermostat variables to enforce two moment conditions taken from the Maxwell-Boltzmann distribution, $\langle\, (p^2/T)\,\rangle = 1$ *and* $\langle\, (p^4/T^2)\,\rangle = 3$, it is practical to use a *single* thermostat variable consistent with *both* conditions. This weak-control idea was proved successful for both the harmonic-oscillator and the pendulum. These successes suggested a similar treatment of the Quartic and Mexican Hat potentials :

$$\phi_{\text{Quartic}} = (q^4/4)\ ;\ \phi_{\text{MH}} = (q^4/4) - (q^2/2)\ .$$

The Mexican Hat potential is the simplest model with two smooth minima, and provides a prototype for the quantitative study of transition states and "rare events".

For the quartic potential I chose 10,000 different combinations of thermostat forces of the form :

$$\ddot{q} = \dot{p} = -q^3 - \zeta[\ \alpha p + \beta(p^3/T) + \gamma(p^5/T^2)\ ]\ .$$

Here the single friction coefficient ζ exerts weak control on three velocity moments :

$$\dot{\zeta} = \alpha[\ (p^2/T) - 1\ ] + \beta[\ (p^4/T^2) - 3(p^2/T)\ ] + \gamma[\ (p^6/T^3) - 5(p^4/T^2)\ ]\ ,$$

with the three parameters (α, β, γ) selected randomly in the range from -1 to $+1$. This approach is successful for the harmonic oscillator and the simple pendulum, suggesting that a similar approach could be applied to the quartic potential, $\phi(q) = (q^4/4)$. But our attempts indicated, wrongly as it turned out, that the quartic potential could not be thermostatted with a single control variable.

Among 10,000 thermostat combinations we selected the 48 which produced the best values for the fourth moment $\langle\, p^4\,\rangle \simeq 3$. But examination of these 48 cross sections with $(p = 0)$ revealed either tori or a chaotic sea containing holes. There were holes in the cross sections in every case I tried. It became evident that the relatively smooth and simple quartic potential is a greater challenge than the harmonic oscillator or the pendulum.

By adding a little more complexity we can get closer to Gibbs. Consider the one-body two-dimensional Hamiltonian system with

$$\mathcal{H} = (1/2)(p_x^2 + p_y^2) + (1/4)(x^2 + y^2)^2\ .$$

By itself this system has a vanishing largest Lyapunov exponent indicating that the motion lies on a torus. The dynamics remains regular if a single friction coefficient ζ is added to the equations of motion. Adding *two* friction coefficients, one controlling p_x^2 and the other p_y^2 can lead to chaos.

Exercise : Consider a doubly-thermostatted quartic oscillator with initial values $(x, y, p_x, p_y, \zeta_x, \zeta_y) = (1, 1, 1, 1, 1, 1)$:

$$\dot{p}_x = -x(x^2 + y^2) - \zeta_x p_x \; ; \; \dot{p}_y = -y(x^2 + y^2) - \zeta_y p_y \; ;$$

$$\dot{\zeta}_x = (1/4)(p_x^2 - 1) \; ; \; \dot{\zeta}_y = (3/4)(p_y^2 - 1) \; .$$

Confirm that the largest Lyapunov exponent is $\lambda_1 = 0.054$ using 2×10^9 timesteps with $dt = 0.001$. Develop and implement a test designed to show whether or not this value depends upon the initial conditions.

Given the difficulty of thermostatting a one-dimensional quartic potential Carol and I decided to set that task as the 2016 Ian Snook Prize problem, with the solution due by the end of 2016. Diego Tapias, Alessandro Bravetti, and David Sanders solved the problem with a "logistic thermostat".[23] A slight variant of their idea[24],

$$\dot{q} = p \; ; \; \dot{p} = q - q^3 - \alpha p \tanh(\alpha \zeta) \; ; \; \dot{\zeta} = p^2 - 1 \; ,$$

provides an ergodic solution for $\alpha = 6.9$ (and not for substantially smaller values). Precisely "why" this hyperbolic form is successful is a good question. This problem is quite stiff with $\alpha = 6.9$, showing spurious structure in $\zeta(q)$ sections from an adaptive Runge-Kutta integrator with errors confined to the range $10^{-13 \pm 1}$. A smooth acccurate Gibbs'-ensemble solution resulted[24] with errors constrained to the smaller range $10^{-16 \pm 1}$.

7.9   The Conservative and Dissipative Baker Maps

The Nonequilibrium Baker Map described in Section 13 of **Lecture 6** has the following form analogous to motion equations :

$$\text{if } (q < p - \sqrt{(2/9)}) \text{ then}$$

$$q = (11q/6) - (7p/6) + \sqrt{49/18} \; ; \; p = (11p/6) - (7q/6) - \sqrt{25/18} \; ;$$

$$\text{or if } (q > p - \sqrt{(2/9)}) \text{ then}$$

$$q = (11q/12) - (7p/12) - \sqrt{49/72} \; ; \; p = (11p/12) - (7q/12) - \sqrt{1/72} \; .$$

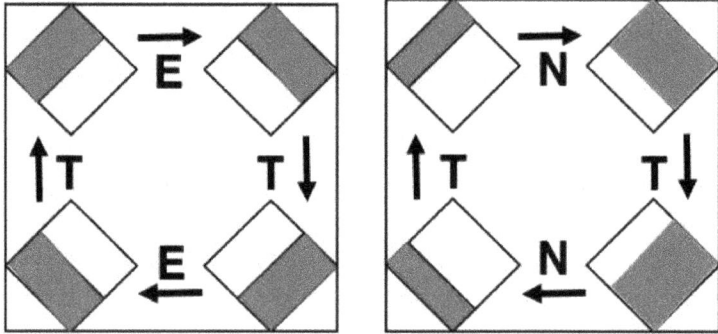

Fig. 7.14: The equilibrium Baker Map "E" conserves area but not length. The nonequilibrium map "N" conserves neither. Both maps are ergodic, covering the 2×2 (q,p) space, E with uniform density and N with a fractal strange attractor. The "T" mapping changes the sign of the vertical coordinate $p : T(q, \pm p) \to (q, \mp p)$. All three maps are time-reversible: $M^{-1} = TMT$. The top and bottom points in the space are (unstable) fixed points of both the E and N mappings.

To what extent do the maps capture the characteristics of continuous flows such as the Galton Board studied here? We see from **Figure 7.14** that the mapping B can be reversed by applying the operations TBT. For flows like the Galton Board this operation is the same as [1] changing the signs of all of the momenta and friction coefficients, $\{ \pm p, \pm \zeta \} \longrightarrow \{ \mp p, \mp \zeta \}$ and [2] integrating for a timestep ; and then [3] repeating the sign-change operation. Evidently in either case there is a flow connecting the repellor "past" (time-reversed attractor) to the attractor in the "future".

In both cases, the Baker Map and the nonequilibrium flows, both the past and the future acquire fractal character as one departs from the present. It is evident that the structures are mirror images so that both are equally rare and comprise zero-volume attractor/repellor "strange" sets in the phase space. The difference between the attractor and repellor can be quantified through the Lyapunov spectrum. Evidently the time-reversal operation changes the signs of the Lyapunov exponents too.

Galton Repellor → Galton Attractor

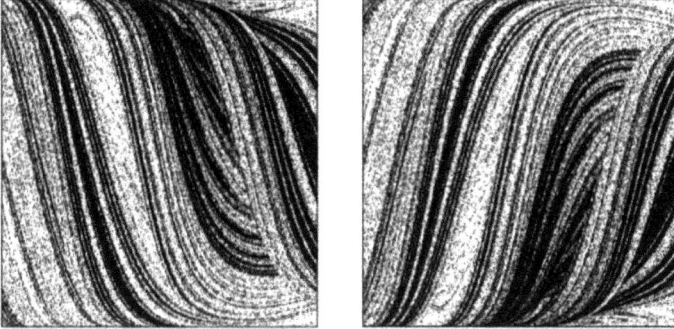

Fig. 7.15: This nonequilibrium Galton Board attractor was constructed with a field strength 3. The repellor is obtained by reversing the sign of the vertical coordinate, equivalent to changing the sign of the velocity or of the time. The abscissa is the angle $0 < \alpha < \pi$, playing the role of a coordinate, and the ordinate is $-1 < \sin(\beta) < +1$, playing the role of a momentum. The repellor source and attractor sink are both ergodic, with density everywhere, but with an information dimension of 1.80. The Galton Board dynamics is fully time-reversible.

Figure 7.15 shows an attractor-repellor pair for the Galton Board flow problem. The information dimension is 1.80. The mechanism driving the motion from repellor to attractor is exactly the same as in the Baker Map. Continuum flow problems all show that the statistical entropy production (which makes it possible for a system to convert an infinite amount of work into heat and increased entropy in a steady state) can be expressed equally well in terms of phase-volume shrinkage or heat transfer divided by temperature.[22]

Thus the time-reversible deterministic models we constructed in order to simulate steady nonequilibrium flows have provided a quantitative explanation for the Second Law of Thermodynamics. The mechanism is the loss of Gibbs' phase-space volume through its steady conversion to external entropy increase.

7.10   Summary of Lecture 7

Although Thermodynamics and Hydrodynamics are macroscopic disciplines we have seen that idealized small systems provide analogs of the larger systems treated with continuum mechanics. The simplest two-disk problem, with a proper choice of boundary conditions [periodic, with an aspect ratio of $\sqrt{3}$] illustrates the sliding mechanism typical of two-dimensional melting. This two-body problem, the correlated-cell model, has also a thermodynamic equation of state displaying a finite-system van der Waals' pressure-volume loop that is reasonably close to the results of modern simulations carried out with 2^{20} hard disks. Only a few dozen disks are enough to illustrate the percolation transition that links low-density extensive free volumes to high-density intensive free volumes. It is interesting that in three dimensions (hard spheres, for example) there are two distinct percolation transitions which occur at two different densities.[25] The volume available to a sphere changes from extensive to intensive on compression. At a *different* density the mean "cluster size" created by overlapping excluded volumes changes in the reverse direction, from intensive to extensive.

Nonequilibrium properties have their small-system analogs too. The Galton Board model, in which gravitational energy is converted into extracted heat is a good model for mesoscopic color conductivity in which half the particles in a thermostatted manybody system are accelerated to the right, and half to the left. Likewise, the Galton Staircase generates another family of field-driven thermostatted steady states : a Nosé-Hoover thermostat maintains the time-averaged kinetic temperature of a particle confined by a sinusoidal field and driven by an additional constant force. In both these cases, a single control variable ζ extracts field energy in the form of "heat" so that a steady nonequilibrium state results.

A single harmonic oscillator in a smooth thermal environment, $T(q) = 1 + \epsilon \tanh(q)$, provides a localized nonequilibrium dynamics. There is a nonvanishing heat current $\langle \, (p^3/2) \, \rangle$ but the mass current is obviously zero. When the oscillator's entropy generation rate is time-averaged, $\langle \, -\dot{Q}/T \, \rangle$, it *exactly compensates* the time-averaged dynamical loss rate of phase volume $\langle \, (\dot{S}_{Gibbs}/k) \, \rangle \; = (\dot{\otimes}/\otimes) \, \rangle$. This equivalence is typical of the deterministic time-reversible thermostats used in nonequilibrium simulations.[22] When such an oscillator is controlled by an ergodic thermostat, such as the 0532 or 1.50 Model, linear-response theory based on Gibbs' canonical ensemble applies for small fields. In this case the thermostats provide a quantitative description of the nonequilibrium linear response to dissipation.

Baker Repellor → Baker Attractor

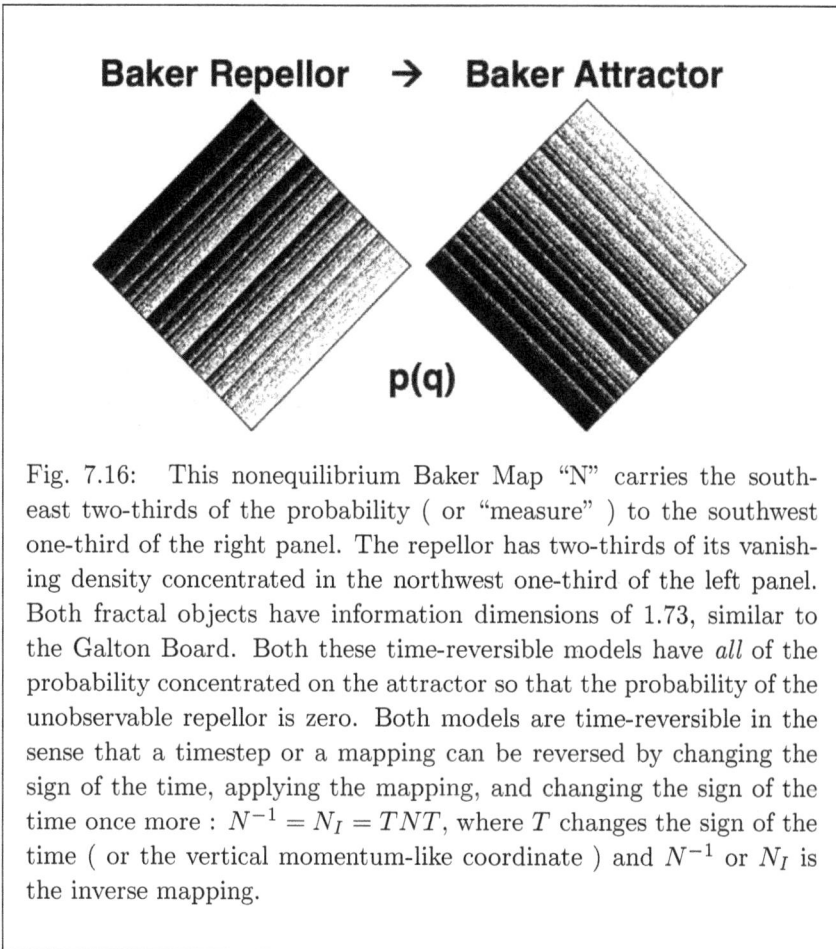

p(q)

Fig. 7.16: This nonequilibrium Baker Map "N" carries the southeast two-thirds of the probability (or "measure") to the southwest one-third of the right panel. The repellor has two-thirds of its vanishing density concentrated in the northwest one-third of the left panel. Both fractal objects have information dimensions of 1.73, similar to the Galton Board. Both these time-reversible models have *all* of the probability concentrated on the attractor so that the probability of the unobservable repellor is zero. Both models are time-reversible in the sense that a timestep or a mapping can be reversed by changing the sign of the time, applying the mapping, and changing the sign of the time once more : $N^{-1} = N_I = TNT$, where T changes the sign of the time (or the vertical momentum-like coordinate) and N^{-1} or N_I is the inverse mapping.

These small-system models can be idealized further to the time-reversible nonequilibrium Baker Map "N" with the repellor and attractor shown in **Figure 7.16**. Replacing the time integration with a simple "nonequilibrium" mapping operation N and introducing the time-reversal mapping T such that

$$(q,p) \xrightarrow{N} (q',p') \; ; \; TNT(q',p') = N^{-1}(q',p') \to (q,p)$$

provides distributions analogous to those from time-reversible thermostatted nonequilibrium flows. The distributions are fractal. Many applications of the mapping (twice as many for quadruple precision as for double) converts the fractal repellor structure to its mirror-image, the fractal strange attractor. At the beginning and at the end of this process the phase volumes of the fractal structures are zero. At the same time the structures are ergodic, with a natural measure everywhere. This paradoxical coexistence of zero volume and ubiquitous density is hard to grasp, but real. The measure of the cells which contain the majority of the density, or "natural measure", approaches zero as the cell size is reduced. A mathematical proof of these properties would seem unconvincing for the reversibility and recurrence reasons stated by Loschmidt and Zermélo, but numerical work provides conclusive evidence that it is true.

Evidently, when an ergodic time-reversible mapping contains regions that change area, so that $dqdp \neq dq'dp'$ the mapping may generate a strange attractor sink for the flow from a mirror image repellor. On the other hand the area changes may represent nothing more than equilibrium fluctuations, as in the case where the distribution is Gibbs' canonical one. The 0532 and 1.50 models are two examples. The crucial property for steady nonequilibrium flows is *net* heat transfer to the surroundings. Whenever this is present the dynamics or the mapping will cause the probability density to condense onto a zero-volume attractor or a stable limit cycle.

It is remarkable to find this simple explanation for the failure of Loschmidt's and Zermélo's objections. Reversing the velocities of a fractal attractor with a negative Lyapunov sum produces the mirror-image fractal repellor with a positive Lyapunov sum. Because the attractor represents a steady state there is no problem with recurrence. Inevitably the collapse onto a fractal attractor indicates the dynamics' obedience to the Second Law of Thermodynamics.

References

1. N. Metropolis, A. W. Rosenbluth, M. N. Rosenbluth, A. H Teller, and E. Teller, "Equation of State Calculations by Fast Computing Machines", The Journal of Chemical Physics **21**, 1087-1092 (1953).
2. B. J. Alder and T. E. Wainwright, "Phase Transition in Elastic Disks", Physical Review **127**, 359-361 (1962).
3. M. Engel, J. A. Anderson, S. C. Glotzer, M. Isobe, E. P. Bernard, and W. Krauth, "Hard-Disk Equation of State : First-Order Liquid-Hexatic Transition in Two Dimensions with Three Simulation Methods", Physical Review E **87**, 042134 (2013) = arχiv 1211.1645.
4. Wm. G. Hoover, N. E. Hoover, and K. Hanson, "Exact Hard-Disk Free Volumes", The Journal of Chemical Physics **70**, 1837-1844 (1979).
5. B. Moran, Wm. G. Hoover, and S. Bestiale, "Diffusion in a Periodic Lorentz Gas", Journal of Statistical Physics **48**, 709-726 (1987).
6. Wm. G. Hoover, H. A. Posch, B. L. Holian, M. J. Gillan, M. Mareschal, and C. Massobrio, "Dissipative Irreversibility From Nosé's Reversible Mechanics", Molecular Simulation **1**, 79-86 (1987).
7. J. C. Sprott, Wm. G. Hoover, and C. G. Hoover, "Heat Conduction, and the Lack Thereof, in Time-Reversible Dynamical Systems: Generalized Nosé-Hoover Oscillators with a Temperature Gradient", Physical Review E **89**, 042914 (2014).
8. Wm. G. Hoover and H. A. Posch, "Chaos and Irreversibility in Simple Model Systems", Chaos **8**, 366-373 (1998).
9. L. Tonks, "The Complete Equation of State of One, Two, and Three-Dimensional Gases of Hard Elastic Spheres", Physical Review **50**, 955-963 (1936).
10. B. J. Alder, Wm. G. Hoover, and T. E. Wainwright, "Cooperative Motion of Hard Disks Leading to Melting", Physical Review Letters **11**, 241-243 (1963).
11. J. C. Sprott, Wm. G. Hoover, and C. G. Hoover, "Heat Conduction, and the Lack Thereof, in Time-Reversible Dynamical Systems : Generalized Nosé-Hoover Oscillators with a Temperature Gradient", Physical Review E **89**, 042915 (2014).
12. Wm. G. Hoover and B. J. Alder, "Studies in Molecular Dynamics. IV. The Pressure, Collision Rate, and Their Number Dependence for Hard Disks", The Journal of Chemical Physics **46**, 686-691 (1967).
13. Wm. G. Hoover and F. H. Ree, "Melting Transition and Communal Entropy for Hard Spheres", The Journal of Chemical Physics **49**, 3609-3617 (1968).
14. F. H. Ree and Wm. G. Hoover, "Fifth and Sixth Virial Coefficients for Hard Spheres and Hard Disks", The Journal of Chemical Physics **40**, 939-950 (1964).
15. N. Clisby and B. M. McCoy, "New Results for Virial Coefficients of Hard Spheres in D Dimensions", Pramana - Journal of Physics **64**, 775-783 (2005) = arχiv 0410511.
16. B. J. Alder, Wm. G. Hoover, and D. A. Young, "Studies in Molecular Dynamics. V. High-Density Equation of State and Entropy for Hard Disks and

Spheres", The Journal of Chemical Physics **49**, 3688-3696 (1968).

17. J. Machta and R. Zwanzig, "Diffusion in a Periodic Lorentz Gas", Physical Review Letters **50**, 1959-1962 (1983).

18. Wm. G. Hoover and B. Moran, "Phase-Space Singularities in Atomistic Planar Diffusive Flow", Physical Review A **40**, 5319-5326 (1989).

19. A. Chhabra and R. V. Jensen, "Direct Determination of the $f(\alpha)$ Singularity Spectrum", Physical Review Letters **62**, 1327-1330 (1989).

20. Wm. G. Hoover, C. G. Hoover, and J. C. Sprott, "Nonequilibrium Systems: Hard Disks and Harmonic Oscillators Near and Far From Equilibrium", Molecular Simulation **42**, 1300-1316 (2016) = arχiv 1507.08302 .

21. P. K. Patra, Wm. G. Hoover, C. G. Hoover, and J. C. Sprott, "The Equivalence of Dissipation from Gibbs' Entropy Production with Phase-Volume Loss in Ergodic Heat-Conducting Oscillators", International Journal of Bifurcation and Chaos **26**, 1650089 (2016) = arχiv 1511.03201 .

22. J. D. Ramshaw, "Entropy Production and Volume Contraction in Thermostatted Hamiltonian Mechanics", Physical Review E **96**, 052122 (2017).

23. D. Tapias, A. Bravetti, and D. Sanders, "Ergodicity of One-Dimensional Systems Coupled to the Logistic Thermostat", Computational Methods in Science and Technology **23**, 11-18 (2017).

24. Wm. G. Hoover and C. G. Hoover, "Singly-Thermostatted Ergodicity in Gibbs' Canonical Ensemble and the 2016 Ian Snook Prize Award", Computational Methods in Science and Technology **23**, 5-8 (2017).

25. L. V. Woodcock, "Percolation Transitions in the Hard-Sphere Fluid", American Institute of Chemical Engineers Journal **58**, 1610-1618 (2012).

Chapter 8

Microscopic Reversibility, Macroscopic Irreversibility

/ Microscopic Reversibility / Macroscopic Irreversibility / Størmer and Yoshida Reversibility / Dettmann's Hamiltonian / Benettin's Lyapunov Exponent Algorithm / Lyapunov Instability of a Bouncing Ball / Macroscopic Irreversibility / Shockwave Modelling / Shockwave Stability / Navier-Stokes-Fourier Modelling / Inelastic Ball Collision / Strong Shockwaves / Shear-Map Stability / Baker-Map Stability / Summary of Lecture 8 /

8.1 Microscopic Reversibility ; Macroscopic Irreversibility

Microscopic time reversibility is generally a property of motion equations in physics. In **Lecture 7** we learned that Maps like the Baker Map B can also share the reversibility property, $B^{-1} = TBT$. Macroscopic *Irreversibility* is often characteristic of "materials in motion" or even static materials just conducting heat. Mechanics or dynamics is typically time-reversible in the sense that a reversed movie obeys the same mechanical or dynamical laws for either direction of projection. Standard classical mechanics (Newton, Lagrange, Hamilton) can be expressed in terms of coordinates and velocities or coordinates and momenta, where the momenta are derivatives of the Lagrangian function $\mathcal{L}$ and used to define the Hamiltonian $\mathcal{H}$:

$$\mathcal{L}(q, \dot{q}) = K(\dot{q}) - \Phi(q) \; ; \; p = (\partial \mathcal{L}/\partial \dot{q}) \; ; \; \dot{p} = (\partial \mathcal{L}/\partial q) = F(q) \; .$$

$$\mathcal{H}(q, p) = K(p) + \Phi(q) \; ; \; \dot{q} = (p/m) \; ; \; \dot{p} = F(q) = m\ddot{q} = -\nabla_q \Phi \; .$$

It is possible to pick and choose among these approaches. Equilibrium constant-energy simulations of simple point-mass fluids are most easily and efficiently "solved" using Størmer's algorithm :

$$q(t + dt) - 2q(t) + q(t - dt) = dt^2 F(q)/m \; .$$

It is evident that the solution can be carried arbitrarily far in either time
direction once two coordinate values have been specified for each Cartesian
degree of freedom. Carol pointed out that the momenta, needed to compute
the pressure and the temperature, can be defined with centered differences
in any one of three ways :

$$p2 = \frac{(q_{n+1} - q_{n-1})}{2dt} \; ;$$

$$p4 = \frac{4}{3}\frac{(q_{n+1} - q_{n-1})}{2dt} - \frac{1}{3}\frac{(q_{n+2} - q_{n-2})}{4dt} \; ;$$

$$p6 = \frac{3}{2}\frac{(q_{n+1} - q_{n-1})}{2dt} - \frac{3}{5}\frac{(q_{n+2} - q_{n-2})}{4dt} + \frac{1}{10}\frac{(q_{n+3} - q_{n-3})}{6dt} \; .$$

In problems with spherical or cylindrical geometry it is simplest to express
the kinetic energy in terms of velocities, differentiating the Lagrangian to
find the equations of motion in a useful form. A mass m with two de-
grees of freedom, for example, has kinetic energy $(m\dot{r}^2 + mr^2\dot{\theta}^2)/2$ in polar
coordinates, from which the motion equations for $\ddot{r}$ and $\ddot{\theta}$ follow easily.

> **Exercise :** Find the equations of motion for a two-dimensional har-
> monic oscillator with Lagrangian $\mathcal{L} = (\dot{r}^2 + r^2\dot{\theta}^2 - r^2)/2$ and find initial
> conditions leading to $r = \sqrt{(x^2 + y^2)} = \cos(t)$.

More complicated situations can arise if constraints are added to the
Lagrangian. The rigid rotor problem can be solved by including the product
of a Lagrange multiplier λ and the "constraint", $(r_{12}^2 - 1)/2$, in the dynamics:

$$\mathcal{L} = (\dot{x}_1^2 + \dot{y}_1^2 + \dot{x}_2^2 + \dot{y}_2^2)/2 + \lambda(x_1^2 - 2x_1 x_2 + x_2^2 + y_1^2 - 2y_1 y_2 + y_2^2 - 1)/2 \; .$$

The idea is that so long as the constraint is satisfied (and λ can be chosen
to enforce it so that the constraint no longer appears in the motion equations) the motion will satisfy the usual unconstrained Lagrange equations
conserving mass, momentum, and energy. The equations of motion are

$$\ddot{x}_i = +\lambda(x_i - x_j) \; ; \; \ddot{y}_i = +\lambda(y_i - y_j) \; ,$$

In order to satisfy the constraint differentiate it twice with respect to time
and solve for $\lambda = -(\dot{r}_{12}^2/r_{12}^2)/2$. The initial conditions must be chosen so
as to satisfy the constraint and its first time derivative. Because these motion equations involve velocities as well as coordinates the simple Størmer
algorithm is not applicable. Instead a fourth-order Runge-Kutta algorithm

would be a good choice. This problem can be solved with as many as eight differential equations, Cartesian or polar, or with as few as one (in a coro-tating frame at the center of mass) and provides a good test of one's skills at implementing integrators and constraints.

Nearly all problems in nonequilibrium molecular dynamics involve differential equations with righthandsides that depend upon velocities as well as coordinates. The Nosé-Hoover thermostat equation exerts "integral control" over the kinetic energy. The mean value over long times, $\langle p^2 \rangle = T$, is guaranteed by the friction coefficient equation. With $\dot{p} = -q - \zeta p$ the evolution of the control variable, $\dot{\zeta} = p^2 - T$, provides a simple example of a velocity-dependent force. Such models reward the policy of keeping modules on hand which implement Runge-Kutta integrators advancing a vector YY(N) from time T to time T+DT with the help of a function RHS(YY,YYP) which provides the righthandsides of the NEQ first-order differential equations. Carol gave several illustrations in **Lecture 2**. So long as the underlying differential equations are "reversible" any reasonable integrator will be able to produce a reversible trajectory provided that the timestep is small.

The apparent accuracy of the integration can easily be deceptive. Conserving energy is far from a foolproof test. In most cases dynamical motion equations are Lyapunov unstable so that roundoff errors of order 10^{-16} for double-precision calculations, will increase exponentially, with no remaining memory of the initial conditions once $e^{\lambda t} = 10^{+16} \rightarrow t \simeq (37/\lambda)$.

The Størmer algorithm can run for billions of timesteps with insignificant energy drift but with the trajectory itself far from accurate due to Lyapunov instability. For smoothly continuous forces and times short enough that the Lyapunov instability is unimportant the reversibility of the program can be checked by integrating several steps forward with DT, followed by the same number of steps with the sign of the timestep negative rather than positive. One should be able to confirm that the error in such a test varies as the fourth power of DT, at a fixed time, for a "fourth-order" integrator. It is easy to confirm that nature, meaning the World and Universe we live in, is *not* reversible. None of us is getting any younger. James Gleick's books, *Chaos* and *Time Travel : A History*, are well worth a look. The notion of "time" itself seems to puzzle philosophers, but not us ! We are happy to follow Newton's model of classical mechanics, thinking, speaking, and writing of time as "what a clock measures".

8.2 Størmer's Reversibility and Yoshida's Hamiltonian

There is an interesting connection between Størmer's centered-difference
"Leapfrog" algorithm and a Hamiltonian discovered by Haruo Yoshida.[1]
The correspondence does not depend upon the value of the timestep dt.
We will choose it equal to unity, $dt = 1$, a choice for which the Størmer
oscillator equations of motion have particularly simple solutions.

$$dt = 1 \longrightarrow q_{n+1} = 2q_n - q_{n-1} - dt^2 q_n = q_n - q_{n-1} .$$

Carol considered a solution initiated at a turning point, generating the
sequence of coordinates $\{ +1, +(1/2), -(1/2), -1, -(1/2), +(1/2) \}$, which
then repeats. Let us instead begin with two equal coordinates $q_n = \pm 1 =
q_{n-1}$, which generates a different series of six repeating coordinates (also
with period 6) :

$$\{ q \} = \ldots \pm 1, \pm 1, 0 , \mp 1, \mp 1, 0 \ldots (\text{repeats}) .$$

If we use the simplest Euler algorithm, $\{ q = q + p ; p = p - q \}$, to find a
sequence of corresponding momenta there are two different solutions, given
an initial coordinate/momentum pair (q, p) . The two correspond to first
incrementing q and then p or the reverse, incrementing p first :

$$\{ Q = Q + P ; P = P - Q \} \text{ or } \{ P = P - Q ; Q = Q + P \} .$$

The results of these two possibilities, the first at the right and the second
at the left, are shown in **Figure 8.1**. They give the two hexagonal orbits
shown there.

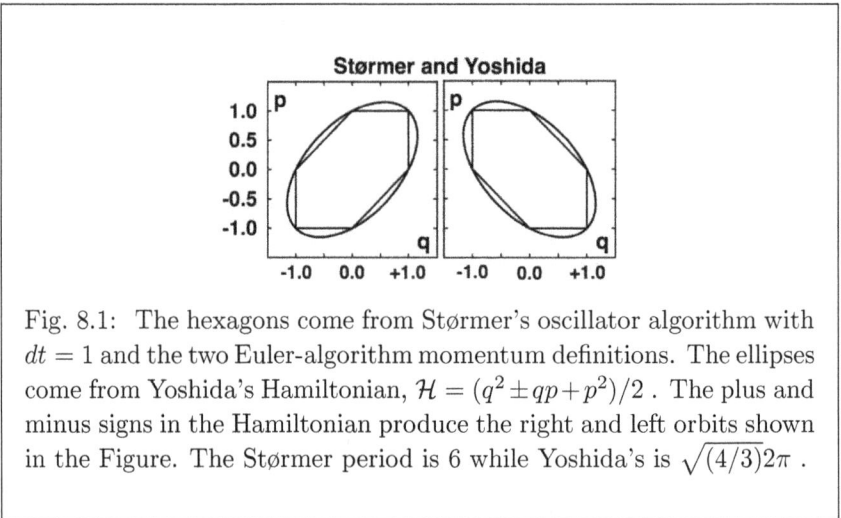

Fig. 8.1: The hexagons come from Størmer's oscillator algorithm with
$dt = 1$ and the two Euler-algorithm momentum definitions. The ellipses
come from Yoshida's Hamiltonian, $\mathcal{H} = (q^2 \pm qp + p^2)/2$. The plus and
minus signs in the Hamiltonian produce the right and left orbits shown
in the Figure. The Størmer period is 6 while Yoshida's is $\sqrt{(4/3)}2\pi$.

Yoshida showed how to find special Hamiltonians which produce continuous differentiable trajectories consistent with the discrete finite-difference coordinates and momenta generated by symplectic integrators like Størmer's.[1] Yoshida's Hamiltonian for the one-dimensional oscillator is particularly simple :

$$\mathcal{H} = (q^2 \pm qpdt + p^2)/2 \text{ with } dt = 1 \rightarrow \{ \dot{q} = p \pm (q/2) \; ; \; \dot{p} = -q \mp (p/2) \} \; .$$

The continuous solution of these two first-order differential equations agrees exactly with the Størmer finite-difference points. Yoshida's Hamiltonian trajectories are the continuous ellipses shown in **Figure 8.1**. Although the trajectory points are identical the *times* are different, reminiscent of Rahman's integrator and Nosé's time scaling. The period of the smooth Yoshida oscillator trajectory exceeds the exact oscillator value of 2π by a factor of $\sqrt{(4/3)}$ while the Størmer finite-difference period is too short, 6 rather than 2π.

The equations of motion from Yoshida's Hamiltonian are certainly not reversible like Størmer's despite their close relationship to a harmonic oscillator. But a 45° rotation in the phase space, choosing a new (coordinate/momentum) pair $\sqrt{(1/2)}(q \pm p)$ diagonalizes the Hamiltonian. In the new variables the motion *is* reversible. Changing the sign of the new momentum produces a trajectory tracing out $q(t)$ in reverse. This is exactly the same change of variables that was required to convert the irreversible square Baker Maps into equivalent time-reversible diamond-shaped maps.

8.3 Reversibility with Dettmann's Hamiltonian

Continuing with the theme of microscopic reversibility let us review Dettmann's formulation of the Nosé-Hoover oscillator. Dettmann used a clever idea,[2] setting a Hamiltonian equal to zero, $\mathcal{H}_D = s\mathcal{H}_N \equiv 0$, so as to avoid the need for time-scaling in deriving a canonical dynamics :

$$2\mathcal{H}_D = s\mathcal{H}_N = (p^2/s) + sq^2 + s\ln(s^2) + s\zeta^2 \equiv 0 \longrightarrow$$

$$\{ \dot{q} = (p/s) \; ; \; \dot{p} = -sq \; ; \; \dot{s} = s\zeta \; ; \; \dot{\zeta} = (p/s)^2 - 1 \} \longrightarrow$$

$$\{ \dot{q} = p \; ; \; \dot{p} = -q - \zeta p \; ; \; \dot{s} = s\zeta \; ; \; \dot{\zeta} = p^2 - 1 \} \; .$$

[Dettmann Oscillator Equations setting $(p/s) \rightarrow p$] .

Sign changes in the two momenta, p and ζ, along with reversing time are quite consistent with a reversed sequence of the coordinates q and s.

Dettmann's equations are perfectly reversible. Numerical reversals of the three treatments of the classical harmonic-oscillator model, Newton's or Yoshida's or Størmer's, are all of them easy to carry out because the motion is microscopically *stable*. Small perturbations to the oscillator equations, $q \to q + \delta_q$ and $p \to p + \delta_p$ oscillate rather than growing exponentially :

$$\{ \, \dot{q} = p \; ; \; \dot{p} = -q \, \} \; \longrightarrow \{ \, \dot{\delta}_q = \delta_p \; ; \; \dot{\delta}_p = -\delta_q \, \} \longrightarrow$$

$$\{ \, \ddot{\delta}_q = -\delta_q \; ; \; \ddot{\delta}_p = -\delta_p \, \} \; .$$

The situation is different with Nosé's or Dettmann's oscillator equations. Let us consider Dettmann's, having made the substitution $(p/s) \to p$ in order to recover the Nosé-Hoover form :

$$\{ \, \dot{q} = p \; ; \; \dot{p} = -q - \zeta p \; ; \; \dot{\zeta} = p^2 - 1 \, \}$$

$$(\dot{\otimes}/\otimes) = (\partial \dot{q}/\partial q) + (\partial \dot{p}/\partial p) + (\partial \dot{\zeta}/\partial \zeta) = -\zeta \; .$$

$$[\, \text{Nosé} - \text{Hoover} \,] \; .$$

Here there is the *possibility* of instability (exponential growth of phase volume $\otimes$ rather than decay or oscillation) because the friction coefficient ζ can be negative. So far as I know one cannot make further progress analytically, so as to determine whether or not the flow actually is unstable. The mathematicians' usual procedure in assessing stability is to expand about the fixed points of the flow. But the Nosé-Hoover oscillator has *no* fixed points. Fortunately it is easy to analyze stability *numerically* using Benettin's algorithm.

8.4   Analyzing Reversibility with Benettin's idea

Giancarlo Benettin[3] generalized Stoddard and Ford's idea[4] of looking directly at stability, following the relative motion of two nearby trajectories in order to see whether or not they would separate exponentially fast. The Nosé-Hoover oscillator provides qualitatively different answers to this separation question in different parts of the phase space. Because a clear rendering of a three-dimensional trajectory is difficult to achieve we begin by exploring two-dimensional Poincaré sections cutting through the three-dimensional phase space. **Figure 8.2** shows two cross sections of the (q, p, ζ) phase space : $\zeta(p)$ where $q = 0$ and $p(q)$ where ζ vanishes. Carol introduced the Poincaré section idea in **Lecture 2** with the $p = 0$ cross section which is common to both the Nosé and the Nosé-Hoover oscillators. The Figure shows clearly that there is a chaotic (Lyapunov unstable) sea interpenetrated with tori and periodic orbits.

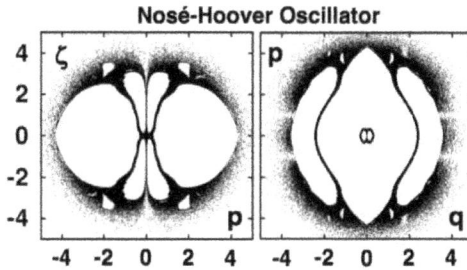

Fig. 8.2: Nosé-Hoover oscillator cross sections with $q = 0$ to the left and $\zeta = 0$ to the right. About half a million plane crossings are shown in these RK4 sections which use two billion timesteps each. Initially $(q, p, \zeta) = (0.0001, 0.0001, 0.0001)$ and $dt = 0.001$.

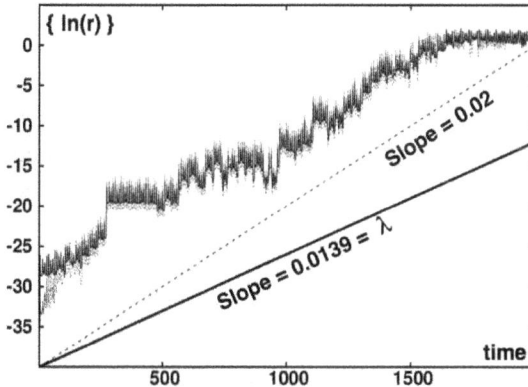

Fig. 8.3: Evolutions of $|r_{12}|$, $|r_{13}|$, $|r_{14}|$ for 2 million fourth-order Runge-Kutta timesteps with $dt = 0.001$. The initial condition of the reference trajectory is $q = p = \zeta = 0.0001$. The three satellite trajectories have initial offsets of length 10^{-16} in the three phase-space directions. The satellite trajectories soon merge. Their growing separation from the reference suggests a Lyapunov exponent of order 0.02. The accurate longtime Lyapunov exponent for this Nosé-Hoover oscillator problem is $\lambda_1 = 0.0139$.

Starting at a reference point in the chaotic sea (q, p, ζ) = $(0.0001, 0.0001, 0.0001)$ I solved the equations of motion for three different perturbations, with an additional 10^{-16} added to q and p and ζ so that there are three nearby satellite trajectories in addition to the reference. **Figure 8.3** shows the logarithm of the separation in phase space of each of the three perturbed trajectories from the reference. In a relatively short run of 2 million timesteps ($dt = 0.001$) the separation rate is roughly $(d\ln(r)/dt) \simeq 0.02$, a bit higher than the longtime averaged value of 0.0139.

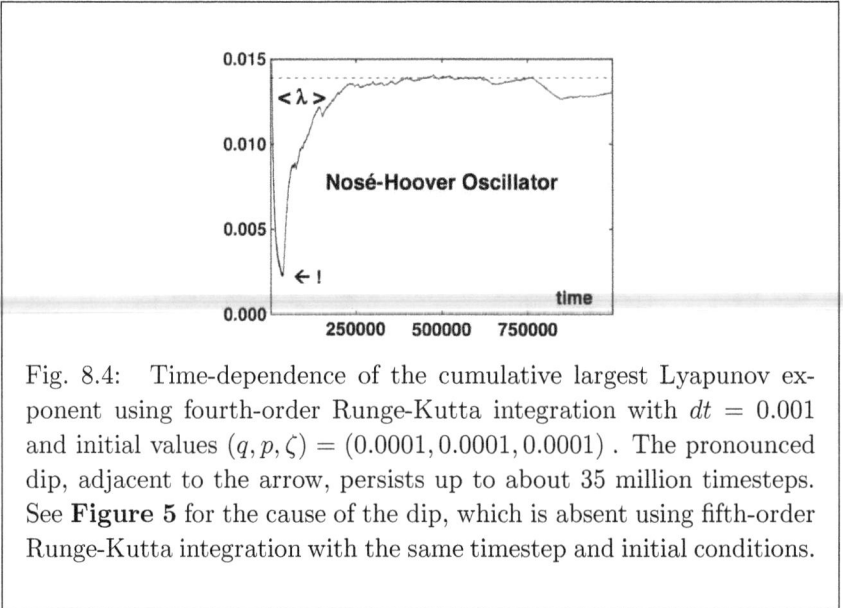

Fig. 8.4: Time-dependence of the cumulative largest Lyapunov exponent using fourth-order Runge-Kutta integration with $dt = 0.001$ and initial values $(q, p, \zeta) = (0.0001, 0.0001, 0.0001)$. The pronounced dip, adjacent to the arrow, persists up to about 35 million timesteps. See **Figure 5** for the cause of the dip, which is absent using fifth-order Runge-Kutta integration with the same timestep and initial conditions.

Figure 8.4 shows an implementation of Benettin's idea[3] for the same problem as that of **Figure 8.3**. Here the simulation is carried to a much longer time interval (one billion timesteps, to $t = 10^6$) . There is only a single satellite trajectory and a single separation r. That separation is maintained nearly constant at $r = 0.000001$. This is achieved by rescaling the distance between the two trajectories at the end of each timestep.

Although both trajectories could be moved we choose to allow the reference trajectory perfect freedom and adjust the satellite trajectory toward or away from the reference at the end of every timestep. The scale factor $\delta/|r|$ and the separation r are related to the largest Lyapunov exponent by $\lambda_1(t) = -\ln(\delta/|r|)/dt$. These simulations leave no doubt that the trajectories in the chaotic sea are indeed chaotic, with an irresistable tendency toward the exponential separation of nearby trajectories. All memory of the initial condition has been lost by a time of about 2000, when $e^{\lambda t} \simeq e^{0.0139t} \simeq e^{28}$. **Figure 8.4** shows an early-time persistent drop in the cumulative average of the local Lyapunov exponent. A followup investigation showed that the trajectory spent a lot of time in the vicinity of the largest torus. See **Figure 8.5**. This attraction persisted for a time of 30,000 or so. It was absent in an otherwise identical RK5 simulation. At long times the RK4 and RK5 estimates of λ_1 were consistent with one another.

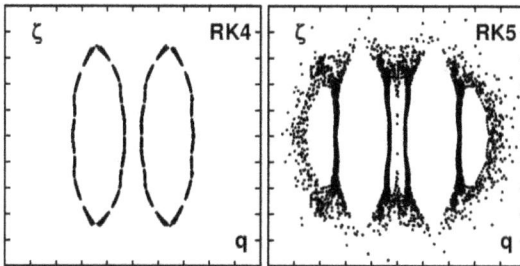

Fig. 8.5: Nosé-Hoover oscillator ($p = 0$) Poincaré sections. Up to a time of about 1.3 million timesteps, $t < 1300$, the RK4 and RK5 trajectories follow each other closely. In the time interval shown in the figure, between 2.5 million and 35 million timesteps, $2500 < t < 35,000$, the RK4 and RK5 trajectories behave very differently. RK4 penetrates the $p = 0$ plane 9161 times, and remains close to a large torus. Meanwhile RK5 explores a larger portion of the chaotic sea with 8201 penetrations. For long times the two integrators both match the chaotic sea value $\lambda_1 = 0.0139$. The ranges shown in the two sections are $-5 < q, \zeta < +5$. The initial condition is $q = p = \zeta = 0.0001$.

In an n-dimensional phase-space flow there are generally n Lyapunov exponents which measure n orthogonal growth rates within a comoving corotating n-dimensional hyperball $\otimes$. In the oscillator case we have just described the calculation of the largest exponent $\lambda_1 = \langle \lambda_1(t) \rangle = 0.0139$. We know also that parallel to the trajectory there can be no longterm growth or decay, so that λ_2 vanishes. Finally, the sum of $\lambda_1 + \lambda_2 + \lambda_3$ is equal to the time-averaged value of $-\zeta$. This follows from the phase-space continuity equation :

$$\lambda_1(t) + \lambda_2(t) + \lambda_3(t) = (\dot{\otimes}/\otimes) = (\partial\dot{q}/\partial q) + (\partial\dot{p}/\partial p) + (\partial\dot{\zeta}/\partial\zeta) = 0 - \zeta + 0 \ .$$

Thus, as the system is known to be Hamiltonian, the mean value of the friction coefficient ζ necessarily vanishes. So in this case we know the whole Lyapunov spectrum, $\{+0.0139, 0, -0,0139\}$. The Nosé oscillator, in its four-dimensional space is likewise Hamiltonian, with an additional Lyapunov exponent and the spectrum $\{+0.0475, 0, 0, -0,0475\}$ as a result of the faster time development of Nosé's stiffer dynamics. We will consider the details of spectrum calculation in much larger phase spaces presently. Let us first turn to a simpler Lyapunov-unstable problem in which the distracting fluctuations induced by manybody mechanics are absent.

8.5   Lyapunov Instability of a Bouncing Ball

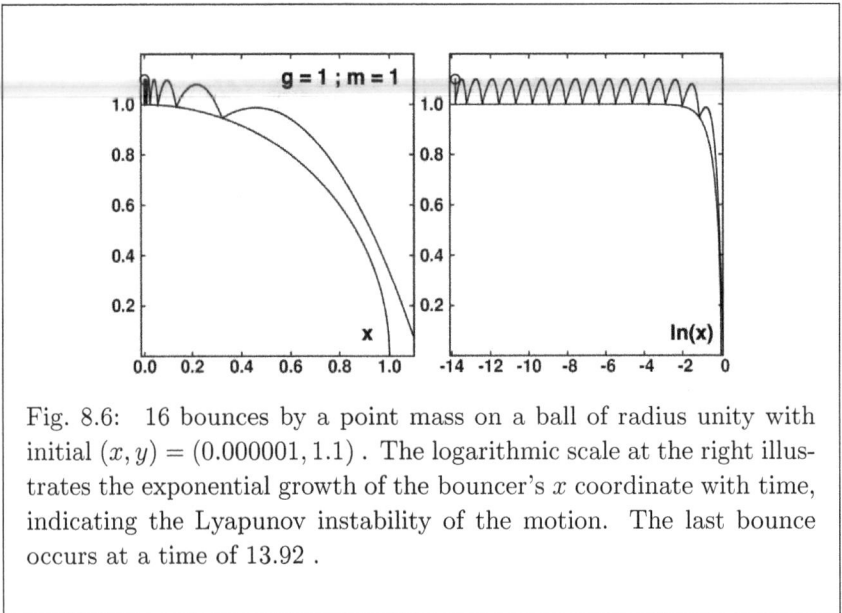

Fig. 8.6:   16 bounces by a point mass on a ball of radius unity with initial $(x, y) = (0.000001, 1.1)$. The logarithmic scale at the right illustrates the exponential growth of the bouncer's x coordinate with time, indicating the Lyapunov instability of the motion. The last bounce occurs at a time of 13.92 .

Consider a mass point in a gravitational field bouncing from the surface of a rigid disk with unit radius. From experience we know that this problem is Lyapunov unstable. The smallest deviation of the bouncing point from the high point of the fixed disk leads exponentially quickly to the mass point's exit from the problem. To make the problem concrete we start with the mass point motionless at $(x, y) = (0.000001, 1.1)$ and we assign the gravitational field a strength of unity. **Figure 8.6** shows the forward bouncing portion of this pedagogical trajectory. The small ball, idealized to a mass point, finally clears the disk after 16 bounces. The largest Lyapunov exponent for this problem is close to unity as the offset increases by a factor of $10^6 \simeq e^{14}$ in a time $\simeq 14$.

This simple mechanics problem, rather like that of a pencil balanced on its point, illustrates the kind of exponential instability which can lead to chaos. Expanding the growth rate of a *small* perturbation δ we need keep only terms through those linear in the perturbation. The linear term can be negative, positive, zero, or imaginary. The nonzero possibilities give exponential decay, exponential growth, and oscillation. Among the many degrees of freedom in a complex dynamical system it is usual to find at least some degrees of freedom with *positive* linear growth, $(\dot{\delta}/\delta) > 0$, the mechanism leading to Lyapunov instability.

Lyapunov instability is computationally irreversible. This key point provides clues leading to a mechanical mechanism for the Second Law of Thermodynamics. Even at equilibrium we cannot recapture the past. Lyapunov instability depends upon this past and not the future. Although there is nothing obvious in the time-reversible equations of motion that would prevent a satellite trajectory from clinging to the reference (so that the exponent would be negative) the Second Law announces that chaos ultimately wins out. Given a chance, a dynamical system moves in the direction for which more choices are available. Let us explore some manifestations of the Second Law for macroscopic systems using molecular dynamics with nonequilibrium boundary conditions.

8.6 Macroscopic Irreversible Flows

The Joule-Thomson experiment of **Figure 8.7** is a steady flow in which high-pressure fluid is forced through a porous plug and leaves with a lower pressure and a higher entropy. In the stationary state the rate at which the internal energy of the fluid increases is equal to the rate at which the pressure difference across the system does work. If we observe a Lagrangian volume V of fluid entering the system the energy increase by the time it exits, $E_{\text{right}} - E_{\text{left}}$ is $(PV)_{\text{left}} - PV_{\text{right}}$. In this steady nonequilibrium flow $(E + PV)_{\text{right}} = (E + PV)_{\text{left}}$ so that the flow occurs at constant enthalpy.

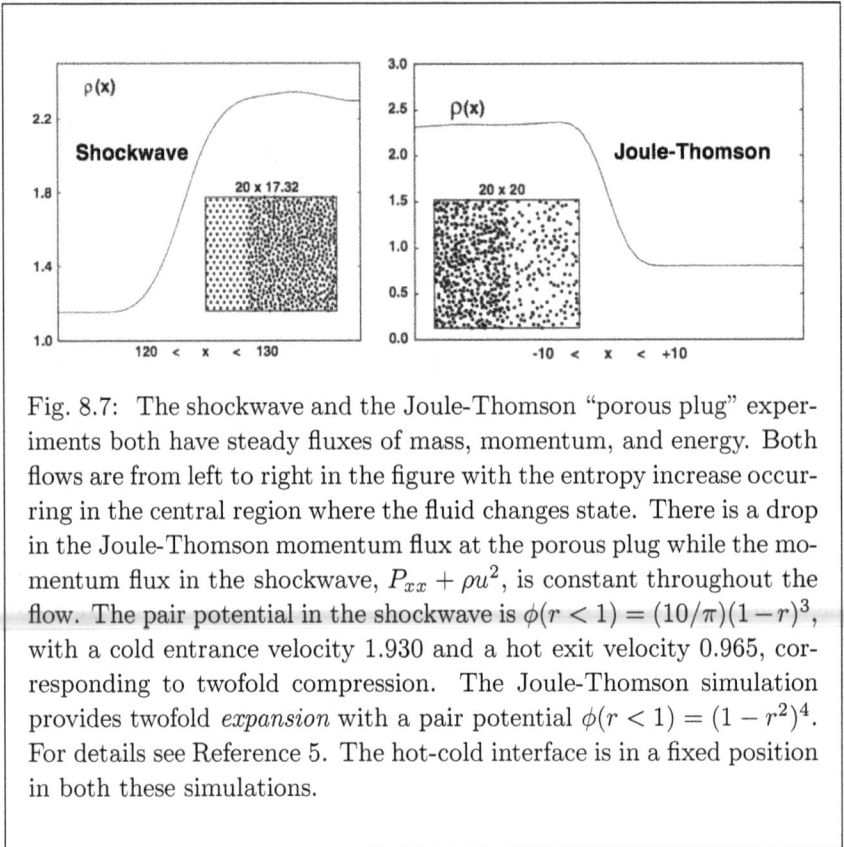

Fig. 8.7: The shockwave and the Joule-Thomson "porous plug" experiments both have steady fluxes of mass, momentum, and energy. Both flows are from left to right in the figure with the entropy increase occurring in the central region where the fluid changes state. There is a drop in the Joule-Thomson momentum flux at the porous plug while the momentum flux in the shockwave, $P_{xx} + \rho u^2$, is constant throughout the flow. The pair potential in the shockwave is $\phi(r < 1) = (10/\pi)(1-r)^3$, with a cold entrance velocity 1.930 and a hot exit velocity 0.965, corresponding to twofold compression. The Joule-Thomson simulation provides twofold *expansion* with a pair potential $\phi(r < 1) = (1 - r^2)^4$. For details see Reference 5. The hot-cold interface is in a fixed position in both these simulations.

The shockwave experiment is likewise a one-dimensional steady flow,[6] but with incoming *low*-pressure fluid and outgoing *high*-pressure fluid. The same arguments as above would lead the tyro to the same conclusion – constant enthalpy. Because the two flows are *both* irreversible it is clear that our analysis is incorrect for one or both of them. In the shockwave problem, the kinetic energy of the moving fluid has been omitted. If we correct for this we have

$$(E + PV)_H - (E + PV)_L = (Nm/2)(u_L^2 - u_H^2) \ .$$

For the shockwave both the mass flux ρu and the momentum flux $P_{xx} + \rho u^2$ are constants throughout the flow, establishing Rayleigh's result that the pressure tensor component P_{xx} varies linearly with volume :

$$P_{xx}(V) = \frac{P_L \times (V_H - V) + P_H \times (V - V_L)}{(V_H - V_L)} \quad [\text{ The Rayleigh Line }] \ .$$

Eliminating the two velocities from the three conservation laws of mass, momentum, and energy shows, after considerable algebra, that the energy change in the fluid is simply related to the high (H) and low (L) pressure forces operating at the boundaries of the flow:

$$E_H - E_L = (P_H + P_L) \times (V_L - V_H)/2 = (V_H + V_L) \times (P_H - P_L)/2$$

[Shock Hugoniot] .

An alternative derivation of this same "Shock Hugoniot" relation appears in **Figure 8** below.

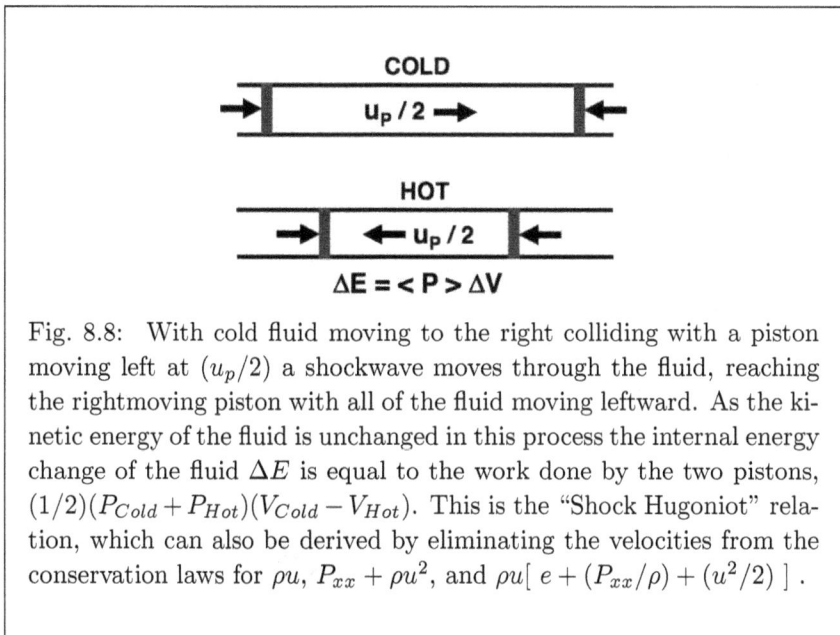

Fig. 8.8: With cold fluid moving to the right colliding with a piston moving left at $(u_p/2)$ a shockwave moves through the fluid, reaching the rightmoving piston with all of the fluid moving leftward. As the kinetic energy of the fluid is unchanged in this process the internal energy change of the fluid ΔE is equal to the work done by the two pistons, $(1/2)(P_{Cold} + P_{Hot})(V_{Cold} - V_{Hot})$. This is the "Shock Hugoniot" relation, which can also be derived by eliminating the velocities from the conservation laws for ρu, $P_{xx} + \rho u^2$, and $\rho u[\, e + (P_{xx}/\rho) + (u^2/2)\,]$.

The phenomenological models used to describe shockwaves are irreversible. Newtonian viscosity is a model describing the extra work done in maintaining a velocity gradient. A Newtonian fluid reacts to a velocity gradient (du/dx) with a linear change in the nonequilibrium pressure tensor induced by viscosity :

$$P_{xx} - P_{eq} = -2\eta(du_x/dx) - \lambda(du_x/dx) .$$

Here η is the shear viscosity and λ is the "second viscosity", related to the "bulk viscosity" by $\eta_v = \lambda + \eta$ in two dimensions and $\eta_v = \lambda + (2/3)\eta$ in three dimensions. For stable fluid flow the shear and bulk viscosities are necessarily positive. Fluid slowing down as it encounters higher density necessarily does extra compressional work, proportional to the square of the strain rate, increasing the entropy. Oddly enough, the expansion undergone in a Joule-Thomson experiment, wherein fluid expands through a porous plug, results in increased entropy as well. How can this be? In order to find out, Karl Travis, Carol, and I set out to simulate the Joule-Thomson experiment using molecular dynamics.[5]

We arranged twin treadmills on the left and right boundaries of the flow controlling the rates of entry and exit of the fluid. At the system's center we installed a potential barrier, $\phi(|x| < 1) = (1/4)(1 - x^2)^4$. **Figure 8.7** shows the steady density profile which results for this simple Joule-Thomson expansion experiment. This simulation uses the short-ranged repulsive potential $\phi(r < 1) = (1 - r^2)^4$. In the center of the system we have our short-ranged potential barrier, our "porous plug", which reduces the momentum flux. In the shockwave the mass, momentum, and energy fluxes are all constant and there is an entropy increase. Evidently the Joule-Thomson experiment cannot conserve all these fluxes without violating the Second Law of Thermodynamics. The lost work done against the potential barrier is evidently the source of entropy that is produced in the Joule-Thomson experiment. The pressure exerted by the plug is a fundamental ingredient of the throttling experiment. The shockwave experiment is governed entirely by its boundaries without the need for any additional internal forces. Let us consider the continuum mechanics of a simple shockwave model.

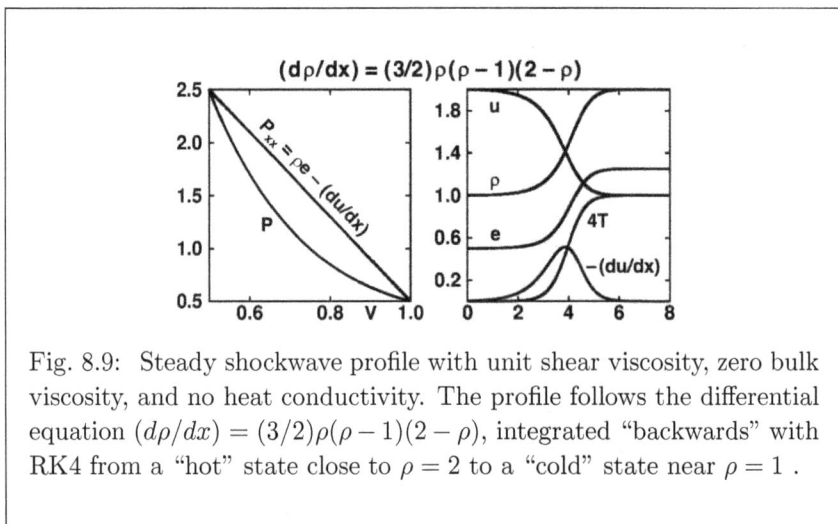

Fig. 8.9: Steady shockwave profile with unit shear viscosity, zero bulk viscosity, and no heat conductivity. The profile follows the differential equation $(d\rho/dx) = (3/2)\rho(\rho - 1)(2 - \rho)$, integrated "backwards" with RK4 from a "hot" state close to $\rho = 2$ to a "cold" state near $\rho = 1$.

8.7 Shockwave Modelling and Irreversibility

The mechanical and thermal equations of state
$$P = \rho e = (\rho^2/2) + \rho T \; ; \; e = (\rho/2) + T \; ; \; \eta = 1 \; ,$$
provide a simple test case to hone our modelling skills. I have chosen it in order to simplify the modelling. This choice of setting a longitudinal viscosity equal to unity corresponds to a vanishing two-dimensional bulk viscosity, $\eta_v = \lambda + \eta \equiv 0$. We will find the internal structure of a shockwave in which the fluxes of mass, momentum, and energy are all constant as the fluid undergoes twofold compression. From the Shock Hugoniot relation $\Delta E = \langle\, P\,\rangle \Delta V$ we find that a compression from $\rho = 1$ with zero temperature to $\rho = 2$ with temperature T gives the following relation :
$$\Delta e = (1+T) - (1/2) = \Delta v \times < P > = [\,(1-(1/2)\,] \times (1/2)[\,(1/2)+2+2T\,]\,.$$
Solving this Hugoniot equation gives $T = (1/4)$ from which the other equilibrium boundary values result :
$$\text{Cold} : \{\, \rho, u, P, e, T\,\} = \{\, 1,\; 2,\; (1/2),\; (1/2),\; 0\,\} \longleftrightarrow$$
$$\text{Hot} : \{\, \rho, u, P, e, T\,\} = \{\, 2,\; 1,\; (5/2),\; (5/4),\; (1/4)\,\}\,.$$
The mass, momentum, and energy fluxes are as follows :
$$\rho u = 2 \; ; \; P_{xx} + \rho u^2 = (9/2) \; ; \; \rho u[\, e + (P_{xx}/\rho) + Q_x + (u^2/2)\,] = 6\,.$$
For simplicity I assume that there is no thermal conductivity so that the heat flux Q_x vanishes, making it possible to eliminate the velocity from the flux equations with the simple result :
$$(d\rho/dx) = (3/2)\rho(\rho - 1)(2 - \rho) \; ; \; \rho(-\infty) = 1 \; ; \; \rho(+\infty) = 2\,.$$
This ordinary differential equation has a positive slope throughout the wave. It can be solved with our fourth-order Runge-Kutta integrator starting "near" the hot boundary, with $\rho = 1.99999$ for instance, and $dx = -0.001$. The results for this choice are shown in **Figure 8.9**.

The lens-shaped area between the equilibrium pressure P and the nonequilibrium P_{xx} can be thought of as the lost work or entropy production due to the shock process. It is easy to see that the entropy production goes smoothly to zero as the compression is reduced, and in fact varies as the cube of the compression as a limiting case. The monotone behavior of the temperature is expected from the Navier-Stokes equations. We will see that strong shockwaves typically show that the longitudinal temperature, like the pressure, can exceed the transverse, by as much as a factor of two. It is also the case that the rise in pressure and in heat flux lag slightly behind the maxima of the velocity and temperature gradients suggesting the need for time delay in the constitutive relations within the shockwave.

8.8   Shockwave Stability

Molecular dynamics made it possible to demonstrate the stability of planar
shockwaves. We could look at wider and wider systems, making sure that
the shockwave profile was insensitive to the width. It is simpler to begin
the simulation with a sinusoidal perturbation in the shape of the boundary
separating the cold and hot fluids (or solids !). By now many molecular
dynamics studies of shockwaves have made it plain that plane shockwaves
are stable in both two and three dimensions. Both these stability checks can
be carried out numerically, as illustrated in **Figure 8.10**. Carol and I and
Francisco Uribe, a colleague from the Universidad Nacional Autónoma de
México, carried out detailed investigations of two-dimensional shockwaves,
presenting our results at meetings in Mexico, Russia, Spain, and the United
Kingdom. Because the proceedings of these meetings are not easily obtained
we have placed all the papers on the Los Alamos arχiv for easy reference.

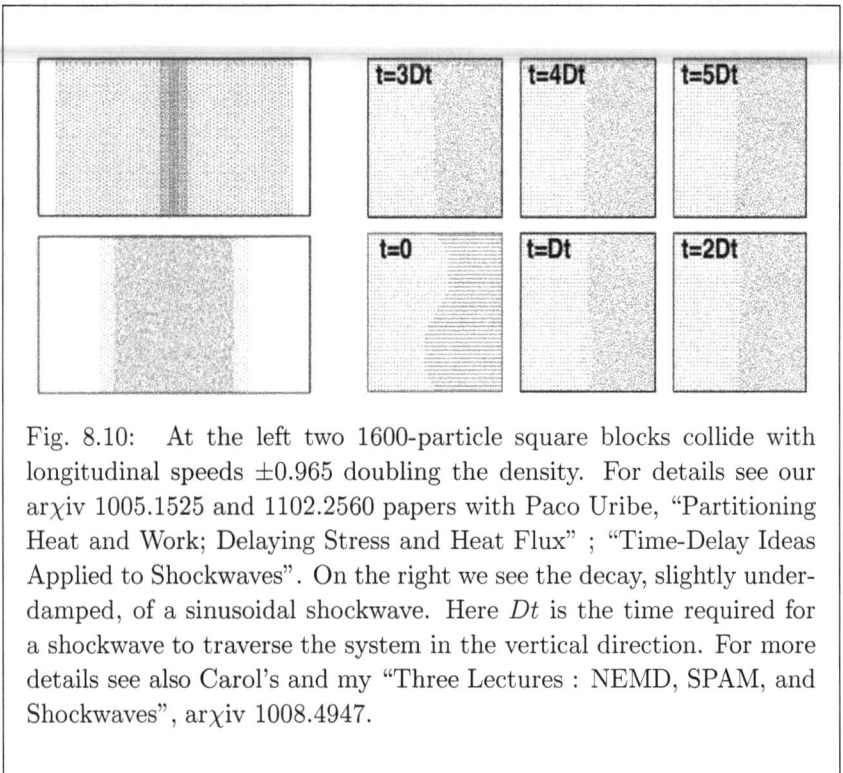

Fig. 8.10:   At the left two 1600-particle square blocks collide with
longitudinal speeds ±0.965 doubling the density. For details see our
arχiv 1005.1525 and 1102.2560 papers with Paco Uribe, "Partitioning
Heat and Work; Delaying Stress and Heat Flux" ; "Time-Delay Ideas
Applied to Shockwaves". On the right we see the decay, slightly underdamped, of a sinusoidal shockwave. Here Dt is the time required for
a shockwave to traverse the system in the vertical direction. For more
details see also Carol's and my "Three Lectures : NEMD, SPAM, and
Shockwaves", arχiv 1008.4947.

In **Figure 8.10** we see six snapshots illustrating the relaxation of a sinusoidal shockwave. The interval separating the snapshots is the *vertical* shock transit time for the system. In this purely-repulsive fluid the wave becomes planar in a time of order one transit time. The snapshot at time Dt reveals that the wave is slightly underdamped, with the amplitude briefly opposite in sign relative to that of the initial sinewave. The small width of shockwaves, just a few atomic spacings, and the simplicity of the conservation laws, make it possible to carry out accurate experiments under high pressure and temperature conditions inaccessible to static experiments. I'll describe such a simulation next.[6]

8.9 Navier-Stokes-Fourier Shockwave Modelling

Shockwaves are supersonic compression waves accelerated by high pressure. A steady wave, travelling at u_s is generated by moving a piston into the material of interest at the steady "particle" or "piston" velocity u_p. In the frame centered on the wave cold material approaches the wave from the left at u_s and departs hot at the right with velocity $u_s - u_p$. Conservation of mass and momentum imply that both ρu and $P_{xx} + \rho u^2$ are constants throughout the wave. Measuring the two speeds gives directly the pressure and the density for the compressed state. The Hugoniot relation then gives the specific energy e.

Within the shockwave the three fluxes can be reduced to two ordinary differential equations for the velocity and temperature gradients. The velocity gradient appears in the longitudinal momentum flux as $-\eta_L(du/dx) = -(\lambda + 2\eta)(du/dx) = -\eta_V(du/dx) - (4\eta/3)(du/dx)$. The energy flux, in addition to the velocity gradient contains the heat-flux vector $-\kappa(dT/dx)$. The two equations can then be solved together as Runge-Kutta integrals of two coupled ordinary differential equations.

From a continuum point of view the conservation relations hold within the shockwave as well as in the initial and final states so that the difference between P_{xx} and $P(\rho, T)$ can be modelled as a longitudinal viscosity coefficient η_L multiplied by the strainrate (du/dx).

These ideas allow laboratory experiments to be carried out and analyzed under conditions which would otherwise be hard to achieve. With explosives it is possible to compress metals to three times their normal density by using pressures of order 70 megabars, comparable to the pressures in the giant planets[7] (Jupiter, Neptune, Saturn, Uranus). Liquid argon reaches such a compression at a pressure of about one megabar.

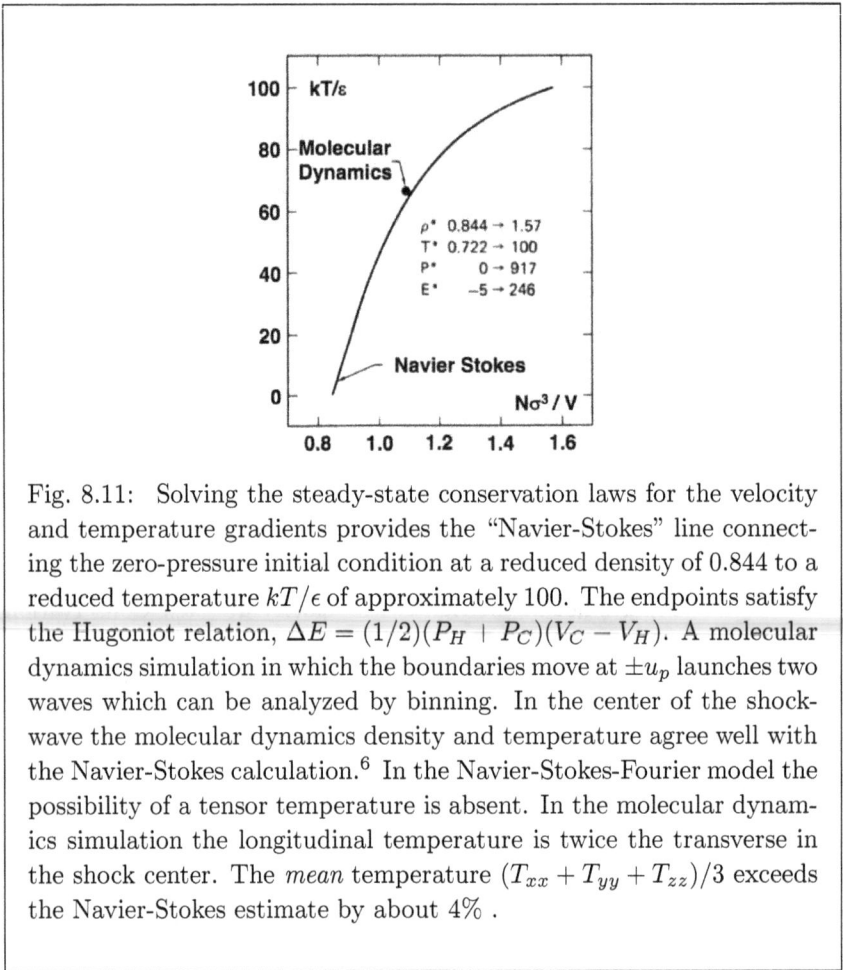

Fig. 8.11: Solving the steady-state conservation laws for the velocity and temperature gradients provides the "Navier-Stokes" line connecting the zero-pressure initial condition at a reduced density of 0.844 to a reduced temperature kT/ϵ of approximately 100. The endpoints satisfy the Hugoniot relation, $\Delta E = (1/2)(P_H + P_C)(V_C - V_H)$. A molecular dynamics simulation in which the boundaries move at $\pm u_p$ launches two waves which can be analyzed by binning. In the center of the shock-wave the molecular dynamics density and temperature agree well with the Navier-Stokes calculation.[6] In the Navier-Stokes-Fourier model the possibility of a tensor temperature is absent. In the molecular dynamics simulation the longitudinal temperature is twice the transverse in the shock center. The *mean* temperature $(T_{xx} + T_{yy} + T_{zz})/3$ exceeds the Navier-Stokes estimate by about 4% .

Because the rare gases are relatively easy to model with pairwise-additive potentials there has been considerable activity confirming the equation of state results of experiments using simple potentials such as Lennard-Jones' :

$$\phi_{LJ}(r) = 4\epsilon[\ (\sigma/r)^{12} - (\sigma/r)^6\ ]\ .$$

For argon at the potential minimum $2^{1/6}\sigma$, the depth of the attractive well divided by Boltzmann's constant k, is equal to 119.8 Kelvins. The characteristic length σ where $\phi_{LJ}(\sigma) = 0$ is 3.405Å, 3.405×10^{-8} cm .

Figure 8.11 shows early (1980) results from a simulation of the shock-wave compression of liquid argon using 4800 Lennard-Jones particles.[6] The boundary conditions in all three spatial dimensions are periodic, but with the separation of the two explicit longitudinal boundaries shrinking to mimic two pistons at opposite ends of the fluid. By summing the properties of particles in "bins" the profiles of density, velocity, stress, and temperature were all determined and compared with a continuum solution of the Navier-Stokes-Fourier constitutive relations. The Figure shows that the molecular dynamics temperature-density states within the shockwave are reproduced rather well with the temperature reaching about 12000 Kelvins. At higher temperatures argon's ionization has to be taken into account and the modelling becomes more challenging.

8.10   Inelastic Ball Collision and Irreversibility

The Lyapunov instability of the bouncing ball problem coupled with analyses of the high-speed deformation of the shock process suggests the simulation of irreversible rapid processes using Hamiltonian mechanics. Despite the time-reversibilty of Hamiltonian motion equations we expect realistic outcomes, in which heat flows from hot to cold and rapid compression increases pressure over its equilibrium value. The difference between Second Law irreversibility and physics-based time-reversible modelling suggests that even relatively-small systems should reveal their mechanisms for irreversibility.

One way to explore this paradoxical problem area further is to consider mesoscopic systems with more than a few particles but with enough that continuum properties are observable. Because the main problem is conceptual (Loschmidt's paradox) we confine our studies to two-dimensional systems where graphical displays of simulation results are simplest.

Carol and I took up a study of "inelastic" collisions of mesoscopic bodies in order to look for what could be called "conservative irreversibility".[8,9] We used Levesque and Verlet's bit-reversible leapfrog integrator[10] in conjunction with fourth-order Runge-Kutta integration to follow the collision of two 400-particle balls. The exact-to-the-last-bit reversibility of the Levesque-Verlet integrator, used for the reference trajectory, made it possible to analyze the merging and separating of the two balls in both time directions, forward and backward.

Two of Milne's fourth-order integration algorithms can be written in a bit-reversible form suitable for molecular dynamics and dynamical systems' studies.[11] See Section 2.9.2 on page 81 for Milne's fourth-order predictor algorithm for second-order equations. The fourth-order predictor for first-order equations, from page 135 of his book[12] is

$$y_{t+dt} = y_{t-3dt} + (4dt/3)[\ 2y'_t - y'_{t-dt} + 2y'_{t-2dt}\ ]\ .$$

As an interesting **Exercise** test this algorithm for stability using the harmonic oscillator and the Nosé-Hoover oscillator with $dt = 0.001$. *Caveat emptor !*

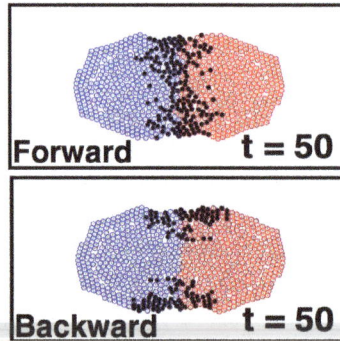

Fig. 8.12: Here we see the "important" particles [those that make above-average contributions to the largest local Lyapunov exponent] evaluated both forward and backward in time at a time of 50. The important particles are blacked in in the figure. The two colliding balls are inverted images of each other. We used the Levesque-Verlet algorithm to reverse the motion at time 100 and analyze the trajectory "backward" in time to the pre-collision state. The bit-reversible algorithm makes it unnecessary to store a forward trajectory in order to process it backward. Notice that in the reversed trajectory, where the coalesced balls are beginning to "neck" in order to separate, that the important particles are on the periphery, where the strain is greatest, rather than at the interface. *Although the motion reverses perfectly the stability does not.* It is remarkable that the two directions of time differ so much (162 *versus* 120 out of the total of 800 particles) in this important-particle measure of chaos. For additional snapshots in this series see Carol's Figures 23-24 on page 89 as well as Reference 9.

All 800 particles interact with a short-ranged soft but repulsive potential, $\phi(r < 1) = (1 - r^2)^4$ and a density-dependent (attractive) manybody potential $(1/2)\sum(\rho_i-1)^2$ where each particle's density is a smooth-particle sum of Lucy's weight functions with a range of 3.5 :

$$\rho_i = \sum_{j=1}^{800} w_{ij} \; ; \; w(r < h = 3.5) = (5/\pi h^2)(1+3z)(1-z)^3; \; z = (r/h) \,.$$

The attractive forces guarantee the stability of the ball's surface and stabilize shapes of minimum surface by providing an effective surface tension.

Initially a single fully-relaxed ball was prepared by adding a constant viscosity to the Runge-Kutta equations of motion for the relaxation of an unstable 20×20 square, $\{ \dot{p} = F - (p/\tau) \}$. After the relaxation process was complete the initial velocities for the collision problem were all chosen equal to $(p_x, p_y) = (+0.10, 0.0)$ in the first ball. Next, a second ball was set next to the first with its particle coordinates and velocities chosen to obey inversion symmetry with the first, $(r_2, p_2) = -(r_1, p_1)$. The leading edges of the balls were separated by just a bit more than the range of the attractive potential, $h = 3.5$. This made it possible to use constant-velocity initial conditions for the first few timesteps. In the Levesque-Verlet simulation of the reference trajectory the momenta needed to initialize the Runge-Kutta simulation of the satellite for the same timestep used the third-order definition[8] with an error of order $+(dt^4/30)\,\dddot{q}$:

$$p_t = -(1/3)(q_{t+2dt} - q_{t-2dt})/(4dt) + (4/3)(q_{t+dt} - q_{t-dt})/(2dt) \,.$$

This approach requires storing the coordinates at five successive times throughout the simulations. In the initial conditions the Levesque-Verlet coordinate values were equally spaced in time corresponding to the constant velocity of the soon-to-collide balls.

The two fully-relaxed zero-temperature balls were then sent off on their collision course with $dt = 0.001$. In order to measure the largest Lyapunov exponent a second two-ball satellite system was launched simultaneously with Runge-Kutta integration. At the end of each timestep the separation between the reference and satellite systems was maintained at $|\delta_{12}| = 10^{-5}$ by rescaling the satellite's coordinate and velocity offsets in the 3200-dimensional phase space. The inelastic deformation of the balls is essentially complete after 10^5 timesteps, at $t = 100$.

The 800 single-particle contributions to the squared offset length ,

$$\delta_{12}^2 = 0.00001^2 = \sum^{800}(\delta_x^2 + \delta_y^2 + \delta_{p_x}^2 + \delta_{p_y}^2) \,,$$

then provide a list of "important" particles, those with above-average contributions to δ_{12}, the separation of the two balls in phase space. We expect to see (and do, in **Figure 8.12**) that the important particles are located in regions with high rates of "plastic strain". Plastic strain is the irreversible deformation of a solid in which vacancies and dislocations make it possible for parallel rows of atoms to undergo relative shearing displacements.

One might well guess wrongly that roughly half the particles would contribute more than the average to the phase-space offset. That guess would be quite incorrect. Half seems to be a great overestimate in inhomogeneous systems where local values of the pressure and temperature are likely to have a strong influence on local contributions to the Lyapunov exponents. At time 50 *forward* in time there are 162 above-average particles. *Backward* in time, with exactly the same coordinates and kinetic energies there are 120 such particles. Both these numbers are well below $N = 800$. Forward in time the important particles lie in the middle third of the composite ball, distributed along the interface where the 400-particle balls have merged. Backward in time the important particles can be found in the necking region, with most of them within three atomic spacings of the composite ball's surface. This asymmetric irreversible measure of instability is simply unavailable from a snapshot $\{ q \}$ of the motion.

Although the complete set of variables at time 50, $\{ q, p \}$, can be processed to evaluate the local stress, strain-rate, temperature, ..., it cannot provide the local Lyapunov vectors or their associated contributions to the exponents. For those a past history of about 10,000 timesteps is required. For this reason I believe that the local Lyapunov vectors will ultimately play a crucial role in explaining the irreversibility of nonequilibrium systems as expressed by the Second Law of Thermodynamics.

This inelastic-collision simulation shows that despite the reversibility of the motion equations the *stability* of the motion is quite different forward and backward in time. The relatively small numbers of above-average particles fits our intuition that the instabilities should be relatively localized to regions that have recently undergone irreversible flows. The local Lyapunov exponents provide an irreversible time dependence making it possible to distinguish the future from the past.

8.11   Strong Shockwaves and Irreversibility

Like the inelastic two-ball collision planar shockwaves include steep gradients of pressure and temperature. Why and how do shockwaves arise? The mechanism is easy to understand. Because sound velocity typically increases with pressure and density the faster denser portion of a wave tends to catch up with the slower less-dense foot of the wave. The result is a steep shockwave, with a local shockwidth on the order of the interparticle spacing. The simplest case, a planar shockwave, can be generated by the symmetric collision of two similar rectangular blocks of material, solid, liquid, or gas. **Figure 8.13** illustrates this idea with a pair of strong shockwaves generated by the collision of two zero-pressure 800-particle slabs.

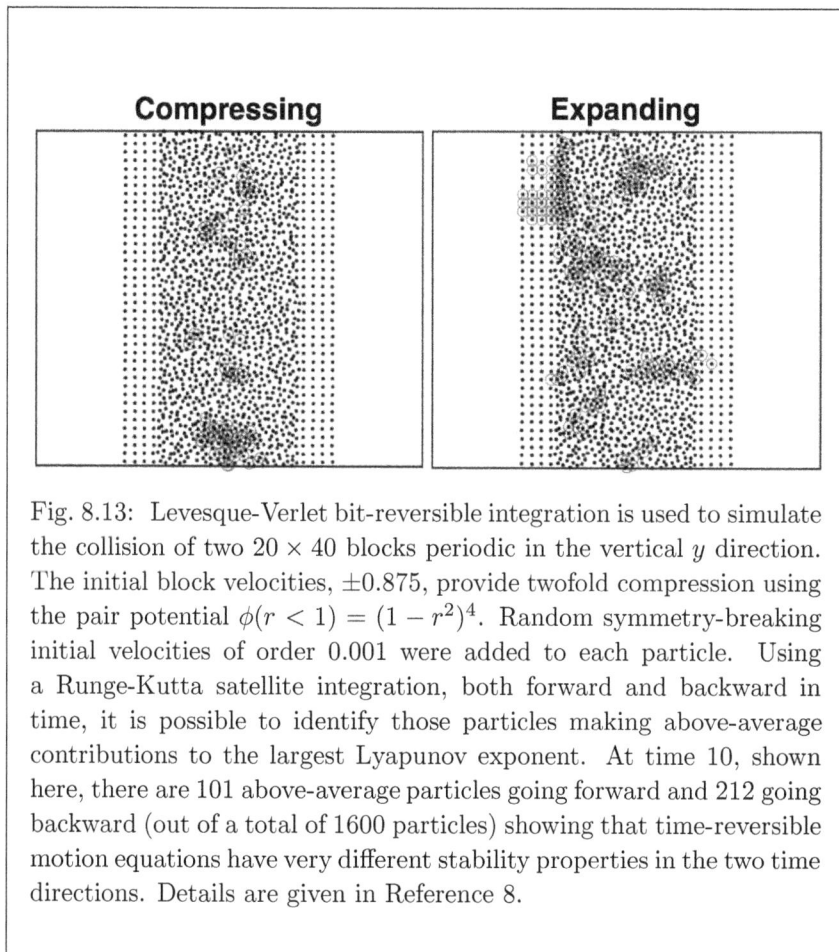

Fig. 8.13: Levesque-Verlet bit-reversible integration is used to simulate the collision of two 20×40 blocks periodic in the vertical y direction. The initial block velocities, ± 0.875, provide twofold compression using the pair potential $\phi(r < 1) = (1 - r^2)^4$. Random symmetry-breaking initial velocities of order 0.001 were added to each particle. Using a Runge-Kutta satellite integration, both forward and backward in time, it is possible to identify those particles making above-average contributions to the largest Lyapunov exponent. At time 10, shown here, there are 101 above-average particles going forward and 212 going backward (out of a total of 1600 particles) showing that time-reversible motion equations have very different stability properties in the two time directions. Details are given in Reference 8.

We choose a square lattice, periodic in the y direction, at unit density. The potential energy is the repulsive pair potential $\phi(r < 1) = (1 - r^2)^4$. The initial potential energy is zero. The two rectangular 20×40 slabs collide with horizontal velocity components of ± 0.875, chosen so as to provide twofold compression of the material. There is no special symmetry in this problem and periodic boundaries are used at the top and bottom, parallel to the x axis. At a time of 10, slightly before the time of maximum compression $t \simeq 12.3$, 101 of the 1600 particles on the left side of **Figure 8.13** make above-average contributions to the phase-space offset vector associated with the largest Lyapunov exponent. At time $t = 30$ the motion is reversed so that at time 50 the configuration (on the right) is identical to the earlier configuration shown in the figure and the momenta are precisely reversed. In this expansion phase 212 particles make above-average contributions to the largest Lyapunov exponent. This is more than twice the number of such particles during the mirror-image compression phase shown at the left of **Figure 8.13**. This example shows that despite the exact reversibility of the motion equations the relative stability in the two time directions is quite different.

We have seen that time-reversible molecular dynamics simulations of strongly "irreversible" processes provide coordinates, stresses, and temperature gradients which are identical in both directions of time (according to molecular dynamics) while the heat flux and the strain rates change sign. Although these differences in sign are inconsistent with Fourier's Law, $Q_x = -\kappa(dT/dx)$, and with viscous or viscoelastic flow laws such as $\sigma = \eta[\ (du_x/dy) + (du_y/dx)\ ]$ there is no doubt that that a molecular dynamics simulation of a shear flow would yield a positive viscosity or a positive shear modulus whether or not the imposed strain rate was positive or negative. So far we have considered models with several hundred degrees of freedom so as to see compelling examples of irreversibility. Next I want to emphasize the value of small system studies to this effort. Let us briefly review two of the smallest, a two-dimensional volume-preserving shear map, a caricature of equilibrium dynamics, and a two-dimensional nonequilibrium Baker Map in which the comoving volume *always* changes by a factor of two, a caricature of nonequilibrium particle dynamics. Both these maps are deterministic and time-reversible.

8.12 Shear-Map Stability

Maps, as opposed to flows, are often studied as prototypes of low-dimensional chaos, with the "Cat Map" and the "Baker Map" classic models for Hamiltonian flows. Because I am particularly fond of time-reversible flows and mappings I devoted some time to their study with an excellent graduate student from Korea, Oyeon Kum, and my Viennese colleague Harald Posch.[13] We pointed out that a toolkit of reversible maps, say A, and B, and C, can be used to generate an infinity of more complicated maps. To accomplish this simply arrange the basic maps in any symmetric combination, such as ACBCAACBCA. The resulting composite maps can then be used to model volume-conserving Hamiltonian flows or nonequilibrium problems like the Galton Board. In these composite symmetric models some parts of the mapping expand while other parts contract in such a way as to fill all of the state space with a singularly variable density.

Let us illustrate these ideas by considering a combination of shear maps in the x and y directions, where the mapping is applied to the unit square, $(-1/2) < (x, y) < (+1/2)$. The X mapping is XOUT = XIN + YIN ; YOUT = YIN and the Y mapping is XOUT = XIN ; YOUT = XIN + YIN . To show the time-reversibility of X let us take the initial point P = $(0.3, 0.4)$ which maps to $(0.7, 0.4)$ [$(-0.3, 0.4)$ when the periodic boundary is applied]. To illustrate that TXT is the inverse of X we apply it to the end product of the X mapping and find that the original point P = $(0.3, 0.4)$ is recovered :

$$TXT(-0.3, 0.4) = TX(-0.3, -0.4) = T(0.3, -0.4) = (0.3, 0.4).$$

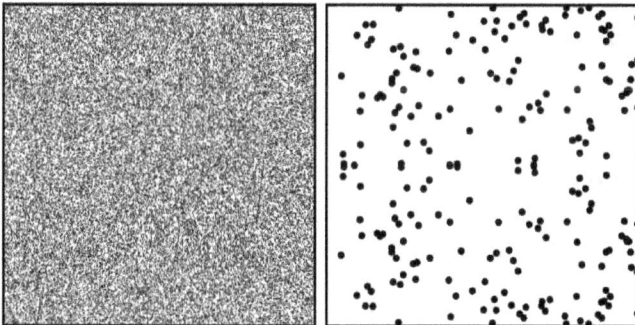

Fig. 8.14: 50,000 points from the single-precision shear map XYYX are shown to the left with $-0.5 < (x, y) < +0.5$. The initial values were $(x, y) = (0.3, 0.4)$. At the right we can see the symmetry exhibited by a complete period of 2,097,152 points. The plotted points occupy a tiny central square $-0.005 < (x, y) < +0.005$, 1/10000 part of the unit square, which should contain about 210 points. The mean values of x^2 and y^2, averaged over the complete period reproduce the expected values of $(1/12)$ with 11-digit accuracy!

Because a single-precision division of the unit interval corresponds to 2^{23} discrete values of x we expect that any of the area-preserving reversible unit-square maps, can access approximately 2^{46} states so that a periodic orbit should result after roughly $2^{23} = 8,389,152$ iterations. To test this idea I iterated the XYYX time-reversible map for 20,000,000 iterations and checked the output file to see where the last data point had occurred previously. The period of the orbit (its "Poincaré cycle" length) is 2,097,152. A look at the distribution of points in the unit square suggests that a more precise application of the map would be ergodic. Further evidence came from the mean values of XOUT*XOUT and YOUT*YOUT. These reproduced the exact value, $(1/12)$, to a surprising 11 significant figures. The uniform coverage of the square suggests that the map could be pressed into service as a pseudorandom number generator. An interesting feature of the XYYX map, in addition to its uniform coverage of the plane is the perfect symmetry of the periodic orbit of length 2,097,152. The central 0.01% of the unit square is shown in **Figure 8.14**. I would guess that it is possible to find a periodic orbit for the map with double-precision arithmetic, but its length, likely of order $2^{52} \simeq 10^{15.65}$ iterations, will require an unusual enthusiasm to take the time and energy to seek it out.

By introducing time-reversible maps based on mirror operations that expand and contract areas we were able to generate fractal strange attractors with fractal information dimensions varying from 2 to 1, with the dimensionality varying close to quadratically in the degree of area change induced by a time-reversible mapping. Improving upon this map problem was set as the Snook Prize Problem in 2015. Puneet Patra[14] won the Prize and provided a variety of colorful maps with beautiful curves, both area-preserving and dissipative, by introducing nonlinearity into his time-reversible and deterministic mappings.

8.13 Baker-Map Stability

The Baker Map is an apt model for stability. It illustrates many of the principles that apply to large nonequilibrium systems. The Baker Map operates in a 2×2 square and on a digital computer this square has only a finite number of states. This number is quite large, roughly $2^{23} \times 2^{23}$ for single precision. Fortuitously, this is just the right magnitude for an understanding of Zermélo's and Loschmidt's Paradoxes. With a finite number of states the motion must eventually repeat.

Repetition is the subject of Zermélo's Paradox : If a trajectory recurs how can it be that entropy increases? Reversal is the subject of Loschmidt's Paradox : If a trajectory is reversible how can it be that entropy increases? These paradoxes represent the tension between time-reversible motion equations and the increase of entropy described in thermodynamics. Let us examine these ideas from the perspective of the time-reversible Baker Map. The number of states for the map using single-precision arithmetic is small enough that we can observe both Zermélo's recurrence and Loschmidt's reversal. Let us recall the nonequilibrium map :

```
IF(Q.LT.P-SQRT(2.0/9)) THEN
        QNEW = +(11.0/ 6)*Q - (7.0/ 6)*P + SQRT(49.0/18)
        PNEW = +(11.0/ 6)*P - (7.0/ 6)*Q - SQRT(25.0/18)
ENDIF
```

```
IF(Q.GT.P-SQRT(2.0/9)) THEN
        QNEW = +(11.0/12)*Q - (7.0/12)*P - SQRT(49.0/72)
        PNEW = +(11.0/12)*P - (7.0/12)*Q - SQRT( 1.0/72)
ENDIF
```

Figure 8.15 shows the 2×2 diamond-shaped region filled in by the map. With the initial condition $(q, p) = (0.6, 0.8)$ the map traces out what looks like a fractal structure. A detailed inspection shows that the number of *pairs* of points within a distance $r = \sqrt{q^2 + p^2}$ varies as $r^{1.59}$. The logarithmic plot shows that this correspondence holds nicely for $e^{-10} < r < e^{-1}$. The "correlation dimension" is the slope of the curves in the appropriate limits, $D_C = 1.59$.

The number of phase points in the 2×2 diamond is about 2^{46}, but the fractal nature of the distribution shows that the number accessible to the Baker Map is much smaller, about $\sqrt{2^{1.59 \times 23}} \simeq 2^{18}$. A random walk on N sites eventually lands on a periodic orbit. In the large-N limit the averaged number of jumps until the periodic orbit is reached and the averaged number of jumps on the orbit, until it returns, are both $J = \sqrt{N\pi/8}$:

$$N \simeq 2^{1.59 \times 23} \simeq 2^{36.57} = 10^{11} \to J \simeq 200,000 .$$

Our Baker Map example, after 243,784 iterations from $(q, p) = (0.6, 0.8)$, reaches a dissipative periodic orbit of length 157,342. In the simpler but related "Birthday Problem" 23 randomly-chosen people have a 50% probability that two of them have the same birthday. $\sqrt{(365\pi/2)} \simeq 23.9$. See Reference 15 for additional example problems.

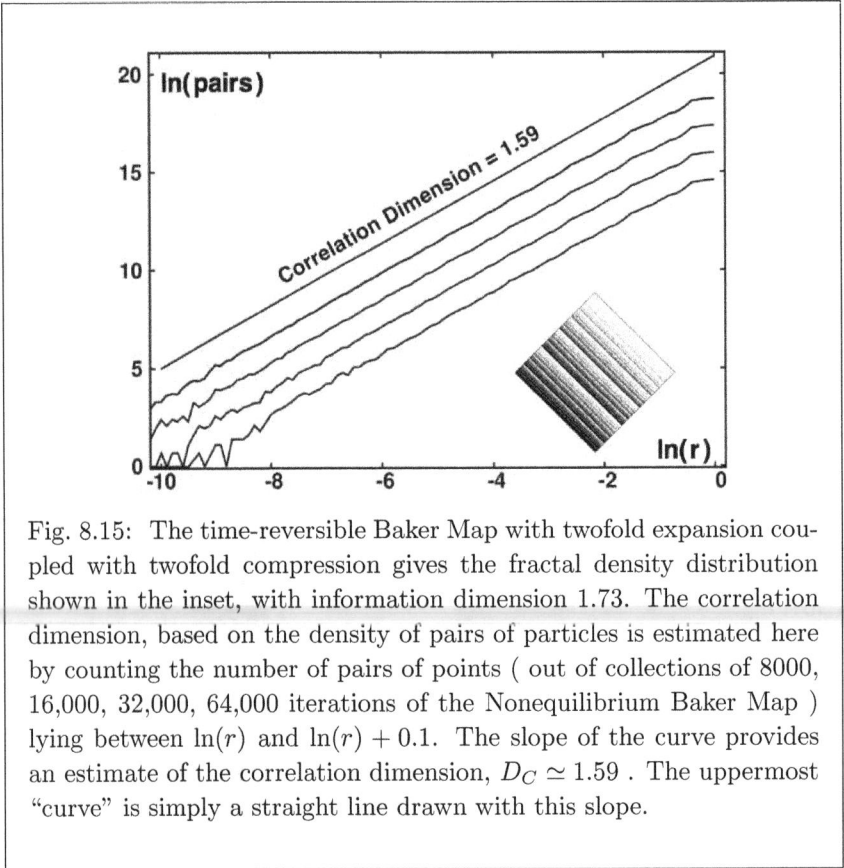

Fig. 8.15:   The time-reversible Baker Map with twofold expansion coupled with twofold compression gives the fractal density distribution
shown in the inset, with information dimension 1.73. The correlation
dimension, based on the density of pairs of particles is estimated here
by counting the number of pairs of points (out of collections of 8000,
16,000, 32,000, 64,000 iterations of the Nonequilibrium Baker Map)
lying between $\ln(r)$ and $\ln(r) + 0.1$. The slope of the curve provides
an estimate of the correlation dimension, $D_C \simeq 1.59$. The uppermost
"curve" is simply a straight line drawn with this slope.

This example problem shows that the repellor is unobservable in practice
due to the many-to-one nature of the mapping. Two-thirds of the iterations
of the mapping reduce a small area $dqdp$ to just half its original size. In a
simulation roundoff error suggests that only a bit more than a tenth of the
time will a jump be reversed. Thus the mathematicians are, from a rigorous
standpoint, correct when they state that the map is reversible. Those that
calculate, on the other hand, are quite correct when they state that the
map is throughly dissipative (failing to reverse about 90% of the time)
and thus obeys the Second Law of Thermodynamics. The time required for
calculators to encounter Poincaré recurrence in a two-dimensional space is
still feasible with double precision ($\simeq 10^{32}$ points) but not with quadruple
precision and its $\simeq 10^{68}$ points.

8.14 Summary of Lecture 8

We have explored the simulation of nonequilibrium irreversible processes with reversible equations of motion. An interesting consequence of these investigations is the discovery of repellor-attractor pairs, often ergodic and typically fractal with the repellor and attractor having probabilities of zero and one respectively despite their nonzero densities throughout the small-system phase spaces. The Baker Map and the Galton Board share these properties with simulations of many particles exposed to temperature and velocity gradients.

Imposing the boundary conditions on these flows is often a challenge. The study of shockwaves (as well as the closely-related Joule-Thomson experiment) with their stationary nonequilibrium flows is made possible by using equilibrium boundary conditions at the system's entrance and exit. The bit-reversible algorithm of Levesque and Verlet[10] and our adaptation of Milne's[11, 12] fourth-order predictor algorithm for second-order differential equations make it possible to explore the time asymmetry of the instabilities which are included in irreversible processes despite the time-reversibility of their equations of motion.

The Lyapunov instabilities of these reversed far-from-equilibrium flows emphasize very different particles forward and backward in time, with the particles in question being the very ones that macroscopic irreversible constitutive equations would suggest. The Lyapunov exponents are particularly useful tools for investigating these problems and will be the focus of our investigations in the next two Lectures.

References

1. H. Yoshida, "Recent Progress in the Theory and Application of Symplectic Integrators", Celestial Mechanics and Dynamical Astronomy **56**, 27-43 (1993).
2. C. P. Dettmann and G. P. Morriss, "Hamiltonian Reformulation and Pairing of Lyapunov Exponents for Nosé-Hoover Dynamics", Physical Review E **55**, 3693-3696 (1997).
3. G. Benettin, L. Galgani, and J-M. Strelcyn, "Kolmogorov Entropy and Numerical Experiments", Physical Review A **14**, 2338-2345 (1976).
4. S. D. Stoddard and J. Ford, "Numerical Experiments on the Stochastic Behavior of a Lennard-Jones Gas System", Physical Review A **8**, 1504-1512 (1973).
5. Wm. G. Hoover, C. G. Hoover, and K. P. Travis, "Shockwave Compres-

sion and Joule-Thomson Expansion", Physical Review Letters **112**, 144504 (2014).

6. B. L. Holian, Wm. G. Hoover, B. Moran, and G. K. Straub, "Shockwave Structure *via* Nonequilibrium Molecular Dynamics and Navier-Stokes Continuum Mechanics", Physical Review A **22**, 2798-2808 (1980).

7. C. E. Ragan III, "UltraHigh-Pressure Shockwave Experiments", Physical Review A **21**, 458-463 (1980).

8. Wm. G. Hoover and C. G. Hoover, "Time's Arrow for Shockwaves; Bit-Reversible Lyapunov and 'Covariant' Vectors ; Symmetry Breaking", Computational Methods in Science and Technology **19**, 69-75 and e1 (Figure 3 erratum) (2013).

9. Wm. G. Hoover and C. G. Hoover, "Time-Symmetry Breaking in Hamiltonian Mechanics", Computational Methods in Science and Technology **19**, 77-87 (2013).

10. D. Levesque and L. Verlet, "Molecular Dynamics and Time Reversibility", Journal of Statistical Physics **72**, 519-537 (1993).

11. Wm. G. Hoover and C. G. Hoover, "Bit-Reversible Version of Milne's Fourth-Order Time-Reversible Integrator for Molecular Dynamics", Computational Methods in Science and Technology **23**, 299-303 (2017).

12. Wm. E. Milne, *Numerical Calculus* (Princeton University Press, Princeton, New Jersey, 1949). Reprinted by Princeton University Press in 2015.

13. Wm. G. Hoover, O. Kum, and H. A. Posch, "Time-Reversible Dissipative Ergodic Maps", Physical Review E **53**, 2123-2129 (1996).

14. P. K. Patra, "Two-Dimensional Time-Reversible Ergodic Maps with Provisions for Dissipation", Computational Methods in Science and Technology **22**, 61-70 (2016).

15. Ch. Dellago and Wm. G. Hoover, "Finite-Precision Stationary States at and Away from Equilibrium", Physical Review E **62**, 6275-6281 (2000).

Chapter 9

Lyapunov Instability, Fractals, Chaos I

/ Lyapunov Exponents / Springy Pendulum / Symmetry Breaking / Calculation of Lyapunov Exponents / Rescaling / Lagrange Multiplier / Linearization / Gram-Schmidt Algorithm / Exponent Pairing / Equilibrium and Nonequilibrium Lyapunov Spectra / Steady Shear Spectra / Equilibrium and Nonequilibrium Baker Maps / Summary of Lecture 9 /

9.1 Lyapunov Exponents

Carol and I have commented on chaos previously, an essential ingredient of any deterministic dynamical models aiming to reproduce Gibbs' smooth many-dimensional unbounded distributions in phase space. The key measure of chaos is the "local" Lyapunov exponent $\lambda_1(t)$, the rate at which neighboring trajectories separate. This one-dimensional growth idea has been generalized to measuring the growth rates of two-dimensional areas $\lambda_1(t) + \lambda_2(t)$, three-dimensional volumes $\lambda_1(t) + \lambda_2(t) + \lambda_3(t)$, . . . all the way up to the rate of change of hypervolumes with the full dimensionality of the phase space itself. The close connection of phase volume to irreversibility and Gibbs' entropy has been tightened up through the development of thermostatted mechanics following the lines pioneered by Ashurst, Woodcock, Nosé and many others in the 1970s and 1980s.

In this Lecture our goal is to illustrate and explain clearly three methods for computing not just the largest Lyapunov exponent $\lambda_1(t)$ and its time average $\lambda_1 = \langle\, \lambda_1(t) \,\rangle$, but *all* of them, and not just for systems with one or two degrees of freedom, but for those with *many*. We begin by setting out the spectrum calculation for a four-dimensional problem : a Hooke's-Law pendulum, the "springy pendulum". With the details of this example problem firmly in hand, I will describe two nonequilibrium "mesoscopic" problem types, small enough for laptop replication while large enough to suggest and illustrate the general principles underlying the microscopic interpretation of the Second Law of Thermodynamics.

The consequences of imposing steady *nonequilibrium* conditions on an otherwise conservative dynamics turn out to be profound. Typically nonequilibrium conditions result in the collapse of the phase-space distributions from Gibbs' smooth equilibrium forms to fractals, distributions of intricate and æsthetic beauty.[1] Fractal distributions display paradoxical properties. For some interesting details we will touch again on the two-dimensional nonequilibrium Baker Map which sheds light on Loschmidt's reversibility paradox as well as Zermélo's recurrence paradox.

9.2 Lyapunov Spectrum for the Springy Pendulum

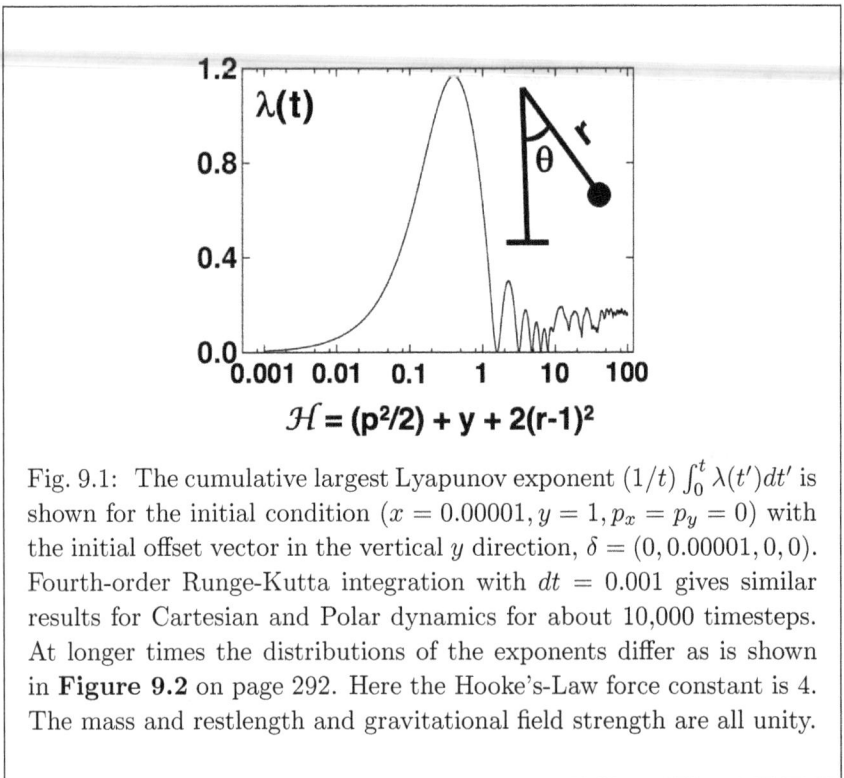

Fig. 9.1: The cumulative largest Lyapunov exponent $(1/t)\int_0^t \lambda(t')dt'$ is shown for the initial condition $(x = 0.00001, y = 1, p_x = p_y = 0)$ with the initial offset vector in the vertical y direction, $\delta = (0, 0.00001, 0, 0)$. Fourth-order Runge-Kutta integration with $dt = 0.001$ gives similar results for Cartesian and Polar dynamics for about 10,000 timesteps. At longer times the distributions of the exponents differ as is shown in **Figure 9.2** on page 292. Here the Hooke's-Law force constant is 4. The mass and restlength and gravitational field strength are all unity.

The springy pendulum of **Figure 9.1** is a good exercise in Hamiltonian mechanics. It has both chaotic and periodic orbits and can be solved in either Cartesian or Polar coordinates. There are many variants in two dimensions or in three. One can, for instance, link pendula together to make a chain with many degrees of freedom.[2] There is an obvious Theorem demonstrating that deterministic flows in a two-dimensional (q, p) bounded state space are not very interesting. This is because the moving phase point (q_t, p_t), unless it stops, is bound to encounter its trajectory again, forming a closed loop, and hence a stable periodic orbit.

With discontinuous maps in two dimensions or with flows in three dimensions things do become much more interesting. We can then have chaos because there is no need for a moving point ever to encounter its past. It must however, come arbitrarily close, as established by a less obvious and more interesting Theorem proving the inevitability of "Poincaré Recurrence". The unstable periodic orbits characterizing chaotic flows inevitably provide near recurrences. The most familiar example of such systems is Lorenz' "Butterfly" attractor.

Exercise : See if you can develop a proof of Poincaré Recurrence for Hamiltonian systems by applying Liouville's Theorem to the future location of phase points initially occupying a closed finite region of constant probability density between the energies of E and $E + dE$, where dE is also finite, but small.

Exercise : Write an RK4 program that generates the two stable normal modes of the springy pendulum. Determine their frequencies.

The springy Hooke's-Law pendulum of **Figure 9.1** is relatively easy to simulate and to visualize. The motion occurs in a two-dimensional coordinate space on a three-dimensional energy surface in a four-dimensional (q, p) phase space, either (x, y, p_x, p_y) or $(r, \theta, p_r, p_\theta)$. It illustrates many typical properties of chaotic Hamiltonian systems. For simplicity we set the mass, restlength, and gravitational acceleration all equal to unity and the spring constant of 4. The Hamiltonian is :

$$\mathcal{H}(x, y, p_x, p_y) = 2(\sqrt{x^2 + y^2} - 1)^2 + y + (p_x^2 + p_y^2)/2 .$$

There are two fixed points where the gravitational force on the bob is offset by the Hooke's Law spring force. We can find them by solving the

Hamiltonian's motion equations for the vertical acceleration with both x and $\ddot{y}$ equal to zero. The field acceleration is alway $\ddot{y}_{\text{field}} = -1$. There are two configurations where the spring force is $\ddot{y}_{\text{spring}} = +1$ leading to the two equilibrium fixed points, with ($\dot{x}, \dot{y}, \dot{p}_x, \dot{p}_y$) all equal to zero :

$$y = +(3/4) \rightarrow \ddot{y} = +1 - 1 \ [\text{ unstable }] \ ;$$
$$y = -(5/4) \rightarrow \ddot{y} = +1 - 1 \ [\text{ stable }] \ .$$

The least-energy static equilibrium at $y = -(5/4)$ is a stable fixed point of the dynamics, with two "normal modes" corresponding to horizontal and vertical oscillations about that point. The unstable fixed point at $y = (3/4)$ has no restoring force for horizontal displacements.

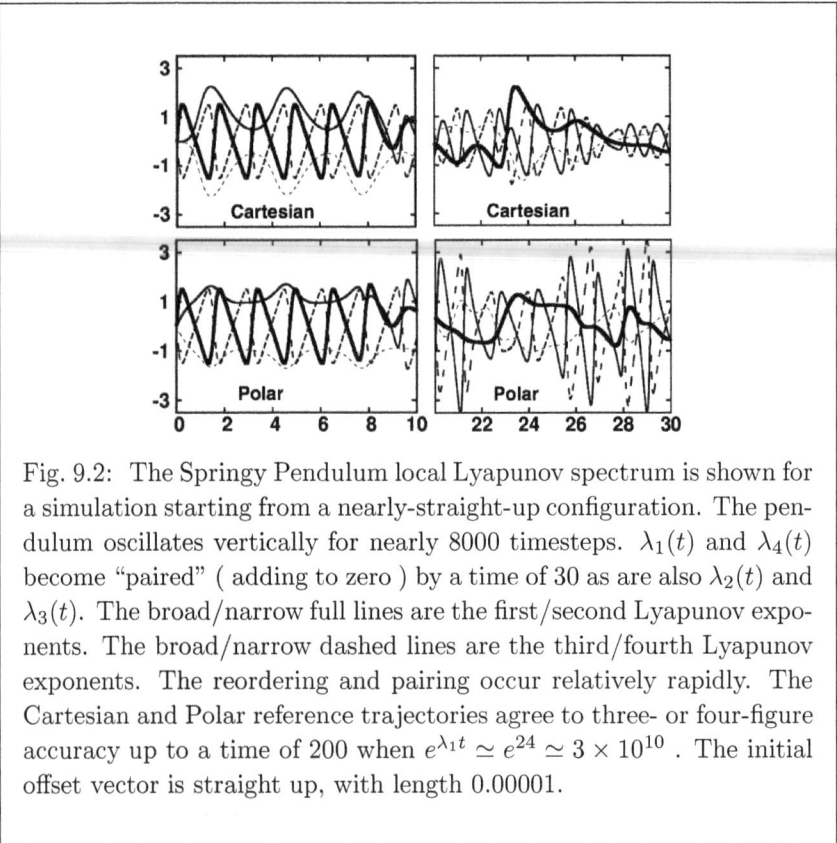

Fig. 9.2: The Springy Pendulum local Lyapunov spectrum is shown for a simulation starting from a nearly-straight-up configuration. The pendulum oscillates vertically for nearly 8000 timesteps. $\lambda_1(t)$ and $\lambda_4(t)$ become "paired" (adding to zero) by a time of 30 as are also $\lambda_2(t)$ and $\lambda_3(t)$. The broad/narrow full lines are the first/second Lyapunov exponents. The broad/narrow dashed lines are the third/fourth Lyapunov exponents. The reordering and pairing occur relatively rapidly. The Cartesian and Polar reference trajectories agree to three- or four-figure accuracy up to a time of 200 when $e^{\lambda_1 t} \simeq e^{24} \simeq 3 \times 10^{10}$. The initial offset vector is straight up, with length 0.00001.

If the displacement of the spring is large (but not too large) we can expect to see chaotic motions like the prototype analyzed in **Figures 9.1- 9.8**. The situation is slightly complicated, with both periodic and chaotic solutions coexisting in the three-dimensional phase volume that corresponds to the initial energy of the spring $E = 1$.

A good starting condition for chaos is $(x = 0.00001, y = 1)$ with an energy just exceeding unity, more than enough to exceed the energy of $(7/8)$ required to pass *over* the pendulum's support. **Figure 9.1** illustrates the time-averaged Lyapunov exponent over the first 10^5 timesteps for a chaotic orbit with an energy barely exceeding unity. Over time the orbit comes close to repeating (typical chaotic seas contain infinitely-many periodic but unstable chaotic orbits). Evidently there are many stable orbits too, like the two least-energy normal modes. We can also imagine periodic orbits at *high* energy $E >> 1$. They correspond to nearly free rapid rotation far from the central support point.

Figure 9.2 shows both the short-term and the intermediate-term Lyapunov spectra for the initial condition of **Figure 9.1**. For a time on the order of ten the motion is nearly periodic. In the polar-coordinate case the kinetic energy is :

$$[\, p_r^2 + (p_\theta/r)^2 \,]/2 = [\, \dot{r}^2 + r^2\dot{\theta}^2 \,]/2 \longrightarrow \dot{\theta} = (p_\theta/r^2) \ .$$

The Lyapunov exponents for the two coordinate systems show the same periodicity but different spectra. For times of order 25 the two different descriptions both lead to exponent pairing, with $\lambda_1 + \lambda_4$ and $\lambda_2 + \lambda_3$ both vanishing. Liouville's Theorem implies that the sum of all four exponents vanishes, as phase volume is conserved by Hamiltonian mechanics. The additional pairings depend upon the directions of the five separating trajectories defining the spectrum $\{\ \lambda_i\ \}$. The generality of this longer-term pairing for more complex situations is unclear, an interesting research topic.

Exercise : Write an RK4 program to solve for the motion of a three-dimensional springy pendulum. Compare the spherical polar coordinates' trajectory to the Cartesian one for the nearly-straight-up initial condition $(x, y, z) = (0.00001, 0, 1)$.

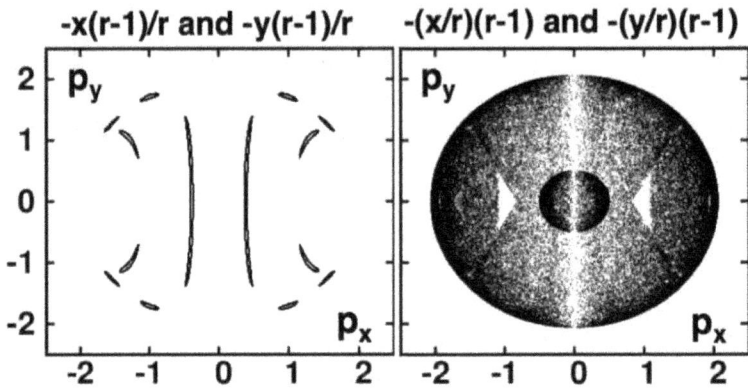

Fig. 9.3: The section on the right shows the chaotic { p_x, p_y } states reached with initial state $(x, y, p_x, p_y) = (0.00001, 1, 0, 0)$. RK4 and RK5 integrators with $dt = 0.001$ were used. RK5 gave $\lambda_1 \simeq 0.12_2$. Our RK4 simulation indicated a vanishing Lyapunov exponent throughout the range $1.45 \times 10^7 < \texttt{time} < 1.85 \times 10^7$. That section is on the left and corresponds to a complex torus. This regular RK4 solution was generated using $[x(r-1)/r, y(r-1)/r]$. With the more-nearly-accurate RK5 algorithm or with a different sequence of arithmetic operations in RK4, $[(x/r)(r-1), (y/r)(r-1)]$, the regular episode disappeared.

Two cross-sectional plots of (p_x, p_y) with $x = 0$ are shown in **Figure 9.3**. The sections accrue points each time the pendulum passes $x = 0$. Two different types of points stand out in the chaotic section. The small dark circle with radius $(1/2)$ represents those kinetic-energy states available near the unstable fixed point $y = (3/4)$. The larger outside circle, with radius $\sqrt{17}/2$ represents the many more numerous states available near the stable fixed point at $y = -(5/4)$. We can see places where stable periodic orbits penetrate the section and we can see the white "nullcline" at $p_x = 0$ where $\dot{x}$ vanishes so that there is no flux through the $x = 0$ plane.

We can easily carry out an identical analysis in polar coordinates and would find the same chaotic sea and corresponding holes in it representing stable periodic orbits. How can we be sure that the motion is chaotic, with small perturbations increasing exponentially fast with time? The problem requires care because Hamiltonian chaos is typically complicated, with a complex fractal boundary between a chaotic sea and infinitely-many stable

periodic orbits. Numerical solutions have a tendency to wander, particularly during relatively long simulations with 10^{11} or 10^{12} timesteps. Recognizing that RK4 integration tends to lose energy while RK5 tends to gain,
one useful approach is adaptive integration in which the difference between
these two integrator types is kept small by adjusting the timestep (the
adaptive integration I talked about in Section 1.7.3, page 23).

The springy pendulum can exhibit some interesting difficulties in long
simulations. See again **Figure 9.3**. That Figure shows that there are stable
periodic orbits quite close to the unstable periodic orbit generated with our
nearly-straight-up initial condition. The springy pendulum is nevertheless
a useful model for generating and interpreting Lyapunov spectra from the
standpoint of dynamical systems theory. With the pendulum support point
at the origin and using a Hookean spring of unit rest length and force
constant 4, the Hamiltonian has the form :

$$\mathcal{H} = 2(\sqrt{(x^2 + y^2)} - 1)^2 + y + (p_x^2/2) + (p_y^2/2) .$$

To solve this same problem in polar coordinates (r, θ) we need to find the
radial and angular momenta (p_r, p_θ). Remembering, or deriving, the Lagrangian kinetic energy, $(\dot{r}^2/2) + (r^2\dot{\theta}^2/2)$ allows us to write the Hamiltonian
for the pendulum in terms of polar coordinates and momenta :

$$\mathcal{L}(q, \dot{q}) = (\dot{r}^2/2) + (r^2\dot{\theta}^2/2) - \Phi(q) \text{ and } \{ p \equiv \partial\mathcal{L}/\partial\dot{q} \} \longrightarrow$$

$$\mathcal{H}(q, p) = (p_r^2/2) + (p_\theta^2/2r^2) + 2(r - 1)^2 - r\cos(\theta) .$$

Like most nonlinear Hamiltonian systems (but unlike the equilibrium
Galton Board problem, which is *always* chaotic at densities greater than
three-fourths of close packing) the springy pendulum has both regular
and chaotic trajectories. If the initial position of the pendulum is close to
straight up, $(x, y) \simeq (0, 1)$, the motion is chaotic and one can define a
Lyapunov spectrum $\{ \lambda_1, \lambda_2, \lambda_3, \lambda_4 \}$. To do so imagine a hypersphere in
the four-dimensional Cartesian or Polar phase space. (The constant energy
"surface" is three-dimensional.) Follow the motion forward or backward
in time. If the hypersphere is small enough it will soon distort into a
hyperellipsoid. The longtime-averaged growth rate of the principal axis
defines $\lambda_1 \equiv \langle \lambda_1(t) \rangle$. The next-highest exponent describes a direction
perpendicular to the first which has the next-highest long-time-averaged
growth rate, $\lambda_2 \equiv \langle \lambda_2(t) \rangle$.

9.2.1   *A Symmetry-Breaking Surprise*

Because running the trajectory backward in time simply interchanges the directions of stretching and shrinking one might justifiably expect that the local Lyapunov spectrum is the same in both directions of time. Although that symmetry satisfies the equations of motion, this expectation is false, as a relatively simple simulation shows. The lack of time symmetry in the local exponents is due to the fact that the Lyapunov exponents can depend only upon the past, not at all on the future.

Figure 9.4 illustrates the time reversibility of the springy pendulum up to a time of 100. We will see that the largest Lyapunov exponent is about 0.122 so that the inevitable double-precision roundoff errors of order 10^{-16} would be amplified by a factor of $e^{\lambda_1 t} = e^{122} \simeq 10^{+53}$ were we to continue the simulation to a time of 1000.

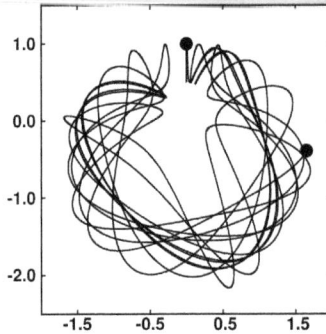

Fig. 9.4:   A chaotic trajectory of the springy pendulum. Integrating either the Cartesian or the Polar equations of motion for the springy pendulum gives essentially the same trajectory for both, ending up at the rightmost filled circle at time 100. Reversing the sign of $dt = 0.001$ in the RK5 integrator and integrating for another 100,000 steps returns to the initial point $(x, y) = (0.00001, 1)$ within visual accuracy.

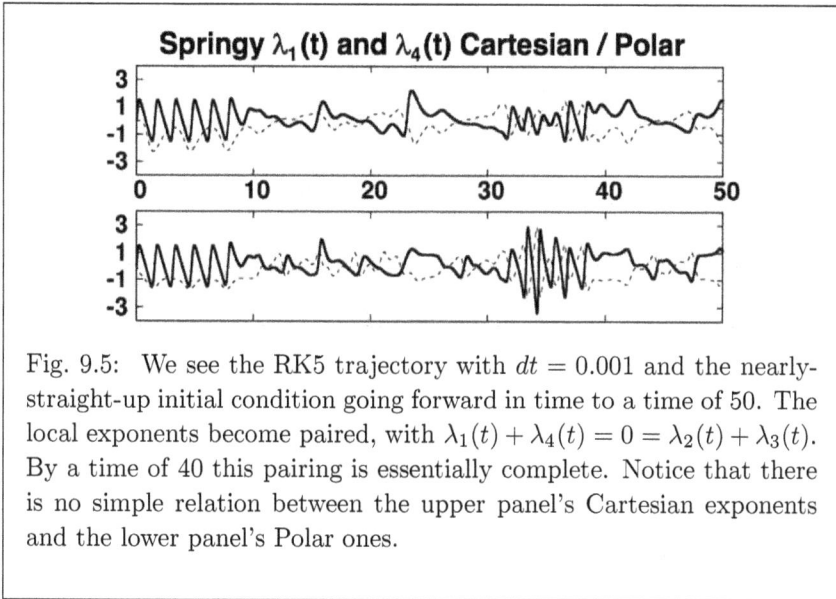

Springy $\lambda_1(t)$ and $\lambda_4(t)$ Cartesian / Polar

Fig. 9.5: We see the RK5 trajectory with $dt = 0.001$ and the nearly-straight-up initial condition going forward in time to a time of 50. The local exponents become paired, with $\lambda_1(t) + \lambda_4(t) = 0 = \lambda_2(t) + \lambda_3(t)$. By a time of 40 this pairing is essentially complete. Notice that there is no simple relation between the upper panel's Cartesian exponents and the lower panel's Polar ones.

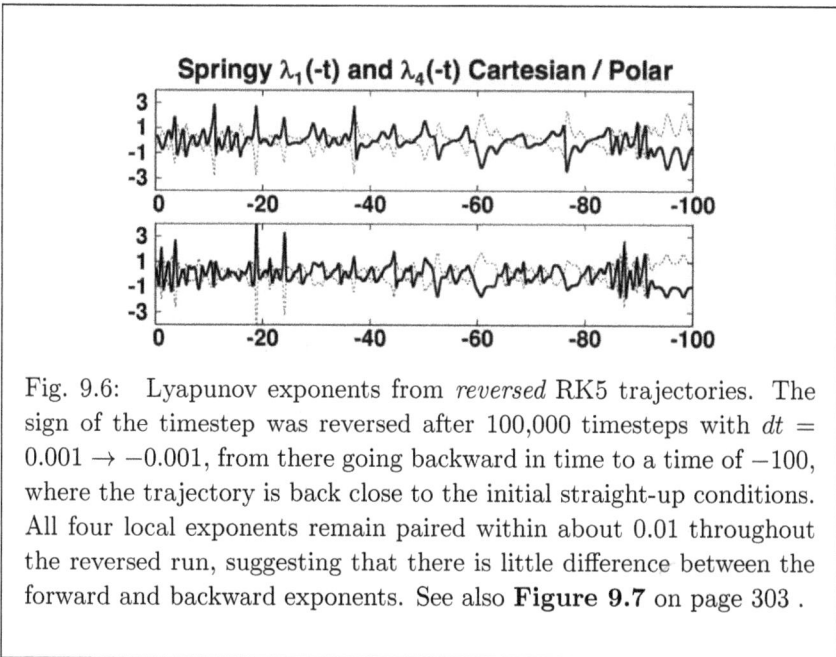

Springy $\lambda_1(-t)$ and $\lambda_4(-t)$ Cartesian / Polar

Fig. 9.6: Lyapunov exponents from *reversed* RK5 trajectories. The sign of the timestep was reversed after 100,000 timesteps with $dt = 0.001 \rightarrow -0.001$, from there going backward in time to a time of -100, where the trajectory is back close to the initial straight-up conditions. All four local exponents remain paired within about 0.01 throughout the reversed run, suggesting that there is little difference between the forward and backward exponents. See also **Figure 9.7** on page 303 .

Because the motion satisfies the Liouville Theorem for Hamiltonian systems the instantaneous summed spectrum $\lambda_1(t) + \lambda_2(t) + \lambda_3(t) + \lambda_4(t)$ necessarily vanishes in the four-dimensional case, but *not* in three (and the motion is on a three-dimensional surface of constant energy) :

$$\dot{f}(x, y, p_x, p_y) \equiv \dot{f}(r, \theta, p_r, p_\theta) = 0 \xleftrightarrow{\mathcal{H}} \otimes(x, y, p_x, p_y) \equiv \otimes(r, \theta, p_r, p_\theta) \equiv 0 ,$$

The theorem isn't quite true (because it doesn't apply) in the three-dimensional case but would become so if the thickness of the three-dimensional energy shell (including all states between E and $E + dE$) were included.

Numerical simulation shows that there is a strong tendency for the first and last Lyapunov exponents to "pair", $\lambda_1(t) + \lambda_4(t) = 0$. See **Figures 9.5 and 9.6**. Whenever the first and last exponents do pair (around time 25 in **Figure 9.5**) the second and third are necessarily likewise paired, $\lambda_2(t) + \lambda_3(t) = 0$ as the sum of all four exponents vanishes :

$$\sum [\, (\partial \dot{q}/\partial q) + \sum (\partial \dot{p}/\partial p) \,] \equiv \lambda_1(t) + \lambda_2(t) + \lambda_3(t) + \lambda_4(t) \equiv (\dot{\otimes}/\otimes) \equiv 0 .$$

Pairing of the instantaneous "local" exponents implies pairing of the "global" long-time-averaged exponents. The growth rate along the trajectory direction can be determined from the motion equations using either Cartesian or polar coordinates :

$$\dot{x} = p_x \; ; \; \dot{y} = p_y \; ; \; \dot{p}_x = -4(x/r)(r-1) \; ; \; \dot{p}_y = -1 - 4(y/r)(r-1) .$$

Here all the derivatives $(\partial \dot{q}/\partial q)$ and $(\partial \dot{p}/\partial p)$ vanish. This is likewise true in the polar-coordinate case :

$$\dot{r} = p_r \; ; \; \dot{\theta} = (p_\theta/r^2) \; ; \; \dot{p}_r = (p_\theta^2/r^3) - 4(r-1) + \cos(\theta) \; ; \; \dot{p}_\theta = -r \sin(\theta) .$$

Exercise : Plot the time-dependence of the speed $|\, v_t \,|$ along the trajectory for the two sets of motion equations for $0 < t < 100$.

Exercise : By storing the trajectory forward in time analyze a reversed satellite trajectory generated by constraining the integrated values to stay within a distance 0.000001 of the stored ones and using a negative timestep $dt < 0$. Show that there is no simple relation between the local values of $\lambda_1(t)$ going forward and backward in time.

9.2.2 Three Methods for Computing $\lambda_1(t)$

The largest Lyapunov exponent is relatively easy to compute, in either Cartesian or polar coordinates, by integrating the same motion equations for a reference and a satellite trajectory. Keeping track of the rescaling needed at the end of each timestep to keep their separation constant provides the local values of the largest Lyapunov exponent. A simulation of one hundred billion timesteps covering a time of one hundred million, with an initial Hamiltonian of unity (start with $x = 0.000001$; $y = 1.000000$) provides an estimate $\lambda_1 = 0.122 \pm 0.001$, significantly below the estimate from 1990, $\lambda_1 = 0.128$, given in Reference 2.

There are three methods for determining the largest exponent $\lambda_1(t)$:

[1] Follow the dynamics of a satellite trajectory constrained to remain a small distance (0.000001 is a good choice) from the reference. Over a single timestep the linear growth equation can be solved to find the local Lyapunov exponent:

$$\dot{\delta} = \lambda\delta \longrightarrow \delta(t) = 0.000001e^{\lambda dt} \longrightarrow \lambda(dt) = (1/dt)\ln\left(\frac{\delta(dt)}{\delta(0)}\right) .$$

[2] Compute the Lagrange multiplier λ that implements a constraint force,

$$F_c(x, y, p_x, p_y) = -\lambda[(x_s - x_r), (y_s - y_r), (p_{x_s} - p_{x_r}), (p_{y_s} - p_{y_r})] ;$$

$$\lambda = [\delta_x\dot{\delta}_x + \delta_y\dot{\delta}_y + \delta_{p_x}\dot{\delta}_{p_x} + \delta_{p_y}\dot{\delta}_{p_y}]/\delta^2 .$$

This Lagrange multiplier is equal to the local Lyapunov exponent ! [3,4]

[3] Linearize the motion equations to compute the Lyapunov exponent in "tangent space" where the length of the offset vector can be chosen equal to unity. This method, though elegant, is tedious to program if the derivatives are complicated. For the pendulum problem we have again :

$$\lambda = [\delta_x\dot{\delta}_x + \delta_y\dot{\delta}_y + \delta_{p_x}\dot{\delta}_{p_x} + \delta_{p_y}\dot{\delta}_{p_y}]/\delta^2 .$$

Now the δ vectors have unit length rather than an infinitesimal length.

$$\dot{\delta}_x = \delta_{p_x} - \lambda\delta_x \; ; \; \dot{\delta}_y = \delta_{p_y} - \lambda\delta_y \; ;$$

$$\dot{\delta}_{p_x} = -4\delta_x[1 - (1/r)] - 4(x/r^3)(x\delta_x + y\delta_y) - \lambda\delta_{p_x} \; ;$$

$$\dot{\delta}_{p_y} = -4\delta_y[1 - (1/r)] - 4(y/r^3)(x\delta_x + y\delta_y) - \lambda\delta_{p_y} .$$

Because all three methods should agree the pragmatic choice is the simplest, the first method, brute-force rescaling of the offset vector δ, which we highly recommend. Methods [2] and [3] need an occasional rescaling of the δ vector to compensate for the effects of computer roundoff error. Each of the three methods can be generalized to yield the complete spectrum. For this the Gram-Schmidt algorithm is both simple and effective. I describe the computation of the spectrum and the detailed structure of the Gram-Schmidt algorithm in the next two Sections.

9.3   Computing the Lyapunov Spectrum $\{\ \lambda(t)\ \}$

Although Liouville's phase-space continuity equation prohibits change in
phase volume for Hamiltonian dynamics there are many nonHamiltonian
situations in which the phase volume does change in response to heat transfer. Overall the longtime-averaged change is either zero (the equilibrium
situation) or negative (in nonequilibrium simulations). In the nonequilibrium case the phase-volume dimensionality is typically fractional. This
means that the occupied phase-space states make up a distribution with
the volume having a fractional dimensionality less than that of the embed-
ding space. The Galton Board and Nonequilibrium Baker Map problems
of **Lecture 7** are good examples of dimensionality loss and fractal distributions. A knowledge of the complete Lyapunov spectrum can be used
to understand and analyze such fractal distributions in more complicated
problems with many degrees of freedom.

   A geometric view of phase-space deformation which can lead to fractal
structures considers the comoving flow and deformation of an orthogonal
set of basis vectors of length δ following the motion. In numerical work
it is convenient to choose the vector length several orders of magnitude
smaller than typical features of the flow. For problems like the harmonic
oscillator or the springy pendulum a convenient choice is 10^{-6}. In Method
[3] of the last Section where the deformation occurs in "tangent space",
responding to linearized equations of motion, the vector length is arbitrary.
It is conventional to choose it equal to unity.

   The comoving corotating flow of a set of basis vectors was implemented
in a stepwise rescaling algorithm that was discovered and implemented by
Shimada and Nagashima[5] as well as Benettin's group.[6] The "first vector"
δ_1 is the same one we have already studied. It keeps track of the constantly
changing and rotating separation of two trajectories (the reference and the
satellite) with the separation restored at the end of each timestep. Harald
Posch and I pointed out that the stepwise adjustment can be replaced by a
continuous constraint, using a Lagrange multiplier,[3] and that this idea can
be extended to the complete spectrum of n exponents by defining an array
of $n(n + 1)/2$ Lagrange multipliers.[4] Let us describe the simplest version
of the orthonormalization procedure.

   Either sequentially or continuously the "first" offset vector δ_1 is scaled
to yield $\lambda_1(t)$ which describes a local one-dimensional growth rate in phase
space. With this accomplished the "second" offset vector δ_2 is constrained
to remain perpendicular to the first. In a sequential algorithm this is a

two-step process: first the projection of δ_2 parallel to δ_1 is removed :

$$\delta_2 = \delta_2 - \left(\frac{\delta_1 \cdot \delta_2}{|\delta_1| \, |\delta_2|} \right) \delta_1 \ .$$

Next, the newly orthogonal δ_2 is rescaled to its original length. This rescaling gives the second local Lyapunov exponent λ_2. Alternatively, these latter two steps can both be accomplished simultaneously with two *continuously varying* Lagrange multipliers, λ_{12} and λ_{22} :

$$\dot{\delta}_2 = D \cdot \delta_2 - \lambda_{12}\delta_1 - \lambda_{22}\delta_2 \ .$$

Here $D \cdot \delta_2$ (where D stands for Dynamical matrix) represents the effect of the unconstrained equations of motion on the vector δ_2.

Adding a third vector to the description should provide practice enough to understand and appreciate the Gram-Schmidt algorithm. In the Lagrange-multiplier approach this requires three more multipliers. In the brute-force approach the projections on δ_1 and δ_2 are both removed :

$$\delta_3 = \delta_3 - \left(\frac{\delta_1 \cdot \delta_3}{|\delta_1| \, |\delta_3|} \right) \delta_1 - \left(\frac{\delta_2 \cdot \delta_3}{|\delta_2| \, |\delta_3|} \right) \delta_2$$

Then the newly-orthogonal vector δ_3 is rescaled and provides $\lambda_3(t)$, the third local Lapunov exponent.

9.3.1 *Gram-Schmidt Orthonormalization Algorithm*

A description of phase-space dynamics is simplified by introducing and maintaining an orthonormal set of basis vectors comoving with the flow. The "Gram-Schmidt" algorithm (credited by Wikipedia to Pierre-Simon Laplace) provides a straightforward approach to maintaining the orthogonality and lengths of a set of NVEX vectors. The algorithm on the next page converts NVEX vectors with NDIM components each to an orthonormal set of unit vectors :

$$\{ \, \mathtt{DALIN(NDIM, NVEX)} \, \} \longrightarrow \{ \, \mathtt{DALOUT(NDIM, NVEX)} \, \} \ .$$

As the Kth vector is rescaled the corresponding local Lyapunov exponent YAPL(K) is computed. Programming the orthonormalization begins by removing the projection of the Kth vector on the preceding K-1, first computing the dot product in the 20 loop below and then subtracting the projection in the 21 loop. Once orthogonality has been achieved for the Kth DALIN the 35 loop computes the natural logarithm of the length of the Kth orthogonal vector. This gives the Kth local Lyapunov exponent YAPL(K) . The Kth vector is then scaled in the 40 loop, producing the Kth orthonormalized vector, DALOUT(NDIM,K).

```
      DO 60 K = 1,NVEX
      IF(K.EQ.1) GO TO 30
      DO 22 J = 1,K-1
      DOT = 0.0D00
      DO 20 I = 1,NDIM
      DOT = DOT + DALOUT(I,J)*DALIN(I,K)
20    CONTINUE
      DO 21 I = 1,NDIM
      DALIN(I,K) = DALIN(I,K) - DALOUT(I,J))*DOT
21    CONTINUE
22    CONTINUE
30    DOT = 0.0D00
      DO 35 I = 1,NDIM
      DOT = DOT + DALIN(I,K)*DALIN(I,K)
35    CONTINUE
      YAP(K) = DSQRT(DOT)
      YAPL(K) = DLOG(YAP(K))/DT
      DO 40 I = 1,NDIM
40    DALOUT(I,K) = DALIN(I,K)/YAP(K)
60    CONTINUE
```

9.3.2 *Pairing of the Local Hamiltonian Exponents*

Remember that the local exponents have a simple geometric interpretation.
$\lambda_1 + \lambda_2$ gives the rate at which the two-dimensional area defined by δ_1 and
δ_2 grows or shrinks in time. $\lambda_1 + \lambda_2 + \lambda_3$ likewise describes the growth
rate of the volume described by the first three vectors, and so on. For
a Hamiltonian system (where the phase volume is preserved precisely)
Liouville's Theorem implies that the sum of the exponents vanishes. Once
the exponents are paired it is not easy to see whether or not they can
"unpair". Much experimentation with the springy pendulum and with the
springy *double pendulum* ,

$$\mathcal{H}_{DP} = (\kappa/2)[\sqrt{(x_1^2 + y_1^2)} - 1]^2 + (\kappa/2)[\sqrt{(x_2 - x_1)^2 + (y_2 - y_1)^2} - 1]^2$$

$$+ gy_2 + (p_{x_1}^2 + p_{y_1}^2 + p_{x_2}^2 + p_{y_2}^2)/2 \ .$$

suggests that once established, pairing is likely to continue. I tried occasional discontinuous sign and magnitude changes of the gravitational constant g as well as jumps in the springs' force constant κ. I found no significant departure from pairing. In every case once pairing had set in there

was never "unpairing" in response to changes of the gravitational and force constants from their initial values ($g = 1$; $\kappa = 4$).

Starting with an arbitrary basis set of orthogonal vectors is easy. In the Cartesian case we simply choose the vectors parallel to the phase-space axes. A certain amount of time (a few multiples of $1/\lambda_1$ is enough) is required for the vectors' directions to converge.

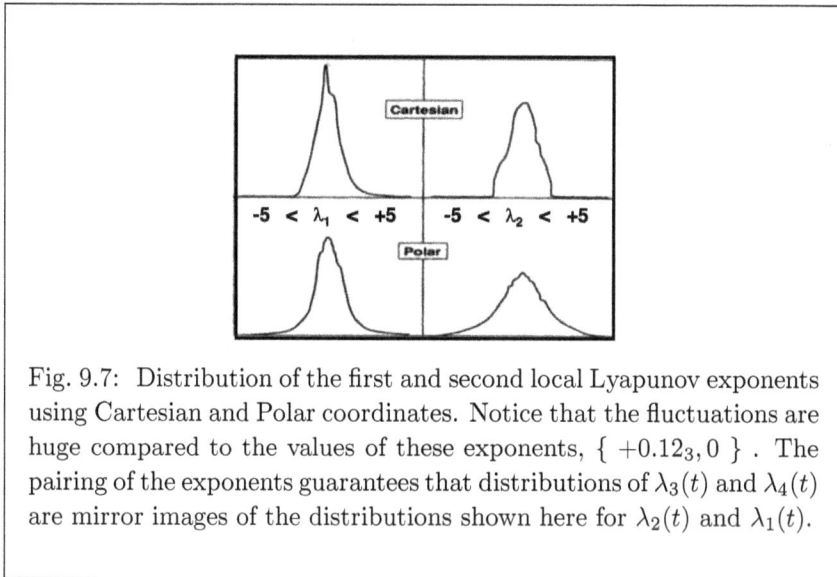

Fig. 9.7: Distribution of the first and second local Lyapunov exponents using Cartesian and Polar coordinates. Notice that the fluctuations are huge compared to the values of these exponents, $\{ +0.12_3, 0 \}$. The pairing of the exponents guarantees that distributions of $\lambda_3(t)$ and $\lambda_4(t)$ are mirror images of the distributions shown here for $\lambda_2(t)$ and $\lambda_1(t)$.

Figure 9.7 shows that the polar versions of the two vanishing exponents, (λ_2, λ_3) can reach local values as distant as ± 5, far from their time-averaged values. As the Figure clearly shows, the distribution of the exponents is very different in the two coordinate systems. In retrospect the spectrum of the springy oscillator is very simple. The instantaneous sum of the exponents vanishes. Empirically this is also true of the pairs :

$$\lambda_1(t) + \lambda_4(t) \equiv 0 \ ; \ \lambda_2(t) + \lambda_3(t) \equiv 0 \ .$$

One could imagine a simple explanation for the pairing : because the Hamiltonian equations of motion are time-reversible any *expanding* direction in phase space becomes a *contracting* direction in the reversed motion. If you picture a sphere becoming an ellipsoid following the dynamics, time-reversal ($dt \to -dt$) simply runs the dynamics backward. This explanation is false ! As we mentioned in 9.2.1 the forward and backward ellipsoids are different.

Because the motion within a comoving sphere of phase points is Lyapunov unstable the direction of the maximum Lyapunov exponent prevails
in the forward time direction, but always lags a bit behind the instantanous
direction of maximum growth.[7] Reversing the motion converts maximum
growth to maximum shrinkage but again there is a lag so that the local
exponents forward and backward in time are not simply related to one another. In order to "see" this it is most convincing to "store" a forward
trajectory long enough for the local vectors to converge. In **Figure 9.8**,
I chose a time of 500, using half a million timesteps, about ten times the
minimum needed. I then processed the stored (q, p) data backward in time
so as to compare the forward and backward Lyapunov exponents and their
associated vectors for the same trajectory, just processed backward.

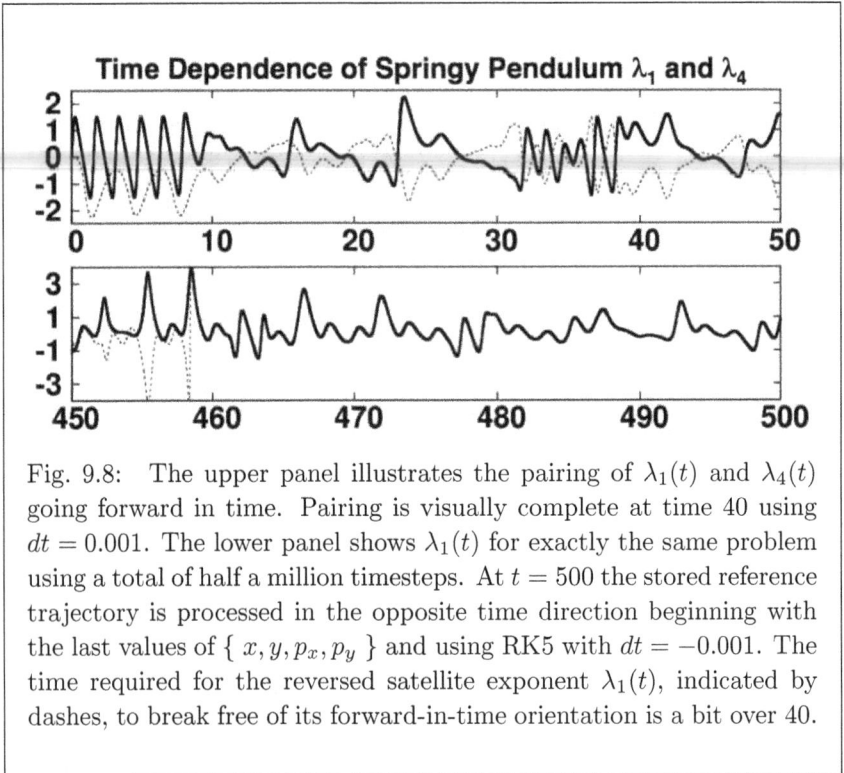

Fig. 9.8: The upper panel illustrates the pairing of $\lambda_1(t)$ and $\lambda_4(t)$
going forward in time. Pairing is visually complete at time 40 using
$dt = 0.001$. The lower panel shows $\lambda_1(t)$ for exactly the same problem
using a total of half a million timesteps. At $t = 500$ the stored reference
trajectory is processed in the opposite time direction beginning with
the last values of $\{ x, y, p_x, p_y \}$ and using RK5 with $dt = -0.001$. The
time required for the reversed satellite exponent $\lambda_1(t)$, indicated by
dashes, to break free of its forward-in-time orientation is a bit over 40.

Exercise : Start with four orthogonal vectors parallel to the (x, y, p_x, p_y) axes and determine the time required for all four of them to converge to a common direction. Use the pendulum equations of motion with the initial conditions $(x, y, p_x, p_y) = (0.00001, 1, 0, 0)$.

This springy pendulum is a good introduction to the theory and practice of Hamiltonian stability studies. It can be extended to manybody systems as we did in analyzing the inelastic collision of two 400-particle balls and two 1600-particle slabs in **Lecture 8**. This analysis can also be extended to apply to continuum problems (such as Rayleigh-Bénard convection) by converting the partial differential equations of continuum mechanics to ordinary differential equations for the evolution of the cell and nodal variables. We will demonstrate this in **Lecture 10**.

Let us go back thirty years to 1987 and the first applications of Lyapunov analysis to stationary nonequilibrium states. 1987 was a significant year for me, the year when I discovered that fractal distributions were *typical* of time-reversible nonequilibrium flows. With Harald Posch and Carol and many other coworkers we made systematic studies of nonequilibrium systems with steady flows of mass, and monentum, and energy, all of them analyzed in terms of their Lyapunov spectra. We began with simple few-body systems with periodic boundary conditions.

9.4 Equilibrium and Nonequilibrium Lyapunov Spectra

After Bill Moran's work on the isokinetic Galton Board problem,[8] with its fractal distributions resulting from impulsive collisions, I carried out, along with the help and support of five colleagues attending a CECAM = **C**entre **E**uropéen de **C**alcul **A**tomique et **M**oléculaire workshop in Orsay in 1986, a study[9] of the phase-space distribution for a particle interacting with a continuous potential, a spatial sinewave plus a constant accelerating field. This continuous-potential problem used isothermal (Nosé-Hoover) mechanics, with the equations of motion :

$$\{ \ddot{q} = \dot{p} = 0.3 - \sin(q) - \zeta p \ ; \ \dot{\zeta} = 0.1(p^2 - 1) \}$$

Both studies, the two-dimensional Galton Board mapping and the three-dimensional Galton Staircase problem, though with qualitatively different collisional forces, showed fractal dimensionality losses consistent with irreversibility and the Second Law of Thermodynamics.

Harald and I developed the Lagrange multiplier approach[3, 4] and began to apply it to systematic studies of Lyapunov spectra. An ambitious analysis of equilibrium simulations with 32 three-dimensional particles (with a 192-exponent spectrum) confirmed our finding for smaller systems that the distribution of the exponents closely resembles the Debye power-law with $\lambda \propto n^{0.38}$, not so different to Debye's sound wave result for the vibrational frequencies of three-dimensional solids $\nu \propto n^{(1/3)}$.

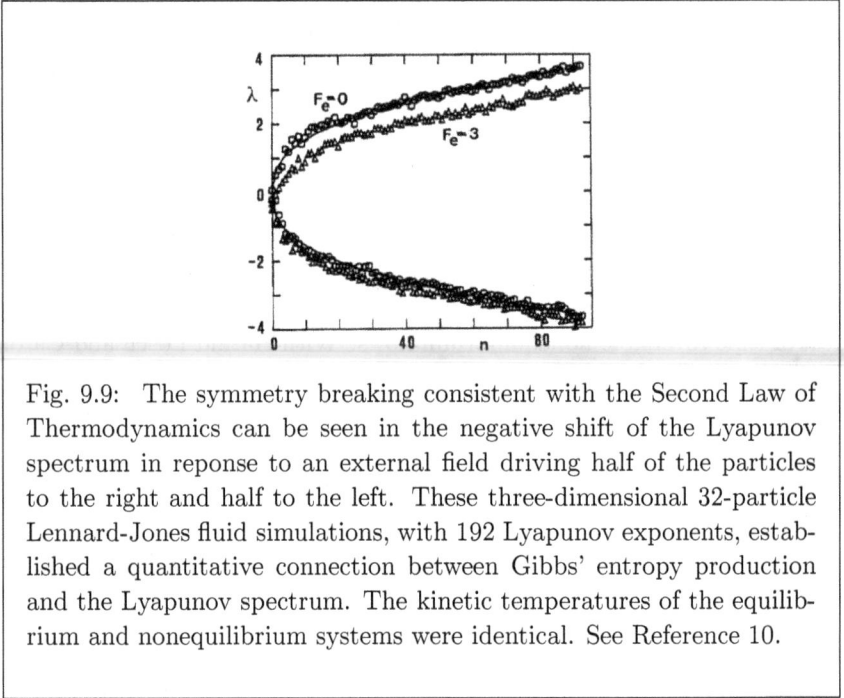

Fig. 9.9: The symmetry breaking consistent with the Second Law of Thermodynamics can be seen in the negative shift of the Lyapunov spectrum in reponse to an external field driving half of the particles to the right and half to the left. These three-dimensional 32-particle Lennard-Jones fluid simulations, with 192 Lyapunov exponents, established a quantitative connection between Gibbs' entropy production and the Lyapunov spectrum. The kinetic temperatures of the equilibrium and nonequilibrium systems were identical. See Reference 10.

It was 1987 when Harald and I carried out what came to be called "color conductivity" studies.[10] Half the particles were accelerated to the right and half to the left using Gaussian isokinetic mechanics with $(kT/\epsilon) = 1.0$ and $(N\sigma^3/V) = 0.5$, using the Lennard-Jones pair potential function, $\phi = 4\epsilon[\ (\sigma/r)^{12} - (\sigma/r)^6\ ]$. **Figure 9.9** compares the distribution of 192 Lyapunov exponents at equilibrium to the nonequilibrium distribution with an accelerating field strength of $3(\epsilon/\sigma)$. Two highly significant findings came from this work. Away from equilibrium the spectrum was shifted uniformly toward more negative values. At a reduced field strength of 3 the sum of the first $\simeq 172$ out of $6N = 192$ exponents is close to zero, showing that the fractal dimension of the distribution is less than the equilibrium value by $\simeq 20$. Because the sum of all the exponents is negative, and equal to the rate of change of Gibbs' entropy,

$$(\dot{S}/Nk)_{\text{Gibbs}} \equiv (1/N)(\dot{\otimes}/\otimes) = -\langle \, \zeta \, \rangle \simeq -(3.5 \times 20/32) \,,$$

we learn that not only is the comoving phase-volume of the distribution zero, the *dimensionality* of that distribution has been reduced by about ten percent relative to the equilibrium one. Beyond the good agreement of nonequilibrium simulations with laboratory experiments, this general feature of time-reversible nonequilibrium steady states, not accessible to direct experimentation, was and is particularly significant. It establishes the great fractal rarity of nonequilibrium phase-space states, showing that a time-reversible thermostatted dynamics accomplishes simulations far from equilibrium in a way impossible with conventional Hamiltonian mechanics.

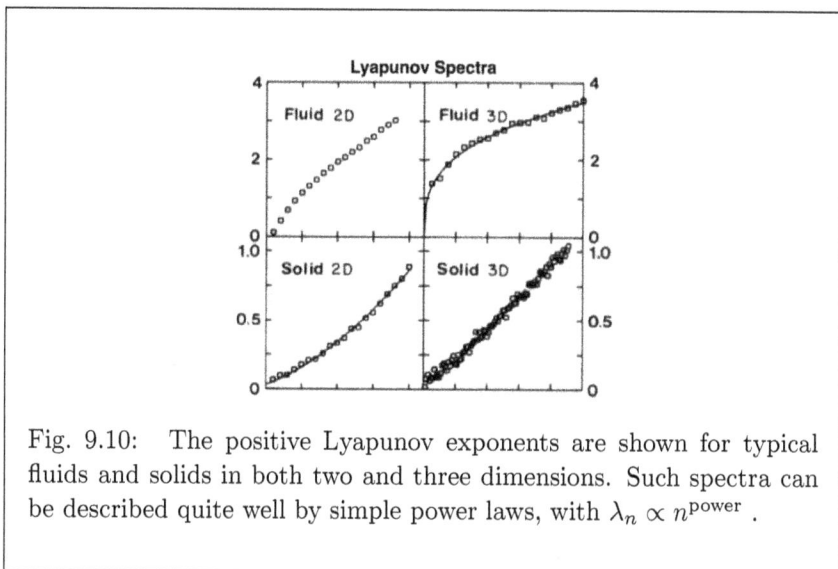

Fig. 9.10:   The positive Lyapunov exponents are shown for typical fluids and solids in both two and three dimensions. Such spectra can be described quite well by simple power laws, with $\lambda_n \propto n^{\text{power}}$.

Further investigation revealed that the Lyapunov spectra for equilibrium systems of just a few particles (8, 9, 12, and 32 particles) showed that equilibrium spectra can usefully be described by smooth power-law curves.[11] The maximum Lyapunov exponent is of key importance in determining the length of time for which the reproducible and reversible trajectories can be generated. **Figure 9.10** shows that the Lyapunov instabilities of chaotic solids and fluids are different in curvature and in magnitude and that the

spectra are simple featureless curves. For much larger systems Harald Posch
and his colleagues pointed out the occurrence of "modes" in the spectrum
with offset vectors that are sinusoidal in the phase space.[12]

9.5   Lyapunov Spectra Under Steady Shear

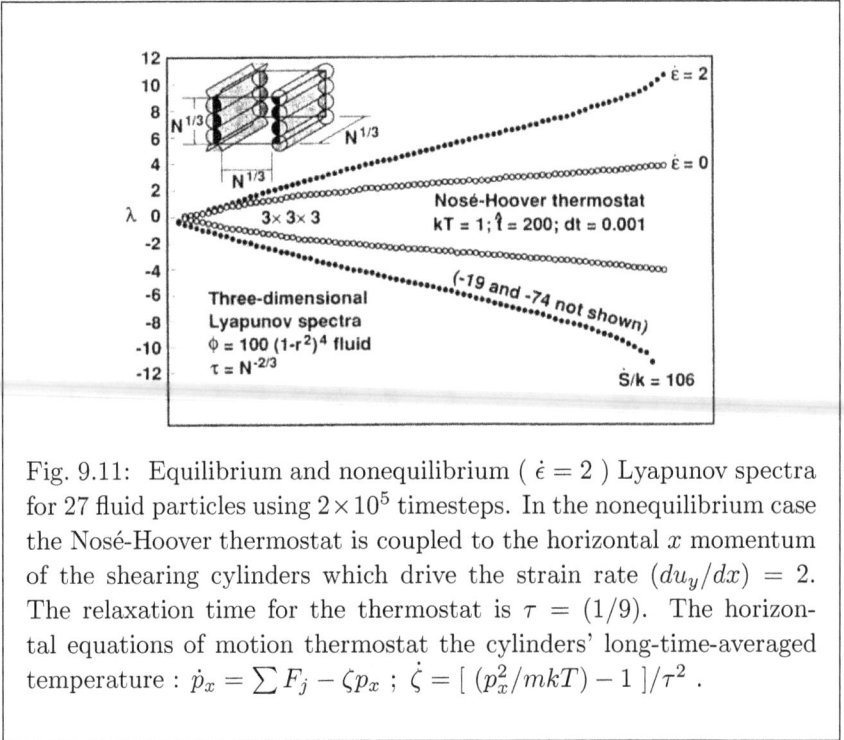

Fig. 9.11: Equilibrium and nonequilibrium ($\dot{\epsilon} = 2$) Lyapunov spectra
for 27 fluid particles using 2×10^5 timesteps. In the nonequilibrium case
the Nosé-Hoover thermostat is coupled to the horizontal x momentum
of the shearing cylinders which drive the strain rate $(du_y/dx) = 2$.
The relaxation time for the thermostat is $\tau = (1/9)$. The horizontal equations of motion thermostat the cylinders' long-time-averaged
temperature : $\dot{p}_x = \sum F_j - \zeta p_x$; $\dot{\zeta} = [(p_x^2/mkT) - 1]/\tau^2$.

The viscous response of fluids to shear can be studied by moving boundary walls on opposite sides of a system in opposite directions. Two and
three-dimensional fluids behave similarly, typically showing enhanced instability and dissipation with a greater separation between positive and
negative exponents as the strain rate is increased. **Figure 9.11** shows
equilibrium and nonequilibrium Lyapunov spectra for a fluid with a strain
rate of $(du_y/dx) = 2$. The cylinders making up the walls normal to the x
axis move coherently, maintaining their wall-to-wall spacing as they move
oppositely in the y direction. The summed-up velocity in the z direction –
into the plane of the paper – is zero, and a constant of the motion.

The Lyapunov spectra for these boundary-driven sheared systems all shared the property that the dimensionality loss was relatively small. Only the last one or two of the negative exponents carried most of the burden of the dissipation. We see that in **Figure 9.11** the last two Lyapunov exponents, -19 and -74, together account for $93/106$ of the dissipation rate $(\dot{S}/k) = 106$. The dimensionality losses for these boundary-driven shear flows never exceeded the extra dimensionality associated with the boundary conditions, $+3$ in two dimensions and $+4$ in three dimensions. In the three-dimensional case the additional phase-space dimensions correspond to the x and y coordinates of the moving cylinders, their fluctuating momentum p_x, and the friction coefficient ζ. The Nosé-Hoover friction coefficient ζ extracts the viscous work done by the boundary forces.

As computer capacity increased, larger and more complex simulations became possible. By 1994, five years later, Harald and I had obtained clearcut evidence that the dimensionality loss at a particular strain rate was *extensive*, proportional to the system size.[13] This demonstrated that our fractal explanation of the Second Law of Thermodynamics was valid for simple shear flows.

The steady flow of heat can likewise be studied with molecular dynamics, sandwiching a Newtonian region between hot and cold boundaries. We will take up such simulations in the next Lecture, illustrating solid-phase heat conduction with the ϕ^4 model Carol discussed in **Lecture 2**. In simulations using that model the particles are tethered to regular lattice sites. Fourier's Law describes their response to temperature gradients quite well.

So far we have seen that nonequilibrium steady states lose dimensionality due to continuing dissipation, converting phase volume to extracted heat, induced by the time-reversible deterministic thermostats. After an equilibrium interlude we will return to the nonequilibrium Baker Map in which similar dissipative phenomena can be seen and understood with a simpler two-dimensional model. We begin by reviewing the equilibrium version, a mapping which preserves area and which is ergodic.

9.6   Equilibrium Baker Map

Remember[14, 15] that it was necessary to rotate the Baker Map 45° in order that it would satisfy our notion of time-reversibility, with $B^{-1} = TBT$. After a [1] mapping forward with B, [2] reversing the time T, [3] again applying the mapping (going backward in time) B, and finally [4] changing the sign of the momentum again, with T, returns us to the initial

(q, p) state. The equilibrium version of the area-conserving Baker Map is as follows :

$$q' = (5q - 3p)/4 \pm \sqrt{9/8} \; ; \; p' = -(3q - 5p)/4 \mp \sqrt{1/8} \; ;$$

with signs for the square roots $(+, -)$ when $(q < p)$ with signs $(-, +)$ when $(q > p)$. **Figure 9.12** shows 500 equilibrium points generated by the Baker Map. The coverage already looks relatively uniform with 250 points started from $(+0.6, +0.8)$ and 250 from $(+0.6, -0.8)$. Notice the inversion symmetry equivalent to watching the development of the map in a mirror (which reflects both q and p).

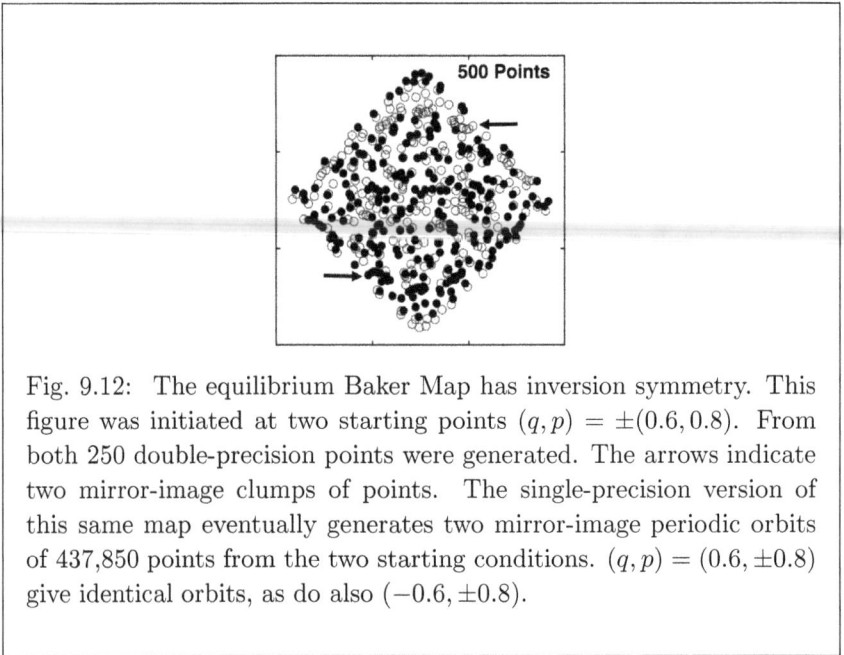

Fig. 9.12: The equilibrium Baker Map has inversion symmetry. This figure was initiated at two starting points $(q, p) = \pm(0.6, 0.8)$. From both 250 double-precision points were generated. The arrows indicate two mirror-image clumps of points. The single-precision version of this same map eventually generates two mirror-image periodic orbits of 437,850 points from the two starting conditions. $(q, p) = (0.6, \pm0.8)$ give identical orbits, as do also $(-0.6, \pm0.8)$.

Let us see if the equilibrium Baker Map really is reversible. Before going on to the next iteration we apply the operations TBT to see whether or not the previous (q, p) is recovered. Because the orbit is in principle and even in practice periodic we need first to compute the single-precision period. Starting with $(q, p) = (0, 0)$ we find that the periodic orbit length is 437,850 of which just 99,602 steps are truly reversible – 22.75 percent of the total. **Figure 9.13** shows the locations of the Baker Map points that *are* reversible in the sense that BTBT really *is* the identity map.

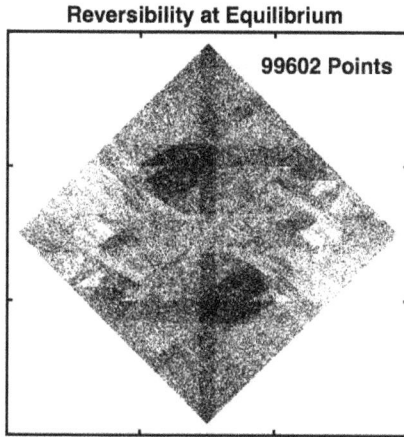

Reversibility at Equilibrium

99602 Points

Fig. 9.13: Starting with $(q, p) = (0.6, 0.8)$ and using single-precision arithmetic, the equilibrium Baker Map is attracted to a periodic orbit of length 437,850. The 99602 points for which the map is reversible, with TBT recovering the previous (q, p) are plotted. The rounding errors vary so as to suggest the inversion symmetry of the map but close inspection shows that the reversible points do not show this symmetry.

Figure 9.13 shows the reversible equilibrium states for the Baker Map. Although the pattern suggests inversion symmetry, as might be expected for the linear mapping, there is no such symmetry when the points are examined in detail. Even for this simplest area-preserving "equilibrium" map only a tiny fraction of the square is sampled and only about one fifth of the sampled points in a representative periodic orbit turn out to be reversible.

Exercise : The equilibrium Baker Map has mirror symmetry, suggesting that the initial point (0,0) would be quite interesting to investigate. Do ! Next consider breaking the symmetry by choosing different paths. Use either $(q > p)$ with $(q \le p)$ or $(q \ge p)$ with $(q < p)$. This exercise shows that simulation is often more reliable than formal theory.

9.7   Nonequilibrium Baker Map

Let us consider again the nonequilibrium, overall compressive Baker Map detailed on page 251. **Figure 9.14** shows the time-reversible points (where TBT recovers the previous point precisely) generated by the nonequilibrium map.

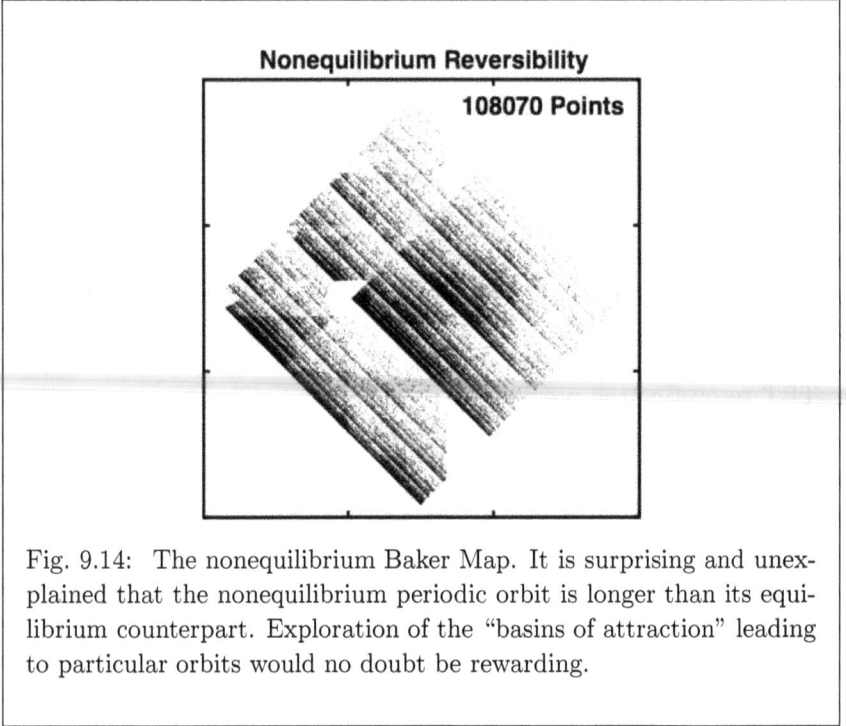

Fig. 9.14: The nonequilibrium Baker Map. It is surprising and unexplained that the nonequilibrium periodic orbit is longer than its equilibrium counterpart. Exploration of the "basins of attraction" leading to particular orbits would no doubt be rewarding.

In the nonequilibrium Baker Map of **Figure 9.14** the overall action of the mapping is to *compress*. Phase volume is reduced, on average, by $2^{1/3}$ with each iteration of the map. Gibbs' entropy is reduced by $(1/3) \ln 2$, on average, through long series of twofold compressions and expansions. The series are typically finite, but long, with the length varying as a fractional power of the total number of phase space states. Using single-precision arithmetic two different initial conditions for the nonequilibrium map, $(+0.6, +0.8)$ and $(-0.6, -0.8)$, eventually lead to exactly the same periodic orbit, a sequence of 157,342 distinct mappings. The asymmetric initial conditions, $(+0.6, -0.8)$ and $(-0.6, -0.8)$, lead instead to a much longer (no doubt the longest for single precision) periodic orbit, of length 1,042,249. This latter length 1,042,249 exceeds the equilibrium Baker one. 108,070 points on the orbit are reversible, 10.37 percent of the total. On the shorter cycle of 157,342 points 16,075 are reversible, **Figure 9.14** shows those points which are time-reversible, 10.22 percent of the total. Remarkably, using the four sign combinations with initial conditions of $(\pm 0.3, \pm 0.5)$ gave three periodic orbits of length 108,070 and one of length 16,075. Evidently the distribution of orbits is relatively complicated.

To investigate this a bit more I considered the 25 periodic orbits reached from initial conditions of { 0.1, 0.2, 0.3, 0.4, 0.5 } for both x and y. 21 of the resulting periodic orbits had the maximum length found, 1,042,249; two had length 157,342; and two had length 76,328. The number of moves required to reach a periodic orbit varied from 12,197 starting at (0.2,0.3) to 1,465,676 starting at 0.5,0.1), in both cases reaching a periodic orbit of length 1,042,249. The Snook Prize problem for 2017 considers a more complex double-pendulum problem in an eight-dimensional phase space with four pairs of exponents.[16]

9.8   Summary of Lecture 9

The chaotic springy pendulum, in both its Cartesian and Polar forms, presents the opportunity for studying pairing in a relatively simple dynamical system. The phase-space distribution is relatively complicated, with tori that pierce the $x = 0$ section many times. The Lyapunov spectrum shows the instability of phase-space volumes of dimensionality from 1 to the full dimensionality of the space. It is noteworthy that the symmetry of the motion equations is broken. We saw that the Baker Maps are seldom reversible for even a single step. Numerical simulations follow models of nature, going in the direction with the greatest number of available states.

Manybody simulations provide Lyapunov spectra which resemble the spectrum of vibration frequencies measured for solids, though with less structure and with exponents closer to the Einstein frequency. The dimensionality loss for color conductivity simulations can be much larger than that found for shear flows. Nevertheless the numerical evidence shows that the loss of dimensionality retains significance for larger systems and provides an explanation for the rarity of nonequilibrium states. The Baker Map provides a simple two-dimensional illustration of fractal distributions and of the lack of reversibility both at and away from equilibrium.

References

1. J. C. Sprott, *Elegant Chaos – Algebraically Simple Chaotic Flows* (World Scientific, Singapore, 2010).
2. Wm. G. Hoover, C. G. Hoover, and H. A. Posch, "Lyapunov Instability of Pendulums, Chains, and Strings", Physical Review A **41**, 2999-3004 (1990).
3. Wm. G. Hoover and H. A. Posch, "Direct Measurement of Lyapunov Exponents", Physics Letters A **113**, 82-84 (1985).
4. Wm. G. Hoover and H. A. Posch, "Direct Measurement of Equilibrium and Nonequilibrium Lyapunov Spectra", Physics Letters A **123**, 227-229 (1987).
5. I. Shimada and T. Nagashima, "A Numerical Approach to Ergodic Problems of Dissipative Dynamical Systems", Progress of Theoretical Physics **61**, 1605-1616 (1979).
6. G. Benettin, L. Galgani, A. Giorgilli, and J-M. Strelcyn, "Lyapunov Characteristic Exponents for Smooth Dynamical Systems and for Hamiltonian Systems; a Method for Computing All of Them", Parts I and II: "Theory and Numerical Application", Meccanica **15**, 9-20 and 21-30 (1980).
7. Wm. G. Hoover, C. G. Hoover, and F. Grond, "Phase-Space Growth Rates, Local Lyapunov Spectra, and Symmetry Breaking for Time-Reversible Dissipative Oscillators", Communications in Nonlinear Science and Numerical Simulation **13**, 1180-1193 (2008).
8. B. Moran, Wm. G. Hoover, and S. Bestiale, "Diffusion in a Periodic Lorentz Gas", Journal of Statistical Physics **48**, 709-726 (1987).
9. Wm. G. Hoover, H. A. Posch, B. L. Holian, M. J. Gillan, M. Mareschal, and C. Massobrio, "Dissipative Irreversibility from Nosé's Reversible Mechanics", Molecular Simulation **1**, 79-86 (1987).
10. H. A. Posch and Wm. G. Hoover, "Chaotic Dynamics in Dense Fluids", Proceedings of the Europhysics Liquid State Conference on Liquids of Small Molecules, editted by M. Nardone, **11G**, 17-18 (1987).
11. H. A. Posch and Wm. G. Hoover, "Equilibrium and Nonequilibrium Lyapunov Spectra for Dense Fluids and Solids", Physical Review A **39**, 2175-2188 (1989).
12. C. Forster, R. Hirschl, and H. A. Posch, "Analysis of Lyapunov Modes for Hard-Disk Fluids", pages 423-431 in Proceedings of the XIVth Conference on Mathematical Physics, editted by J. C. Zambrini (Lisbon, 28 July to 2 August, 2003), (World Scientific Publishing, Singapore, 2005).
13. Wm. G. Hoover and H. A. Posch, "Second-Law Irreversibility and Phase-Space Dimensionality Loss from Time-Reversible Nonequilibrium Steady-State Lyapunov Spectra", Physical Review E **49**, 1913-1920 (1994).
14. Wm. G. Hoover and H. A. Posch, "Chaos and Irreversibility in Simple Model Systems", Chaos **8**, 366-373 (1998).
15. J. Kumičák, "Irreversibility in a Simple Reversible Model", Physical Review E **71**, 016115 (2005).
16. Wm. G. Hoover and C. G. Hoover, "Instantaneous Pairing of Lyapunov Exponents in Chaotic Hamiltonian Dynamics and the 2017 Ian Snook Prizes", Computational Methods in Science and Technology **23**, 73-79 (2017).

Chapter 10

Lyapunov Instability, Fractals, Chaos II

/ Quotes from Alghero, Sardinia, 15 - 17 July 1991 / Galton Board Evolution and Finite-Precision Stationary States / Galton Board Isomorphisms and Fluctuations / Baker Maps at and away from Equilibrium / Baker Map and the Fluctuation Theorem / Dimensionality Loss in 2D Maps and Particulate Flows / 0532 Model Spectra at and away from Equilibrium / ϕ^4 Model Spectra and Dimensionality Loss / Summary of Lecture 10 /

10.1   Setting the Stage

What have we learned about nonequilibrium systems large and small? First of all, we have learned that deterministic thermostats make it possible to model temperature and thermodynamic dissipation is an efficient way, without the need for *random* numbers. We recommend a deterministic approach. Determinism facilitates both portability and reproducibility, qualities essential to the reproducible transmission, maintenance, and reproduction of knowledge. As recently as 1991 the idea that phase volume could collapse onto strange attractors was viewed as damningly artificial by those whose background knowledge predated computers and was based on the Boltzmann equation and its most famous consequence, the H Theorem.

Here is a quote from my remarks at a summertime 1991 discussion meeting on "Microscopic Simulation" in Alghero, on the island of Sardinia[1] : "In the steady state it should be obvious that the [comoving] phase volume, if it is changing, can only get smaller ... One of the characteristics of the attractors is that Gibbs' entropy always diverges in the nonequilibrium case, always minus infinity." Eddie Cohen's response "I do not think so" and Joel Lebowitz' "Boundary conditions [are] only effective at the boundary", were symptomatic of a common reaction to new ideas. Leo Tolstoy, discussing art, expressed this difficulty well (at least in an English translation) :

> "I know that most men, including those at ease with problems of the greatest complexity, can seldom accept even the simplest and most obvious truth if it be such as would oblige them to admit the falsity of conclusions which they have delighted in explaining to Colleagues, which they have proudly taught to others, and which they have woven, thread by thread, into the fabric of their lives."

In 1987 it had become clear to me and many of my own colleagues that the best approach to temperature control was to control the kinetic temperature with feedback. Gauss' thermostat control does this with "differential feedback" — forcing the time derivative of the kinetic energy to vanish. Nosé-Hoover thermostat control uses "integral feedback" — the integrated time-averaged kinetic energy is specified. The fact that both the Gauss and the Nosé-Hoover motion equations are time-reversible was an unexpected educational bonus. Reversibility meant that the old paradoxes generated by Boltzmann and his colleagues now had a real chance of resolution. *Zermélo's recurrence objection* [that irreversibility precludes recurrence] disappears once nonequilibrium steady states are considered. We will see that despite the time-reversibility of the equations of motion the repellor violating the Second Law has a probability so small as to make it unobservable. We have seen that the reversal of just a single Baker Map step only succeeds ten or twenty percent of the time. It is for this very same reason that *Loschmidt's reversal objection* is defeated. It is true that Levesque and Verlet's bit-reversible algorithm, as well as Milne's fourth-order predictor for Newtonian motion equations,[2] allow *perfect* reversibility at equilibrium. However there is no such algorithm, nor any future prospect for one, where *nonequilibrium* simulations are concerned. Nonequilibrium steady states are inherently constrictive many-to-one mappings which are truly irreversible due to the loss of information.

The effective *irreversibility* of dissipative motion equations means one cannot go backward for very long. Precise reversal of even a single Baker-Map step is highly improbable. The 1987 fractal distributions[3] showed that a typical nonequilibrium phase space has three parts: [1] a fractal attractor made up of a negligible fraction of the states but accounting for nearly all the probability; [2] a mirror-image repellor, looking just like the attractor but with *zero probability*, in fact only attainable by an artificial storing and reversal of the attractor ; and [3] the overwhelming remainder, "equilibrium states", useful from the standpoint of linear-response theory and occupying most of the space, but negligible from the standpoint of nonequilibrium measure. Equilibrium states are useless in efforts to understand irreversibility beyond linear response in view of the singular fractal nature of nonequilibrium states.

To enhance our understanding of nonequilibrium systems models like the Baker Map (which illustrates the repellor/attractor mirror-image structure in its most transparent form), the heat-conducting oscillator (which is a flow rather than a map), and the ϕ^4 model (which shows how the boundary flows of heat do indeed penetrate into the interior to generate extensive dimensionality loss) illustrate the general principles in a transparent way. I will describe them all in this Lecture.

Although Hamiltonian systems are incapable of steady-state dissipation[4] Shuichi Nosé used a Hamiltonian basis to construct his thermostat idea. Further, about ten years later Carl Dettmann was able to show that a clever restriction of Hamiltonian mechanics, setting the Hamiltonian to zero, $\mathcal{H} \equiv 0$, produced the equivalent of Nosé-Hoover mechanics. And Nosé-Hoover mechanics *can* model steady-state dissipative systems.

Much more recently the Martyna-Klein-Tuckerman,[5] Hoover-Holian,[6] 0532,[7] and Logistic-thermostat models[8,9] have provided understanding of ergodic nonequilibrium models with three- or four-dimensional phase spaces which can readily be visualized. This understanding completes one project, producing an ergodic canonical dynamics, set in motion by Nosé's 1984 work. Although there is still much more to know, in particular the details of the mechanisms through which attractors attract, the outlines are at present pretty clear. In this Lecture we will reconsider the Galton Board and its relation to Hamiltonian mechanics. We will reexamine the Baker Map and the complexity of its nonequilibrium structure. We will see how the ϕ^4 model for heat conduction leads naturally to the measurement and analysis of substantial dimensionality loss in nonequilibrium phase-space distributions. We will also see how the newer thermostat ideas have enhanced our ability to visualize and understand the structure of nonequilibrium systems. Finally I will summarize what we know of the nature of nonequilibrium states and their simulation.

10.2   The Isokinetic Galton Board *as* a Hamiltonian System

Although the fractal structures generated by the Galton Board might seem "artificial" at first glance, it is possible to understand them from a basis

in Hamiltonian mechanics. If one asks for the form of external field which
causes a mass point to follow the same trajectory as a Gaussian isokinetic
particle "it can be shown" that an *exponential potential* does this.[10] The
exponential form makes it possible for the field strength to increase in just
such a way as to keep the ratio of the field's potential energy to the particle's
kinetic energy constant. The isokinetic ($p_x^2 + p_y^2 \equiv 1$) motion equations in
a field of strength unity are as follows :

$$\dot{x} = p_x \; ; \; \dot{y} = p_y \; ; \; \dot{p}_x = 1 - \zeta p_x \; ; \; \dot{p}_y = -\zeta p_y, \; [\text{ with } \zeta \equiv p_x] \; .$$

Exactly the same trajectory results (see **Figure 10.1**) from the exponentially increasing Hamiltonian $\mathcal{H} = (p^2/2) - (1/2)e^{2x}$:

$$\dot{x} = p_x \; ; \; \dot{y} = p_y \; ; \; \dot{p}_x = e^{2x} \; ; \; \dot{p}_y = 0 \; .$$

In both cases, isokinetic and exponential, the impulsive momentum changes
caused by hard-disk scatterers can be appended to the equations of motion.

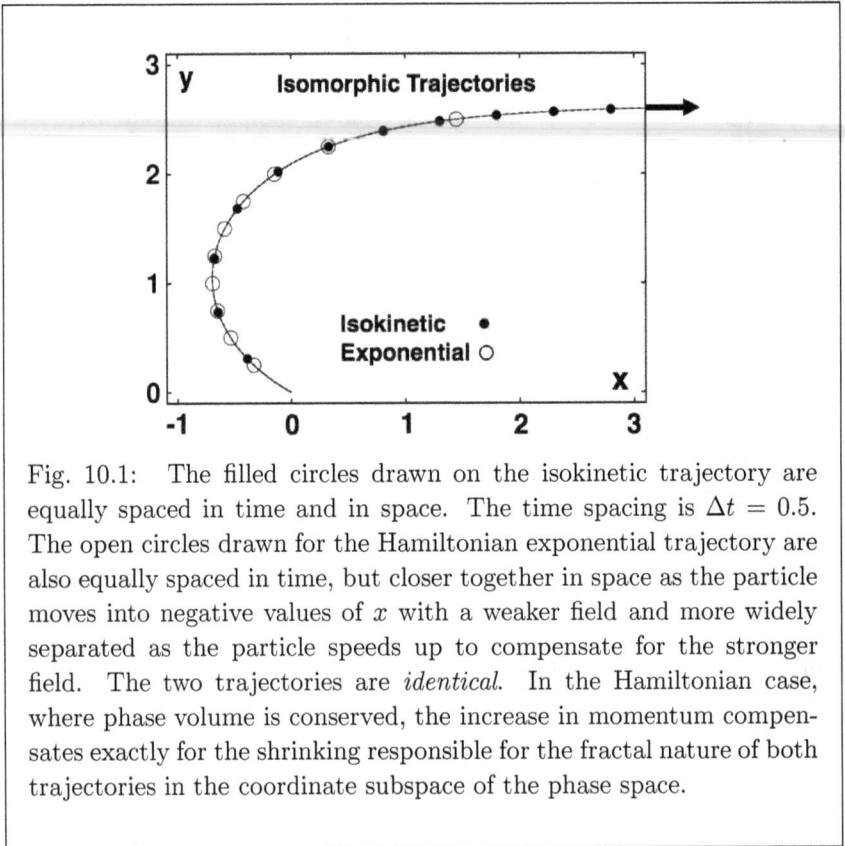

Fig. 10.1: The filled circles drawn on the isokinetic trajectory are
equally spaced in time and in space. The time spacing is $\Delta t = 0.5$.
The open circles drawn for the Hamiltonian exponential trajectory are
also equally spaced in time, but closer together in space as the particle
moves into negative values of x with a weaker field and more widely
separated as the particle speeds up to compensate for the stronger
field. The two trajectories are *identical*. In the Hamiltonian case,
where phase volume is conserved, the increase in momentum compensates exactly for the shrinking responsible for the fractal nature of both
trajectories in the coordinate subspace of the phase space.

Exercise : Show, analytically as well as numerically, that a Hamiltonian particle with the equation of motion $\ddot{x} = e^{2x}$ of **Figure 10.1** arrives at $x = +\infty$ at a finite time.

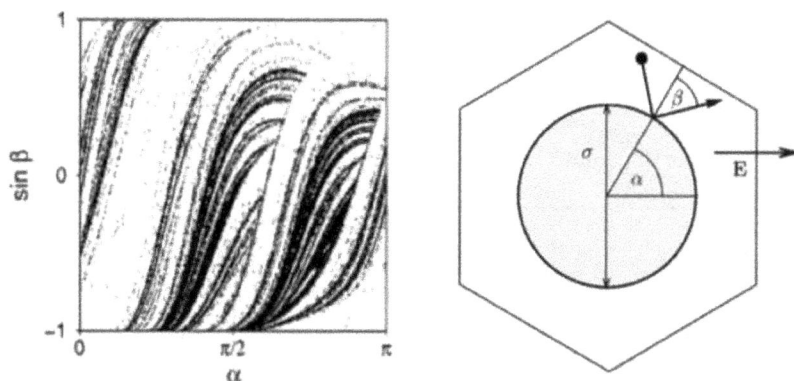

Fig. 10.2: A Galton Board cross section of the type used to study periodic-orbit lengths corresponds to the relatively strong field, $E = 4(p^2/m\sigma)$. The rectangular cells used to implement finite precision are rectangles subdividing the complete cross-sectional area into an $n \times n$ gridwork. The sidelengths of the complete section are $0 < \alpha < \pi$ for the abscissa and $-1 < \sin(\beta) < +1$ for the ordinate. In these simulations carried out with Christoph Dellago the external field is *parallel* to a row of scatterers rather than perpendicular, as was assumed in simulations whose cross sections are illustrated in **Figure 7.7**, page 240.

In Sections 5.9.1 of our two 1999/2012 Time Reversibility books,[11] we showed how the Galton Board attractive section gains in detail as digits are added to the resolution. Finite precision guarantees that any trajectory will end up in a recurring periodic orbit. With Christoph Dellago, who had done his 1995 Ph D work with Harald Posch on Galton-Board Lyapunov exponents, I studied the systematic dependence of the periodic-orbit length as well as the length of the transient approach to it on the finite precision of the simulation program.[12]

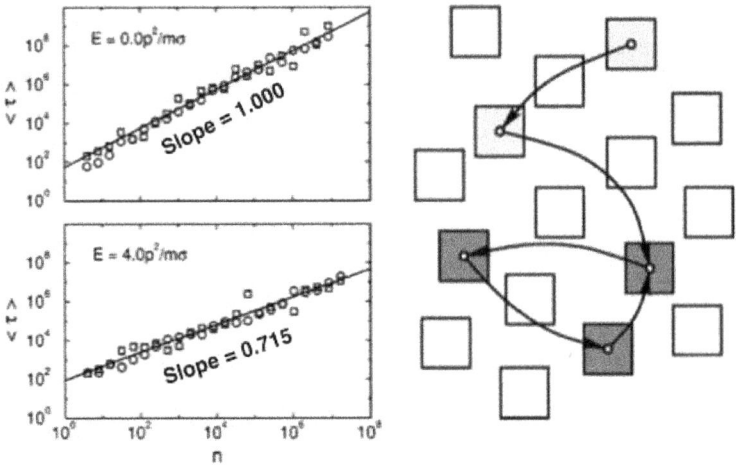

Fig. 10.3: To the left we see that the lengths of transient (circles) and periodic-orbit (squares) trajectories τ vary as the square root of the number of states available. We divide the Galton-Board cross section of **Figure 10.2** into $n \times n$ cells. All of them are equally accessible at equilibrium so that the transient and periodic orbits have a length proportional to $\sqrt{n \times n}$ corresponding to a slope of unity on the log-log plot. At a field strength of $4(p^2/m\sigma)$ the attractor's correlation dimension is $D_C = 1.43$ so that the number of states is of order $n^{1.43}$, corresponding to a slope of 0.715 for that nonequilibrium log-log plot. The largest number of cells in these year-2000 simulations was 2^{46}. To the right is a small caricature of a "statistical model" for understanding the length of the transients and periodic orbits in a space with a finite number of states. The light grey squares represent the transient states, just two in number in our illustration, while the three darker squares represent the ultimate periodic orbit. This seems an apt model for systems that are both chaotic and ergodic, like the Galton Board and the Baker Map. It fits the numerical simulation data very well.

Our first observation was that both these lengths had a power-law relationship to the resolution, the number of digits used in the cross-sectional plots of $\sin(\beta)$ *versus* α. See **Figure 10.2** for the definitions of these two angles and **Figure 10.3** for straight-line plots of the orbit lengths (transient and periodic) with no field ($E = 0$) and with a strong field, $E = 4(p^2/m\sigma)$. The two lengths varied inversely with the cell size at equilibrium and as the cell size to the 0.715 power with the strong nonequilibrium field. Our "discovery" fitting both results, was that the power-law linking the inverse cell size to the trajectory lengths was precisely half the correlation dimension of the fractal section. Our discovery was actually a repeat of Grebogi, Ott, and Yorke's discovery a decade earlier.[13] These results follow from an elaboration of the "Birthday-Problem" reasoning.

Going back to the "Birthday Problem", where the number of pairs of people must be of order 365 for it to be likely to find two with the same Birthday, exactly the same reasoning leads to the conclusion that a typical periodic-orbit length will be of order the square root of the number of pairs of phase-space points. The number of pairs of points within a radius r defines the correlation dimension, which simulations show is 1.43 for the relatively strong field of $4(p^2/m\sigma)$. The chaotic cross section for this field is shown in **Figure 10.2**. It is interesting to see that the entropy of an equilibrium manybody system, $k \ln \Omega$, where Ω is the accessible phase volume or the accessible number of states, is actually much less than that following Gibbs' classical analysis. The number of states on an equilibrium periodic orbit varies as $\sqrt{\Omega}$, where Ω is the actual number of plausible phase-space states. This difference looks like an appropriate topic for pedagogical explorations of the applicability of ergodic dynamics concepts to equilibrium [and nonequilibrium] statistical mechanics.

10.2.1   *A Statistical Model for the Generation of Fractals*

A statistical *model*[12,13] consistent with the slopes in **Figure 10.3** is based on random jumps from one cell to another with a total of N cells. The matrix M describing the jumps { $I \rightarrow J$ } has one random entry per row, and can include the possibility that the jumper is "stuck", $M(I, I) = 1$. The mean value of the trajectory length for this model approaches $\sqrt{(N\pi/2)}$, equally divided between the jumps included in the transient approach, and those of the ultimate stable periodic orbit. In such a large-N limit the probability of finding n trajectories, all of length N, is e^{-n} times that of finding the first, so that it is unlikely to find even a second trajectory of the same length as the first (unless, of course, the problem has two- or more-fold symmetry).

10.3   Variations on the Galton Board Problem

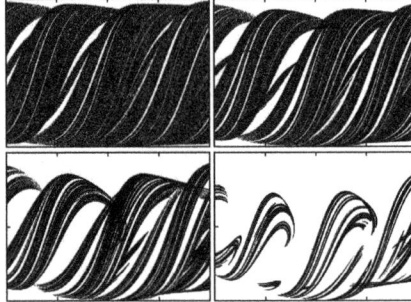

Fig. 10.4:   Here, the magnitude of the velocity in a Galton Board
is unconstrained so that the phase space is four-dimensional rather
than three. The field here and the viscosity are equal, with values
(1/8), (1/4), (1/2), and 1, with 300,000 successive collisions plotted.
The lowest field and viscosity combination are in the upper left corner
with the greatest in the lower right corner. Projections are shown of
the four-dimensional trajectories onto the two-dimensional ($\alpha, \sin(\beta)$)
plane.

Variations on the isokinetic Galton Board problems can be implemented
by using variations of the one-dimensional thermostats, controlling $(p_x^2 + p_y^2)$
and $(p_x^2 + p_y^2)^2$ with integral feedback for example. A constant viscous drag
force, $-\zeta p$, can provide the simplest [though irreversible] control type :

$$\dot{p}_x = F_x + E - \zeta p_x \; ; \; \dot{p}_y = F_y - \zeta p_y \longrightarrow (\dot{\otimes}/\otimes) = -2\zeta \ .$$

See **Figure 10.4**. The four-dimensional phase space, reduced to three by
considering the collisional [$\sin(\beta) \times \alpha$] cross section appears to have at least
one positive Lyapunov exponent for the combinations of field and viscosity
shown in the Figure. There is certainly a zero exponent, corresponding to
the direction of the trajectory. It appears that the fractals (if they *are*
fractal) are definitely not ergodic. Investigation of the details looks like
an interesting computational project, requiring, as it does, the treatment
of the impulsive collisions as a map, with the remainder of the analysis
applying to a flow in the four-dimensional (x, y, p_x, p_y) space.

10.4 The Baker Map, Nosé-Hoover, and The Second Law

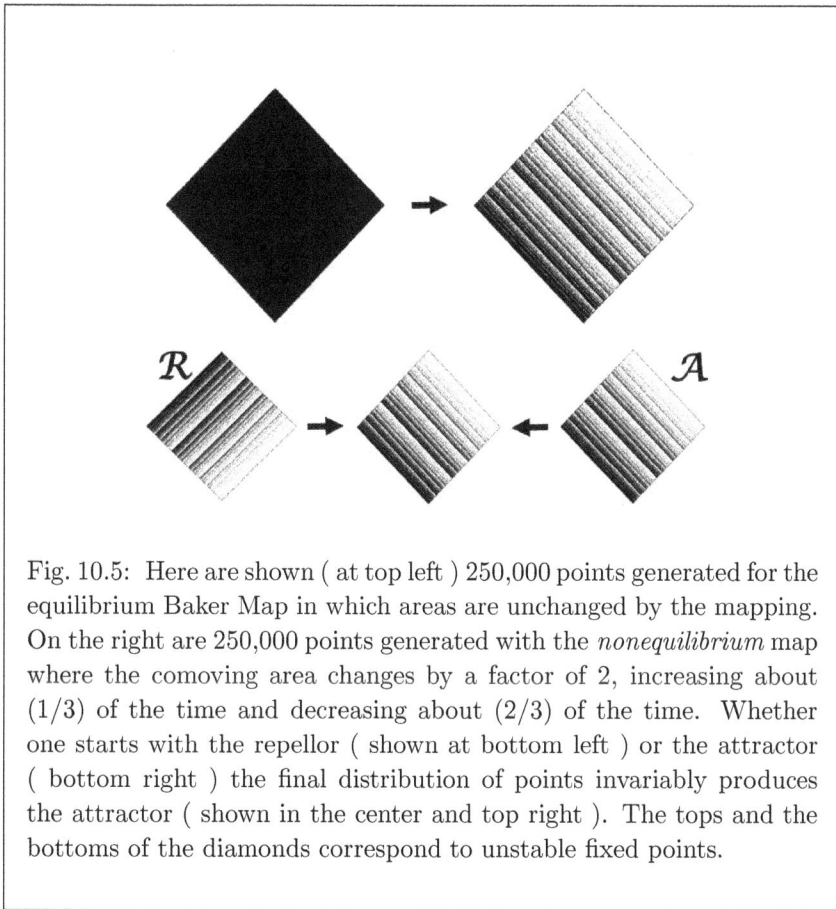

Fig. 10.5: Here are shown (at top left) 250,000 points generated for the equilibrium Baker Map in which areas are unchanged by the mapping. On the right are 250,000 points generated with the *nonequilibrium* map where the comoving area changes by a factor of 2, increasing about (1/3) of the time and decreasing about (2/3) of the time. Whether one starts with the repellor (shown at bottom left) or the attractor (bottom right) the final distribution of points invariably produces the attractor (shown in the center and top right). The tops and the bottoms of the diamonds correspond to unstable fixed points.

The equilibrium Baker Map has a chaotic dynamics of the simplest kind, with infinitesimal areas stretching, and doubling in size, parallel to $q+p = 0$ and halving parallel to $q - p = 0$. A constant density within the 2×2 diamond-shaped domain of **Figure 10.5** is unchanged by the equilibrium mapping. The mapping is apparently ergodic, apart from specially-selected points. Just as in the Galton Board the actual trajectory consists of a transient followed by a periodic part.

If the Map starts at (0.6,0.8) the period is 437,850 using single-precision arithmetic and the following equilibrium form of the map :

```
IF(Q.LT.P) THEN
    QNEW = +1.25*Q - 0.75*P + SQRT(1.125)
    PNEW = -0.75*Q + 1.25*P - SQRT(0.125)
ENDIF

IF(Q.GT.P) THEN
    QNEW = +1.25*Q - 0.75*P - SQRT(1.125)
    PNEW = -0.75*Q + 1.25*P + SQRT(0.125)
ENDIF
```

With (0.6,0.8) point number 1, points 774,913 and 1,212,763 are the first pair to agree and are separated by a periodic orbit with a cycle length of 437,850 points. The nonequilibrium Baker map has changes in area with the southeastern two-thirds of the diamond compressing and the northwestern one-third expanding, in both cases by factors of 2. *A priori* it isn't obvious whether or not the result will be a strange attractor (it is). We saw in **Lecture 9** that the *non*equilibrium Baker Map has a cycle length of 157,342 using the (0.6,0.8) initial condition but that a much longer cycle with length 1,042,249 is actually much more likely.

The Nosé-Hoover oscillator is a similar case. In half of the (q, p, ζ) space, wherever ζ is positive, the flow is contracting. In the other negative-ζ half space the flow expands. Liouville's continuity equation shows that a simple Gaussian is a stationary solution. In addition to Gibbs' Gaussian distribution in (q, p) the friction coefficient ζ likewise has a Gaussian distribution. The "catch" is that there is no mixing between the chaotic sea, which is a three-dimensional "fat fractal" and the infinitely many tori which surround stable periodic orbits and penetrate the sea. By incorporating a temperature gradient in this model the fat-fractal sea is converted into a fractal strange attractor of the type shown in **Figure 10.6**.

The difficulty of predicting whether or not such a "strange attractor" forms just by looking at motion equations is apparently insuperable. Consider the heat-conducting Nosé-Hoover oscillator,

$$\dot{q} = p \; ; \; \dot{p} = -q - \zeta p \; ; \; \dot{\zeta} = p^2 - T(q) \text{ with } T = 1 + \epsilon \tanh(q) \; .$$

In this case, where there is a maximum temperature gradient ϵ there is the possibility for heat flow with a nonzero current $\langle (p^3/2) \rangle$. What actually happens depends entirely on the initial conditions. With a maximum temperature gradient $(dT/dx)_{max} = \epsilon$ and an initial condition within the chaotic sea (see **Figure 10.6**) there *is* a nonzero current (on the average in

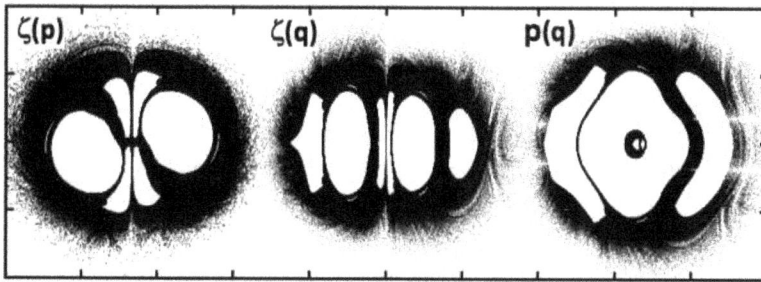

Fig. 10.6: Nosé-Hoover conducting oscillator sections where the temperature is $1 + (1/4)\tanh(q)$, giving rise to a heat current $\langle\,(p^3/2)\,\rangle =$ -0.0016 in the chaotic sea. The initial conditions for these chaotic sections were $(q, p, \zeta) = (0.0001, 0.0001, 0.0001)$. The spacing of the ticks is 3, with the first and last sections displaced by ∓ 10. The "holes" contain conservative tori which disappear when the hyperbolic tangent multiplier exceeds 0.4054.

the direction hot-to-cold), a fractal phase-space distribution, and an overall positive conductivity. For moderate values of ϵ the dynamics produces two different kinds of results : [1] a vanishing mean current with the dynamics restricted to a conservative torus or [2] a hot-to-cold current with a positive conductivity and with the dynamics restricted to the fractal chaotic sea. The boundary between these two solution types, insulating and conducting, is highly complex. There simply is no way to know what will happen without actually looking. Careful inspection[14] shows the persistence of the chaotic heat-conducting sea for maximum temperature gradients in the range $0 < \epsilon < 0.4054$. For larger gradients there is a one-dimensional periodic limit-cycle solution. The mean current on the limit cycle at $\epsilon = 1$ is $\langle\,(p^3/2)\,\rangle = -0.7438$.

10.4.1 *More Details of Nonequilibrium Baker Maps*

It is easy to see that the stretching in the nonequilibrium Baker Map gives a probability of $(2/3)$ for the occupation of the southwestern third of the square. To see the source of the next band of states in the map, $-(1/3) < x < +(1/9)$ we can map backwards using $B^{-1} = TBT$. The source of the band one step backward in time is a diamond-shaped square of sidelength $(2/3)$ with its westernmost point at $(q, p) = (-\sqrt{2}, 0)$. Two steps back the source is simply the set of repellor states $\{q, -p\}$ corresponding to the $\{q, +p\}$ band. The topology of the Baker Map is not complicated. The "same" steady-state distribution can alternatively be generated by a *stochastic* dynamics for (x, y) where R = RUND(INTX,INTY) and Y = 2*RUND(INTX,INTY) - 1. Each pair of (x, y) points in **Figure 10.7** was generated by using "random" numbers from our simple generator RUND(INTX,INTY) which we described on page 116. Values of x are generated by the "stochastic dynamics" :

```
if(R.GT.(2/3)) XNEW = (2*X+1)/3
if(R.LT.(2/3)) XNEW = (X - 2)/3
```

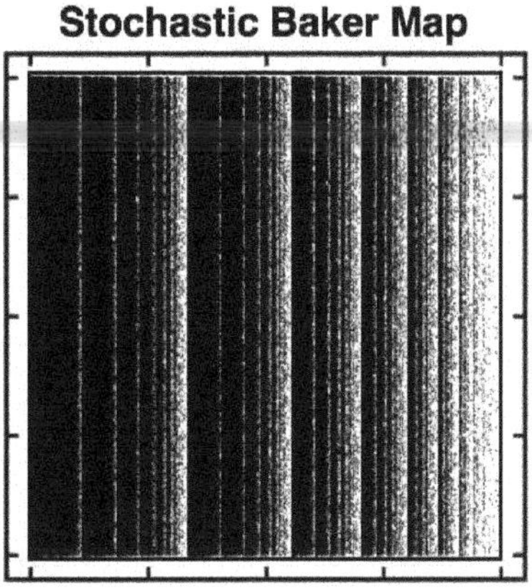

Stochastic Baker Map

Fig. 10.7: Here we see 2^{21} points from the *stochastic* Baker Map with the initial condition $x = 0.6$. We use the random number generator from **Lecture 3**, RUND(INTX,INTY) beginning with INTX = INTY = 0. The points could equally-well have been generated from the deterministic Baker map. The horizontal and vertical coordinates here, $-1 < (x, y) < +1$, are related to those in the diamond configuration through the definitions $\{q \equiv (x - y)/\sqrt{(2)} \; ; \; p \equiv (x + y)/\sqrt{(2)} \}$.

I choose random y values in order to avoid the hassle of setting out a histogram (and to facilitate a comparison with the deterministic version of the Map). **Figure 10.7** shows 2^{21} points in this (x, y) version of a "stochastic" nonequilibrium Baker Map. Using our random number generator provides a cycle length of 2^{21} as our generator produces 2^{22} random numbers before repeating. Two of these numbers are used for each (x, y) point. Joseph Ford was fond of pointing out that "deterministic chaos" is indistinguishable from stochastic chaos. The Baker Map example demonstrates this point.

No doubt the Baker Map could (with considerable effort) be made fully-reversible by incorporating the reversible random number generator discovered by Federico Ricci-Tersenghi and described in a pair of arχiv papers.[15, 16] **Figure 10.8** demonstrates the validity of a probabilistic analysis of the Baker Map where (2/3) of the jumps are compressive and only (1/3) are expansive.

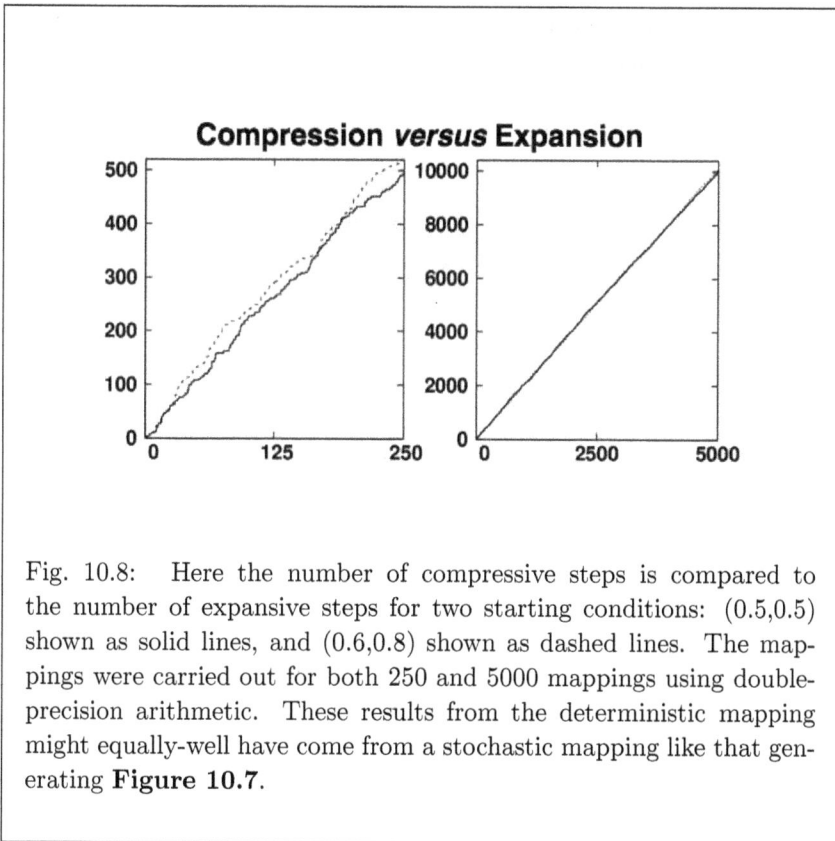

Fig. 10.8: Here the number of compressive steps is compared to the number of expansive steps for two starting conditions: (0.5,0.5) shown as solid lines, and (0.6,0.8) shown as dashed lines. The mappings were carried out for both 250 and 5000 mappings using double-precision arithmetic. These results from the deterministic mapping might equally-well have come from a stochastic mapping like that generating **Figure 10.7**.

10.4.2   *Deterministic Map – Convergence and Reversibility*

The strange attractor generated by the nonequilibrium Baker map owes its simplicity to the longtime two-to-one ratio of compressions to expansions. In **Figure 10.9** I compare the reversibilities of the Single-, Double-, and Quadruple-precision maps, for times of 25, 50, and 100 respectively. Because 2^{25}, 2^{50}. and 2^{100} correspond roughly to the 8-digit, 15-digit, and 30-digit limits of the three algorithms' abilities to discriminate, it is expected to see reversal fail near the end of the three reversed sections of the simulations. The Lyapunov exponent for the mappings should provide quantitative estimates of reliability. In the nonequilibrium map $(1/3)$ of the measure is expanded threefold parallel to $q+p=0$ while $(2/3)$ is expanded by a factor of $(3/2)$. The largest Lyapunov exponent is the weighted average $\lambda_1 = (1/3)\ln(3) + (2/3)\ln(3/2) = (1/3)\ln(27/4) = 0.63651$. In double precision, with 52-bit accuracy we can make an estimate $2^{52} = e^{0.63651n}$ for the number of mappings required to destroy all memory of the initial state. The resulting estimate, $n = 57$ is overly optimisitic.

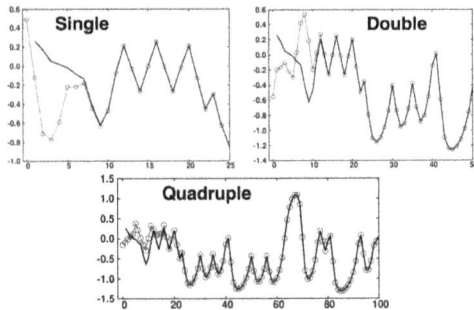

Fig. 10.9:   Here the forward and reversed Baker Map momenta are plotted as functions of the iteration number for single-, double-, and quadruple-precision arithmetic. In the three cases the initial condition was (0.6,0.8). The maps were run forward for 25, 50, and 100 iterations. Reversal was carried out by changing the sign of the momentum and proceeding "backward". The forward momenta (heavy lines) and the reversed backward momenta (light lines) are compared for the three precisions. They are consistent with the expected scaling of the reversal limit (for 18, 39, and 85 mappings respectively).

10.5 Baker Map Dynamics and the Fluctuation Theorem

In 1993 Denis Evans, Eddie Cohen, and Gary Morriss discovered an interesting symmetry, by now "well-known", in their isoenergetic studies of the periodic shear flow of 56 hard disks.[17] They kept track of the time-averaged Gibbs' entropy changes associated with the flow as a function of the averaging time τ. At a strainrate $\dot{\epsilon}$ and over a time window τ the distribution of entropy production rates approaches a smooth curve with a mean value, $\langle\,(\dot{S}/k) = (V/T)\eta\dot{\epsilon}^2\,\rangle_\tau$. The fluctuations about this mean necessarily satisfy the Central Limit Theorem for large τ and vary as $\sqrt{(1/\tau)}$.

Evans, Cohen, and Morriss stressed that both positive and *negative* values of the entropy production can be observed if [1] the system is not too large (56 soft disks in their case) and [2] τ is not too large (a few collision times). *At* equilibrium the positive and negative values even out over time. *Away from* equilibrium the positive values win out. The relatively few time intervals with negative values correspond to periods of overall entropy decrease. These unlikely fluctuations are reflected in the title of their paper, "Probability of Second Law Violations in Shearing Steady Flows". Their key insight was the definition of a nonequilibrium measure in terms of the local Lyapunov exponents.

I expected that an analog of their shear-stress fluctuations could be found in the qualitatively simpler dynamics of a nonequilibrium Baker Map. The simplicity of the map provides a worthwhile pedagogical approach to understanding the Evans-Cohen-Morriss "Fluctuation Theorem". The "Theorem" is not quite an identitity. It relates the probability of phase-space trajectory fragments forward-in-time for a time interval τ to the generally less-likely probability of observing these same fragments with the time order reversed :

$$[\,\mathrm{prob}_{\mathrm{forward}}(\tau)/\mathrm{prob}_{\mathrm{backward}}(\tau)\,] \simeq e^{\tau\dot{S}/k} \simeq e^{\Delta S/k}\,[\,\text{Fluctuation Theorem}\,].$$

I have used $\simeq$ rather than $=$ as reminders that the equalities describe limiting cases and are not valid for short sampling times, $\tau \simeq 0$.

For long times the theorem corresponds to a comparison of the zero-probability unstable repellor to the probability-one strange attractor generated by the forward mapping. It is interesting that Reference 7 of the Evans-Cohen-Morriss' paper describes the measurement of fluctuations for a (fractal) "Skinny Baker Map".[18]

The classic Baker Map maps the unit square $(0 < x, y < 1)$ onto itself with no change in area (a caricature of Liouville's phase volume conservation at equilibrium). A 45° rotation along with a fourfold expansion of a *generalized* map (in which the comoving area changes) gives a fractal distribution. The rotation, together with the definition of (q, p) coordinates relative to the center of the 2×2 square, gives a diamond-shaped time-reversible ergodic dissipative map B :

$$B(q, p) = (q_{new}, p_{new}) \overset{\text{reversible}}{\longleftrightarrow} B(q_{new}, -p_{new}) = (q, -p) \ .$$
$$q \propto (dx + dy) \ ; \ p \propto (dx - dy) \longrightarrow B(q, p) \ .$$

I prefer to consider a particular special case different to the Skinny and Classic maps. In our Baker Map case the comoving density is doubled two-thirds of the time and halved one-third. See again **Figure 10.5** on page 323. The single-step dynamics of the time-reversible nonequilibrium Baker Map B converts a (q, p) phase point to a new one, (q_{new}, p_{new}) with a factor-of-two change in the comoving area. The FORTRAN programming necessary to each iteration of the map $B(q, p \to q_{new}, p_{new})$ is as follows :

```
IF(Q.LT.P - SQRT(2.0/9)) THEN
    QNEW = +(11.0/6)*Q - (7.0/6)*P + SQRT(49.0/18)
    PNEW = +(11.0/6)*P - (7.0/6)*Q - SQRT(25.0/18)
ENDIF
```

```
IF(Q.GT.P - SQRT(2.0/9)) THEN
    QNEW = +(11.0/12)*Q - (7.0/12)*P - SQRT(49.0/72)
    PNEW = +(11.0/12)*P - (7.0/12)*Q - SQRT( 1.0/72)
ENDIF
```

In 1991, Bill Vance was interested in computing the relative probabilities of forward and reversed trajectories using unstable periodic orbits.[19] He showed me that it is easy to confirm the time-reversibility of the "rotated" Baker map $B(q, p \to q_{new}, p_{new})$. First, choose the point (q, p) within the 2×2 diamond-shaped domain of **Figure 10.10** and compute (q_{new}, p_{new}) using B. Then compute $B(q_{new}, -p_{new}) \to (q, -p)$ and confirm that the resulting point $(q, -p)$ is indeed the time-reversed image of the original (q, p) point.

Another point of view from which to analyze the rotated map B is "statistical", based on distributions rather than a single long dynamical trajectory. The statistical viewpoint describes the evolution of the phase-space probability density due to transformations of the comoving area, $dqdp \overset{B}{\to} dq_{new}dp_{new}$. The map *doubles* the comoving area whenever $q < p - \sqrt{(2/9)}$. B *halves* the area when $q > p - \sqrt{(2/9)}$ as can be seen directly from the mapping equations :

$$(11/6)^2 - (7/6)^2 = 2 \ ; \ (11/12)^2 - (7/12)^2 = (1/2) \ .$$

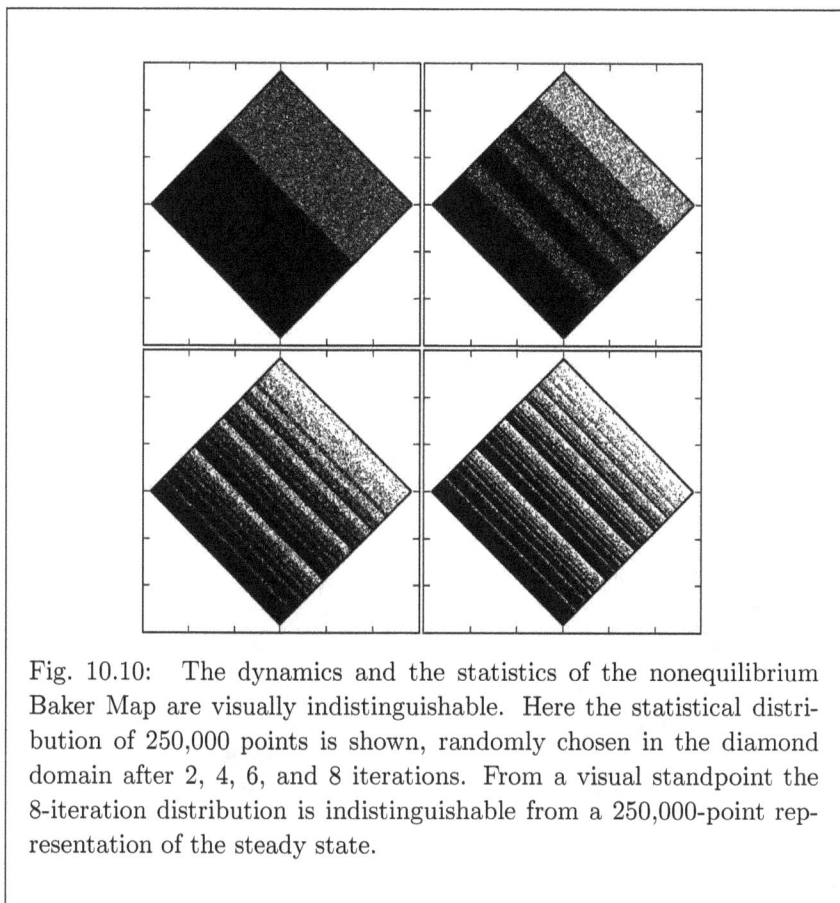

Fig. 10.10: The dynamics and the statistics of the nonequilibrium Baker Map are visually indistinguishable. Here the statistical distribution of 250,000 points is shown, randomly chosen in the diamond domain after 2, 4, 6, and 8 iterations. From a visual standpoint the 8-iteration distribution is indistinguishable from a 250,000-point representation of the steady state.

The probability of observing k expansions with $N - k$ contractions (in any order) in a total of N iterations, with 2^N possible outcomes using the map, is the same as the biased random walk probability :

$$(1/3)^k (2/3)^{N-k} \binom{N}{k} .$$

To illustrate: for the most likely 12-step example the relatively large probability is $(1/3)^4 (2/3)^8 \binom{12}{4}$ while the relatively small probability of the *reversed* walk, $(1/3)^8 (2/3)^4 \binom{12}{4}$ is sixteen times smaller.

If we associate the logarithm of the phase-space area with Gibbs' entropy the entropy changes by $\pm k \ln(2)$ with each iteration of the map. Because the region subject to halving is twice as large as that subject to doubling, the *net* effect is an entropy *decrease* averaging $-(k/3) \ln(2)$ per iteration.

Now consider the steady state. Evidently the steady state must have a steady value of the entropy [or so the usual argument goes]. In order to be consistent with this expectation we must legislate an entropy production of $(k/3)\ln(2)$, on average. We must also imagine a heat reservoir that extracts the entropy produced or transmits the entropy consumed by each iteration of the map and note that in the steady state the reservoir entropy external to the map *increases* on average at a constant rate $(k/3)\ln(2)$ per iteration. Although this idea (based on continuity) might seem a bit suspect for a fractal distribution it appears to be fully consistent with, and nicely illustrates, the ideas and results of Evans, Cohen, and Morriss.

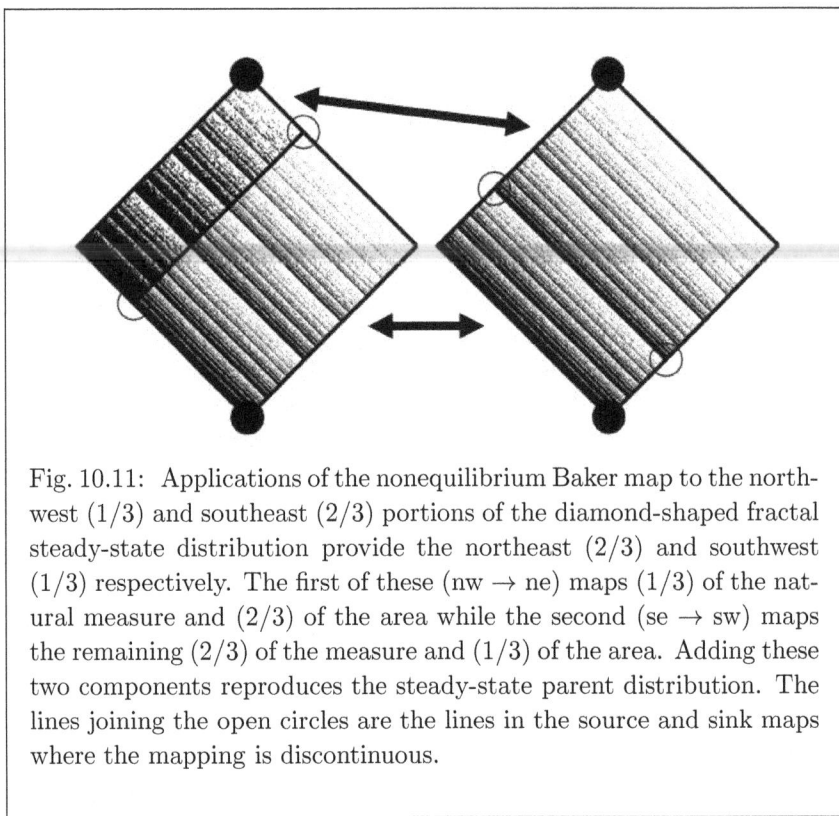

Fig. 10.11: Applications of the nonequilibrium Baker map to the northwest (1/3) and southeast (2/3) portions of the diamond-shaped fractal steady-state distribution provide the northeast (2/3) and southwest (1/3) respectively. The first of these (nw → ne) maps (1/3) of the natural measure and (2/3) of the area while the second (se → sw) maps the remaining (2/3) of the measure and (1/3) of the area. Adding these two components reproduces the steady-state parent distribution. The lines joining the open circles are the lines in the source and sink maps where the mapping is discontinuous.

In the nonequilibrium Baker Map the probability density is asymptotically uniform in the (unstable growth) direction parallel to the line $(q+p) = 0$ and becomes fractal in the (collapsing dissipative) perpendicular direction, parallel to the 45°-degree line $(q-p) = 0$. The southwest third of the diamond-shaped domain necessarily comes to include (2/3) of the natural measure. The southwest third of the remainder (see **Figure 10.11**) is a scale model of the larger northeast third and contains (2/3) of the remaining measure, that is (2/9) of the total. This scaling in the northeast direction continues on with each image of the original smaller by a factor of (2/3) while containing (1/3) of the remaining measure. Thus the total area of $2 \times 2 = 4$ is divided up as follows :

$$(4/3) + (4/3)(2/3) + (4/3)(2/3)^2 + (4/3)(2/3)^3 + \cdots = \frac{(4/3)}{1 - (2/3)} \equiv 4 \ .$$

The total "natural measure" is unity and corresponds to the following sum where the first term corresponds to the southwest third of the domain :

$$(2/3) + (2/3)(1/3) + (2/3)(1/3)^2 + (2/3)(1/3)^3 + \cdots = \frac{(2/3)}{1 - (1/3)} \equiv 1 \ .$$

In our Baker Map (2/3) of the measure expands to the northwest, with a Lyapunov expansion of (3/2), while being squeezed threefold in the perpendicular direction. Simultaneously (1/3) of the measure expands to the southeast, moving mostly east, with a Lyapunov expansion of 3. Because both "new" locations of the measure are perfect scale models of the "old", the coupled bilinear mappings both generate and preserve the stationary fractal distribution shown in **Figure 10.11**.

The fluctuation theorem for a window of τ steps states that the probability of converting work to heat, divided by the (illegal, according to the Second Law) reversed process, is the exponential of the entropy produced going forward in time :

$$\ln [\ \text{prob}_{\text{forward}}(\tau)/\text{prob}_{\text{backward}}(\tau) \ ] = e^{+\tau(\dot{S}/k)} = e^{(\Delta S/k)} \ .$$

For the Baker Map the probabilities $\text{prob}_{\text{forward}}(\tau)$ and $\text{prob}_{\text{backward}}(\tau)$ can be worked out analytically by noting that a sequence of steps forward in time is more probable than its reverse by a factor of $e^{+\tau(\dot{S}/k)}$. There is an isomorphism linking the comoving expansions and contractions of the Baker Map to a random walk in which steps to the right (corresponding to compression) are twice as likely as steps to the left (corresponding to expansion). Let us look at the details of a single example.

Consider the case $\tau = 6$ with 2^6 outcomes for the entropy production ranging from $-6\ln(2)$ with probability $(1/3)^6$ to $+6\ln(2)$ with probability $(2/3)^6$. Because this six-step mapping is equivalent to a random walk problem with left and right probabilities of (1/3) and (2/3) the resulting binomial distribution approximates a Gaussian with mean value

$\langle \sigma \rangle = (\tau/3)\ln(2)$ and mean squared value $\langle \sigma \rangle^2 + (8/3)\langle \sigma \rangle \ln(2)$:

$$P_{-6} = \frac{1 \cdot 1}{729} \ ; \ P_{-4} = \frac{6 \cdot 2}{729} \ ; \ P_{-2} = \frac{15 \cdot 4}{729} \ ; \ P_0 = \frac{20 \cdot 8}{729} \ ;$$

$$P_{+2} = \frac{15 \cdot 16}{729} \ ; \ P_{+4} = \frac{6 \cdot 32}{729} \ ; \ P_{+6} = \frac{1 \cdot 64}{729} \ ,$$

the binomial distribution for a biased random walk. In the general case with a window τ the mean entropy production is $\langle \sigma \rangle = (\tau/3)\ln(2)$ and the mean of the squared entropy production turns out to be

$$\langle \sigma^2 \rangle = [\ (\tau/3)\ ]^2[\ \ln(2)\ ]^2 + (8\tau/9)[\ \ln(2)\ ]^2 = \langle \sigma \rangle^2 + (8/3)\langle \sigma \rangle[\ \ln(2)\ ] \ .$$

From the Central Limit Theorem we see that for sufficiently large values of the sampling time τ the binomial random-walk distribution approaches a Gaussian distribution with the following form :

$$\mathrm{prob}(\sigma) \propto e^{(-3/16)(\sigma - \langle \sigma \rangle)^2/\langle \sigma \rangle \ln(2)} \ .$$

The "Fluctuation Theorem" or "Fluctuation Relation" came from the analysis of dense fluid shear flows carried out in 1993 by Evans, Cohen, and Morriss.[17] Although the flows were nominally steady, maintained at a constant internal energy by ergostat forces, the shear stress fluctuated about its average value, $\eta\dot{\epsilon}$, occasionally exhibiting relatively rare negative trajectory segments. Denis, and Eddie, and Gary noticed that the relative probabilities of positive and negative stresses were correlated with the entropy production during segments of the same length :

$$\ln[\ \mathrm{prob}(-P_{xy})_\tau / \mathrm{prob}(+P_{xy})_\tau\ ] = (\tau\dot{S}/k) \ [\ \text{Fluctuation Relation}\ ] \ .$$

The more probable sign agrees with the Second Law of Thermodynamics and the less-probable one does not – it is instead an unlikely fluctuation. For "long times", but not so long as to make the unlikely sign impossible to observe, the two probabilities are related to this "Fluctuation Relation" which is itself closely related to Liouville's Theorem. The impact of this work has been tremendous and is the subject of many reviews.[20]

A simpler illustration of this correlation between fluctuations and entropy production can be based on the nonequilibrium Baker Map provided the time window τ is not too large. The nonequilibrium Baker Map is clearly equivalent to a random walk with steps to the right twice as likely as steps to the left. These steps correspond to area shrinkage (two-thirds of the time) and area expansion (one-third of the time). The Fluctuation Theorem is said to apply when the window τ is sufficiently large. For $\tau = 3$ the longtime average result is two compressive steps and one expansive step while the reversed dynamics gives two expansions and one compression. The logarithm of the ratio of the two probabilities is $\ln[\,(2/3)(2/3)(1/3)/(1/3)(1/3)(2/3)\,] = \ln(2)$, the left side of the Fluctuation Theorem. The entropy production forward in time is evidently $k\ln(2)$, reflecting the additional compressive step resulting from the twofold larger regions of the space leading to compression. Evidently the Fluctuation Theorem *is an identity* for the time-averaged nonequilibrium Baker Map.

10.6 Dimensionality Loss in Maps and Flows

The Baker Map can teach us a great deal about the differences between equilibrium and nonequilibrium systems. As one departs from equilibrium he might imagine that phase volume shrinks, indicating fewer available states. The time-reversible Baker Map, like other more complicated maps of the type I investigated with Harald Posch and Oyeon Kum[21] shows otherwise. The maps demonstrate that it is the phase-space *dimensionality* rather than the area itself that decreases away from equilibrium. See the nonequilibrium Baker-Map data in **Figure 10.12**. A late colleague from Slovakia, Juraj Kumičák (1941-2017), considered generalizations (a usual practice of mathematicians) of the nonequilibrium Baker Map[22] where the compressed rectangular area in a unit square takes up a fraction $(1/w)$ of the space. $w = 2$ corresponds to the "equilibrium" situation where the natural measure is homogeneous and constant, everywhere the same. Kumičák found for the generalized Baker map that the information dimension is :

$$D_I(w) = 1 + \frac{(w-1)\ln(w-1) - w\ln(w)}{\ln(w-1) - w\ln(w)} \ .$$

The information dimension for our map, with $w = 3$, is 1.73368, quite consistent with our numerical work.

Area-conserving time-reversible shear maps, such as the "simple" (linear) shears :

$$X(x,y) = (x+y,y) \ ; \ Y(x,y) = (x,x+y) \ ,$$

resemble the well-known "Cat Map" and can be concatenated into a variety of symmetric derivative maps like $XYYX$ or $XXXYXXX$. These symmetric derivative maps maintain the time-symmetry of their components and serve as models for equilibrium phase-volume-conserving flows.

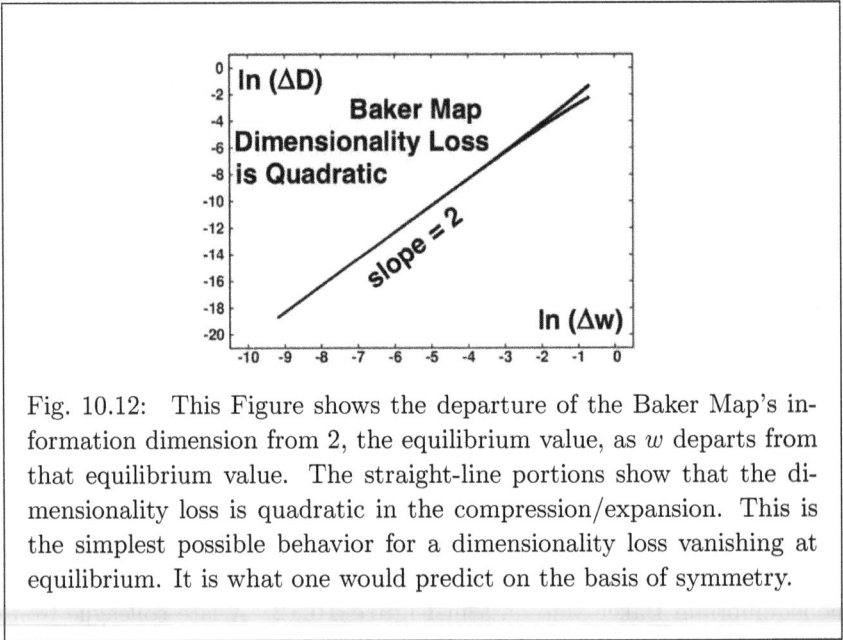

Fig. 10.12: This Figure shows the departure of the Baker Map's information dimension from 2, the equilibrium value, as w departs from that equilibrium value. The straight-line portions show that the dimensionality loss is quadratic in the compression/expansion. This is the simplest possible behavior for a dimensionality loss vanishing at equilibrium. It is what one would predict on the basis of symmetry.

Such maps, even when ergodic and chaotic, lack the possibility for changes of area needed to describe nonequilibrium systems. For nonequilibrium applications a simple "mirror" operation can serve as a model. Consider the upper right quadrant $[\ 0 < x, y < (1/2)\ ]$ of the unit square centered on the origin. With "mirrors" at $x = m$ and $y = m$ such a map $M(x, y) = (x', y')$ gives the new primed coordinates :

$$\frac{x' - m}{m - x} = \frac{m}{(1/2) - m} \ ; \ \frac{y' - m}{m - y} = \frac{m}{(1/2) - m} \ .$$

Figure 10.13 shows fractal attractors generated with the symmetric time-reversible map $XYMYX$. You can compare these to the cruder versions we generated in 1996.[21] The mirror map depends upon the position of the mirrors $(\pm m, \pm m)$ and returns an ergodic distribution with uniform density if the mirror locations divide the square symmetrically with mirrors at $x = \pm(1/4)$ and $y = \pm(1/4)$. Away from these symmetric locations the maps for positive and negative displacments of the mirrors yield exactly the same fractal attractors when the periodic boundaries are taken into account. For mirror positions within a tenth of the symmetric location 0.25 the information dimension of the fractals varies quadratically with the deviation from equilibrium, $D_I \simeq 2 - 31\Delta m^2$, exhibiting the same simplicity as does the nonequilibrium Baker Map.

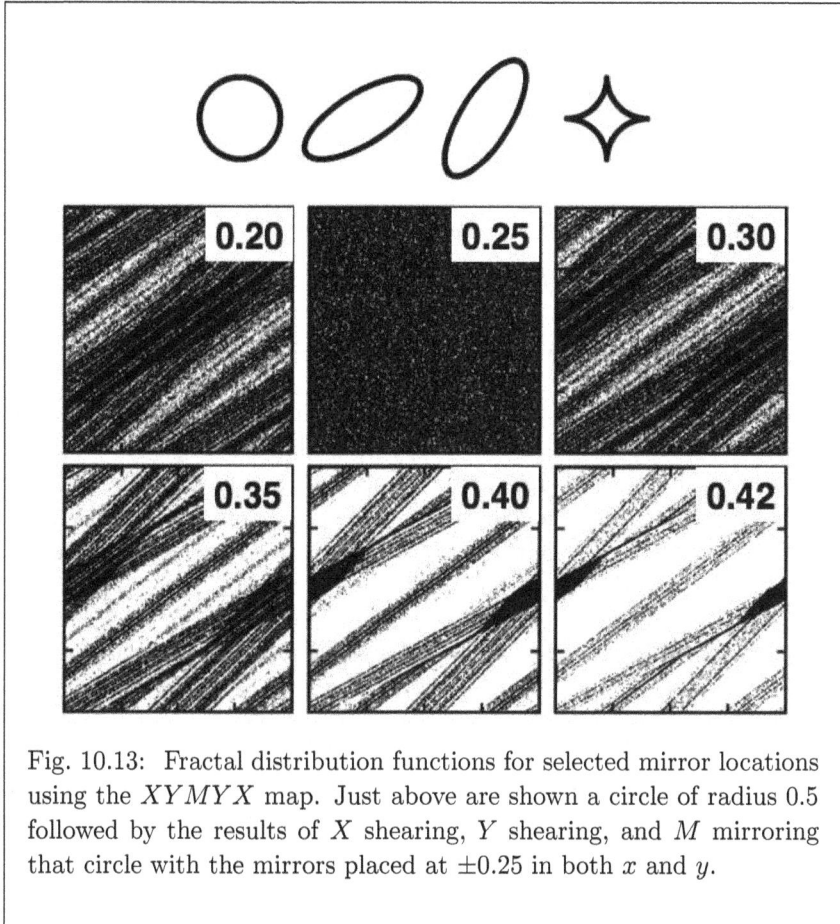

Fig. 10.13: Fractal distribution functions for selected mirror locations using the $XYMYX$ map. Just above are shown a circle of radius 0.5 followed by the results of X shearing, Y shearing, and M mirroring that circle with the mirrors placed at ± 0.25 in both x and y.

Flows likewise show a phase-space dimensionality loss proportional to the square of the departure from equilibrium. In a shear flow thermostatted to maintain the temperature or the energy constant the equations of motion include a friction coefficient ζ tasked with maintaining a steady state.

$$\{ \dot{x} = p_x + \dot{\epsilon}y \; ; \; \dot{y} = p_y \; ; \; \dot{p}_x = F_x - \dot{\epsilon}p_y - \zeta p_x \; ; \; \dot{p}_y = F_y - \zeta p_y \} \; .$$

The friction coefficient ζ generates a steady state by constraining either the internal energy or the kinetic temperature of the system. The resulting change of phase volume with time, the irreversible work done by the shearing boundary conditions, and the mean value of the friction coefficient ζ are all connected as would be expected from thermodynamics :

$$\langle\ \zeta = -\dot{\epsilon}(P_{xy}V/NkT) = +\eta\dot{\epsilon}^2(V/NkT) = -(\dot{\otimes}/\otimes)\ \rangle\ .$$

Evidently the thermostatting frictional forces $-\{\zeta p\}$ extract the heat generated by the shear flow at exactly the rate required by irreversible thermodynamics (as in the Navier-Stokes equations). This approach also furnishes, at small strain rates, a derivation of the Green-Kubo expression for viscosity in terms of shear-stress fluctuations :

$$\eta = (V/kT)\int_0^\infty\langle\ P_{xy}(0)P_{xy}(t')\ \rangle_{\rm eq}dt\ .$$

10.7 Linear Response with the 0532 Oscillator

The somewhat surprising failure of Nosé's mechanics to provide the complete canonical distribution for simple problems like the harmonic oscillator launched an effort to find ergodic thermostats, still with only a few parameters and still time-reversible. This effort involved several dozens of scientists over a period of thirty years. Eventually the incremental progress due to this activity led to two novel thermostat types. "Weak control" of not only the kinetic temperature but also its higher moments, and sometimes those of the configurational temperature, $\langle\ (\nabla\mathcal{H})^2/\nabla^2\mathcal{H}\ \rangle$, was one approach. An example successfully applied to the harmonic oscillator is the 0532 model :

$$\dot{q} = p\ ;\ \dot{p} = -q - \zeta(0.05p + 0.32p^3)\ ;\ \dot{\zeta} = 0.05(p^2 - 1) + 0.32(p^4 - 3p^2)\ .$$

This approach, controlling a linear combination of two velocity moments rather than just one, turned out to be unsuccessful when applied to the quartic and Mexican Hat potentials, inspiring an alternative hyperbolic-tangent "logistic" thermostat[8,9] :

$$\dot{q} = p\ ;\ \dot{p} = q - q^3 - \alpha p\tanh(\alpha\zeta)\ ;\ \dot{\zeta} = p^2 - 1\ .$$

This idea successfully generates the canonical distribution for the quartic oscillator as well as the Mexican Hat potential, $(x^4/4) - (x^2/2)$. In the latter case ergodicity results for α in the neighborhood of 7.

The main advantage of ergodic thermostats is consonance with Gibbs' ensembles. For instance, it is possible to take advantage of the linear-response theory that describes nonequilibrium systems close to equilibrium. Because the 0532 model was successful in thermostatting the harmonic oscillator it was logical to look for heat transport in that case, where linear-response theory could be applied. We chose two smooth variations of temperature $T(q)$ with a relatively small gradient, $\epsilon = 0.1$, well-suited to linearization :

$$0.9 < T(q) = [\,1 + 0.1\tanh(q)\,] < 1.1 \text{ and}$$

$$(10/11) < T(q) = 1/[\,1 - 0.1\tanh(q)\,] < (10/9) ,$$

with the good agreement of the two establishing that both simulations were in the linear regime. Puneet Patra, Clint Sprott, Carol, and I were able to calculate the nonequilibrium heat current which responds to these perturbations, $\simeq 0.1\tanh(q)$, of the canonical distribution. We used a grid of one million ($100 \times 100 \times 100$) equiprobable cells in the equilibrium (q, p, ζ) phase space.[23] This same approach worked well (although with a much more time-consuming calculation, with 100 million cells) in the four-dimensional case of the Hoover-Holian thermostat). In both cases, and also in the logistic-thermostat case, the equilibrium canonical distribution is known, including the distribution(s) of the thermostat variables, Gaussian in the HH case and a hyperbolic function in the logistic case.

10.8 Dimensionality Loss with the ϕ^4 Model

In any event conservation of probability means that the comoving product $(f\otimes)$ flows through phase space with its magnitude unchanged. At equilibrium the accessible phase volume $\otimes$ and its relative probability f are both constant. The time-averaged friction coefficient sum necessarily vanishes at equilibrium, $\langle\,\sum\zeta\,\rangle = 0$. When Liouville's Theorem is extended to nonequilibrium steady states described by thermostatted models the time-averaged result,

$$\langle\,(\partial f/\partial t)\,\rangle = -\sum[\,\partial(f\dot{q})/\partial q\,] - \sum[\,\partial(f\dot{p})/\partial p\,] - [\,\partial(f\dot{\zeta})/\partial\zeta\,]\,\rangle > 0 ,$$

establishes a continuous *loss* of phase volume and a continuous increase of probability density through the mechanism of the thermostat forces,

$$\{\,F = F_{\text{Newton}} - \zeta p\,\} \rightarrow (\dot{f}/f) = -\sum(\partial\dot{p}/\partial p) \longrightarrow$$

$$\langle\,(\dot{f}/f)\,\rangle \equiv \sum\zeta\,\rangle > 0 ;\ \langle\,(\dot{\otimes}/\otimes)\,\rangle \equiv -\sum\zeta\,\rangle < 0\ (!) .$$

Such a loss is reminiscent of a damped harmonic oscillator's inevitable end, the fixed point at the origin, $(q = 0, p = 0)$.

Exercise : Show numerically that the kinetic energy of a harmonic chain can have a nonzero steady-state average if one particle in the chain, "j", has the equation of motion $\ddot{q}_j = q_{j+1} - 2q_j + q_{j-1} - \dot{q}_j$.

If the steady-state average of $(\partial f/\partial t)$ is not zero then it *must* be positive. This is because an overall negative friction would imply phase volume growth and such a *steady-state* growth is impossible without eventual divergence and computational instability. Evidently the only possibility away from equilibrium is a steadily decreasing phase volume.

What is the significance of a steady-state *decreasing* phase volume? At first this idea seemed or seems ridiculous, but soon few-dimensional nonequilibrium toy-problem simulations revealed that the motion equations, although strictly time-reversible, displayed a steady-state symmetry breaking, choosing fractal solutions with *zero* phase volume. Zero phase volume means that the fraction of cells covering the attractor and containing most of the measure (a fixed percentage) approaches zero as the cell size vanishes. This zero-volume-attractor fringe benefit of Nosé's 1984 work was a real surprise in 1985. It led to the Galton staircase and Galton Board simulations of nonequilibrium steady states in 1987.[24, 25] It led to an understanding of the irreversible nature of reversible flows in nonequilibrium steady states. These 1987 results were subsequently accepted and blessed by the mathematicians in 1993.[26]

A great deal of effort went into the diagnosis of phase-volume loss through Lyapunov spectra. The sum of the exponents, starting at the positive end of the spectrum, changes from positive to negative at some stage, indicating the collapse of the phase volume onto a fractal strange attractor. The number of terms in the vanishing sum gives the dimensionality of the phase-space distribution. Though the occupied phase volume vanishes and the probability density diverges it is still possible to compute the ("coarse-grained") "natural measure" by binning trajectory points. Recall the Galton Board Poincaré sections that show limit cycles for strong fields, where the dimensionality of the phase-space distribution has been reduced from three-dimensional to one. Work on periodic two-dimensional shear flows revealed dimensionality losses as large as 15% for systems of 100 soft-disk particles.[27] Of course thermostats are a bit artificial so that a convincing demonstration of dimensionality loss needs to focus on situations in which only a few surface particles are directly affected by the nonequilibrium constraints.

We could glimpse a way forward to a better understanding in 2002. In that year Ken Aoki and Dimitri Kusnezov helped to popularize the very useful ϕ^4 model for heat conduction. In their ϕ^4 model each particle is bound to its own lattice site by a quartic potential while it interacts with its nearest neighbors through Hooke's-Law springs.[28] In the simplest one-dimensional case the Hamiltonian is :

$$\mathcal{H} = \sum^N (p_i^2/2) + \sum_{\text{pairs}} (1/2)(\Delta q_{ij})^2 + \sum^N (q_i^4/4) .$$

The resulting Hamiltonian equations of motion involve forces between nearest-neighbor pairs as well as cubic "on-site" forces:

$$\{ \dot{q}_i = p_i \; ; \; \dot{p}_i = q_{i+1} - 2q_i + q_{i-1} - q_i^3 \} .$$

Boundary forces and thermostat forces used to control the temperature at the boundaries need to be discussed separately and added to these Hamiltonian $\{ \dot{p} \}$ equations. Aoki and Kusnezov used two Nosé-Hoover particles at the boundaries of their one-dimensional simulations. The highlights of Aoki and Kusnezov's work included [1] finding the insensitivity of the sound velocity to temperature ; [2] formulating a good powerlaw representation of the thermal conductivity, $\kappa = (2.83/T^{1.35})$ in one dimension ; [3] assessing the usefulness of local equilibrium as measured by the difference between $\langle p^4 \rangle$ and $3\langle p^2 \rangle^2$; and [4] describing the jumps and the curvatures found near the boundaries, particularly at the cold end, at low temperatures. Aoki and Kusnezov mention the relationship of the Lyapunov spectrum to the fractal dimensionality of the nonequilibrium strange attractors but did not explore the spectra numerically. Harald Posch and I were quick to seize on their ϕ^4 model as a candidate for nonequilibrium dimensionality-loss simulations.

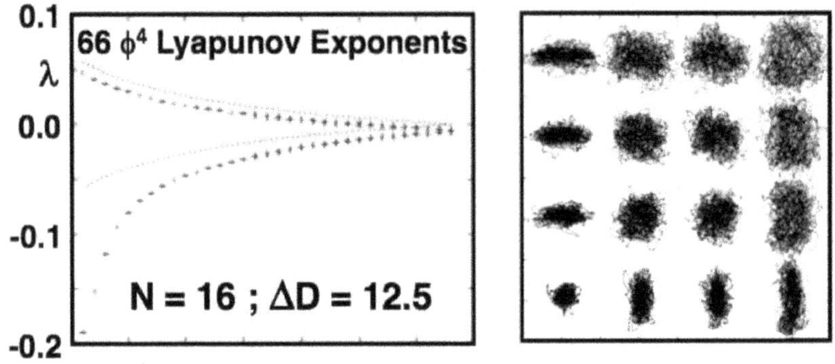

Fig. 10.14:   On the right is a trajectory fragment for 16 ϕ^4 particles with an overall heat flow from the upper right corner (where a Nosé-Hoover thermostat maintains the temperature $T = 0.009$) to the lower left corner where a second thermostat maintains the lower temperature ($T = 0.001$).   On the left are the $66 = 4 \times 16 + 2$ time-averaged Lyapunov exponents for this system. Adding the exponents, beginning with 24 positive λ and including the single vanishing exponent, then adding on successive more-negative exponents, the exponent sum changes from positive to negative between the 53rd and 54th terms, indicating a dimensionality loss of about $66 - 53.5 = 12.5$, well in excess of the number of dimensions (10) { x, y, p_x, p_y, ζ } associated with the two thermostatted particles. [66 includes the two friction coefficients in addition to the $4N$ coordinates and momenta].

Later in that same year we got together with Aoki and Kusnezov and published a paper describing steady-state simulations of heat flow in the 16-particle ϕ^4 system of **Figure 10.14**.[29]   With Ken Aoki, Carol, and Tony De Groot's daughter Stephanie I carried out studies of seven different thermostat types. The systems we investigated were rectangular, 6×4, 10×4, and 18×4, crystals of ϕ^4 particles.[30]

At Livermore we used both differential and integral feedback with differing combinations of $\langle p^4 \rangle$ and $\langle p^2 \rangle$ control, finding a well-behaved heat flux in every case. In the longest systems the time-averaged heat fluxes varied from 0.086 to 0.094 (corresponding to a conductivity around 1.8) with boundary temperatures 0.5 and 1.5 imposed upon the first and last columns of 4 particles each. We reported on these results at a meeting in Dresden.[30] We concluded our comparison of all seven thermostat types with an overall assessment favoring the Nosé-Hoover thermostat :

> "The simplicity of thermostats based on the second moment of the velocity distribution and simply connected to irreversible thermodynamics recommends the use of Nosé-Hoover thermostats whenever possible".

It seemed to us that thermostats exhibiting little number dependence and minimizing deviations from local equilibrium would, other things being equal, be the best choice in simulations. In addition to performing well by those criteria the Nosé-Hoover thermostat is certainly the simplest to implement, explaining our favoring that approach.

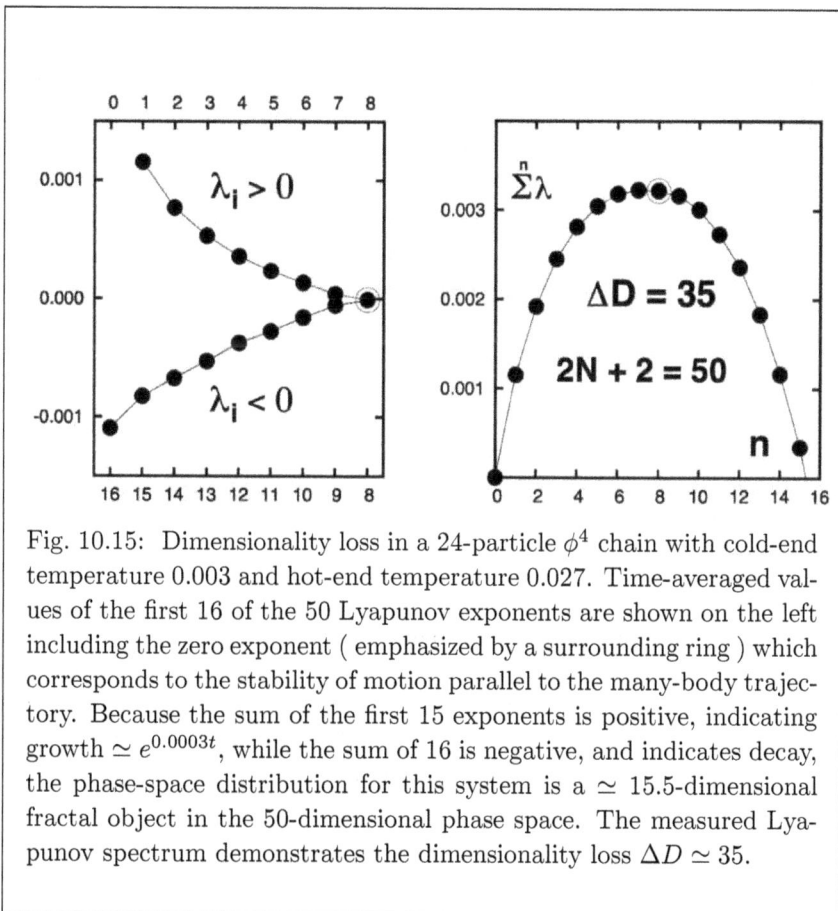

Fig. 10.15: Dimensionality loss in a 24-particle ϕ^4 chain with cold-end temperature 0.003 and hot-end temperature 0.027. Time-averaged values of the first 16 of the 50 Lyapunov exponents are shown on the left including the zero exponent (emphasized by a surrounding ring) which corresponds to the stability of motion parallel to the many-body trajectory. Because the sum of the first 15 exponents is positive, indicating growth $\simeq e^{0.0003t}$, while the sum of 16 is negative, and indicates decay, the phase-space distribution for this system is a $\simeq$ 15.5-dimensional fractal object in the 50-dimensional phase space. The measured Lyapunov spectrum demonstrates the dimensionality loss $\Delta D \simeq 35$.

Much more recently, in our 2015 book,[31] Carol and I decided to revisit the one-dimensional ϕ^4 problem sufficiently far from equilibrium that the nonequilibrium dimensionality loss would overwhelm any reasonable doubts of that concept. We considered a 24-particle chain with Nosé-Hoover thermostats at the ends set to temperatures of 0.003 and 0.027. The system has 50 Lyapunov exponents, with the largest 16 (eight of which are negative) displayed to the left in **Figure 10.15**. The sums of the first n exponents are shown to the right with the 15th sum positive and the next one negative and not shown in the Figure. The dimension of the strange attractor for this ϕ^4 chain is about 15.5, giving a dimensionality loss of about 34.5, a clear demonstration that thermostatting just two particles can result in the loss of phase-space dimensions much greater than the number associated with the two "cold" and "hot" particles. We mention in the book that a 36-particle system with a 74-dimensional phase space shows an information dimension of about 30.5 so that the dimensionality loss in that case is 43.5, about 60% of the total phase-space dimensions. Evidently boundary conditions *can* be effective *throughout* a system.

These simulations are easy to program in a variety of interesting geometries and are perfectly suitable for laptop computers. Exploring the details of the dimensionality loss and its size-dependence on a variety of crystal structures, with or without defects, would open up many interesting research areas.

Exercise : Explore the final steady states of one-dimensional thermostatted rings of harmonic and ϕ^4 particles when one or more particles in the ring have a frictional force $-p$ in their equations of motion.

In retrospect the phase-space dimensionality loss of fluids or solids undergoing nonequilibrium flows could equally well be controlled by a constant friction chosen to place the system in a steady state. In such a case it is logical to expect a Gibbs'-volume loss of dimensionality equal to the rate of work done on the system, divided by the imposed temperature and the largest Lyapunov exponent. That Lyapunov exponent has the magnitude of the sound velocity divided by the interparticle spacing, and corresponds to an effective collision rate and to an Einstein vibrational frequency. I believe that the explanation of the Second Law irreversibility seen in nature is best explained by models where Lyapunov instability is responsible for steering an evolving system toward the more probable portions of phase space. It is not surprising that such portions are exceptional and occupy only a negligible fractal portion of the space with an information dimension less than that of the whole space. The link between the fractal dimension and the Lyapunov spectrum is a conceptual breakthrough in the understanding of the Second Law of Thermodynamics. This understanding quantifies the great rarity of steady nonequilibrium microstates.

10.9 Summary of Lecture 10

By extending Nosé's deterministic and time-reversible thermostat idea to systems away from equilibrium we have found a simple explanation of the way in which irreversible processes result from time-reversible nonequilibrium molecular dynamics. The earliest such simulations, with thermal constraints added to the equations of motion, turned out to be time-reversible as well as fully consistent with the Green-Kubo linear-response predictions for the transport coefficients governing diffusion, viscous flows, and heat conduction. The simplest special case of thermostatted dynamics, rescaling the velocities to maintain the same kinetic energy at every step, turned out to be equivalent to applying *Gauss' Principle of Least Constraint* to the problem of implementing isokinetic dynamics.[32] This link to classical mechanics provided differential-feedback constraint forces linear in the momentum about a decade earlier than Nosé's pioneering Hamiltonian constraint forces of the same form, $\{-\zeta p\}$, but based on integral feedback, with $\dot{\zeta} \propto (p^2/T) - 1$.

The simplest Nosé-Hoover problem, the canonical-ensemble harmonic oscillator, illustrates the difficulty mathematicians have in contributing to an understanding of dynamical systems. The oscillator has a chaotic sea, six percent of the Gaussian distribution in the simplest case,[23]

$$\{ \dot{q} = p \; ; \; \dot{p} = -q - \zeta \; ; \; \dot{\zeta} = p^2 - 1 \} \longrightarrow$$

$$f(q,p,\zeta) \propto e^{-(q^2/2)} e^{-(p^2/2)} e^{-(\zeta^2/2)} \, .$$

The only way to determine whether or not a given point is "in" that sea, or not, is to integrate the equations of motion and look for chaos. The distinction between toroidal periodicity and chaos is clear enough from computation but seems immune to mathematical investigations. For this reason I wish to stress that particular examples, rather than Theorems and Corollaries, provide the shortest path to progress, understanding, and the truth. Feynman regularly stressed this point.

With Christoph Dellago I verified that digital simulations of steady states do indeed repeat (as they must, given that the number of phase-space states is finite). This repetition illustrates exact "Poincaré recurrence" rather than approximate and establishes that the time required to see such a repetition depends upon the two-point correlation dimension D_C of the fractal phase-space distribution, $\tau_{\text{Poincaré}} \propto \epsilon^{-D_C/2}$ where ϵ is the phase-space bin size. Because these results are for *nonequilibrium* steady states recurrence is perfectly fine.

What about Loschmidt's reversibility objection? It is certainly true that reversing a dissipative Nosé-Hoover trajectory gives a trajectory violating the Second Law but obeying the same motion equations as the forward version. This is not troublesome because there is no way to reverse a dissipative trajectory. *There exists no bit-reversible algorithm for steady-state trajectories obeying the Second Law of Thermodynamics.* We found that the reversed nonequilibrium Baker Map, TBT, where T is the time-reversal map changing the sign of the momentum, reproduces the previous state only about ten percent of the time. Reversal for a single step is highly unlikely, even for just one degree of freedom. Further, the slightest error in an attempted reversed trajectory immediately begins to grow. Lyapunov instability guarantees that one will leave the repellor and return to the attractor in a time of order $(37/\lambda_1)$ (double precision) or $(74/\lambda_1)$ (quadruple precision).

The time-reversible deterministic Baker Map, which incorporates the possibility of dissipative area changes, despite ergodicity, is a shorthand illustration of the Second Law. Half a dozen iterations are enough to see a developing fractal attractor from which we can construct the otherwise unobtainable fractal repellor. The Baker Map model is isomorphic to a biased one-dimensional random walk and provides a useful illustration of the Fluctuation Theorem, including an exact calculation of the relative probabilities of forward and reversed trajectory fragments. This model teaches us that the complexities of chaotic flows can often be understood well with low-dimensional models, in this case just two-dimensional, (q, p).

Similarly, the 0532 model provides an ergodic flow in the (q, p, ζ) phase space of the simple oscillator, making it possible to demonstrate linear-response theory exactly and completing the search for a single-thermostat-variable dynamics consistent with Gibbs' canonical ensemble.

The general nature of the dimensionality reduction seen in the Galton investigations of 1987 was confirmed nearly 30 years later through simulations of Aoki and Kusnezov's ϕ^4 model for heat flow. Not only can dissipation and fractal formation penetrate throughout the system due to the rapid rotation of the Lyapunov vectors; but also the resulting dimensionality loss can evidently cover the whole range of possible dimensionalities from the one-dimensional limit cycle to the ergodic coverage of the entire phase space characteristic of Gibbs' ensembles.

The two different views of dissipative dynamics, analysis of a single long trajectory or that of an evolving distribution function have yet to be related in a quantitative way. One reason for this lack is the *unobservable* nature of the Lyapunov spectrum. Although the spectrum is easily accessible from simulations of the equations of motion, it is not readily measurable for "real" systems in the laboratory. The symmetry-breaking displayed by the spectrum, with the exponents and their associated vectors forward and backward in time, can be analyzed to distinguish forward and backward trajectories. The spatial variation of the largest exponent is definitely fractal, though it appears that the forward and backward local exponents at any point can be determined by storing the forward trajectory and processing it backward. I know of no other such simulations than those Carol and I have done on inelastic collisions of many-body drops or periodic slabs.[33] Because a reversed molecular dynamics trajectory, when analyzed on the atomistic scale, will turn out to have negative diffusive, viscous, and thermal transport coefficients there are plenty of possibilities for making interesting movies.

The steady shockwave problem is a good candidate for the better understanding of irreversibility. My most-influential mentor Berni Alder has termed the shockwave the "most irreversible" process. Cold fluid enters on the left and hot fluid leaves on the right. These boundary conditions are purely equilibrium ones and the dynamics is purely Hamiltonian. Liouville's [$\dot{f} \equiv 0$] certainly applies throughout. Nevertheless there is no doubt that the exit entropy is always larger (infinitely larger if the inflow is at zero temperature). Surely this problem deserves more than a cursory study !

References

1. *Microscopic Simulations of Complex Hydrodynamic Phenomena*, editted by M. Mareschal and B. L. Holian, NATO Advanced Science Institute at Alghero, Sardinia, 15-17 July 1991 (Plenum Press, New York, 1992).
2. Wm. G. Hoover and C. G. Hoover, "Bit-Reversible Version of Milne's Fourth-Order Time-Reversible Integrators for Molecular Dynamics", Computational Methods in Science and Technology **23**, 299-303 (2017).
3. Wm. G. Hoover, "Reversible Mechanics and Time's Arrow", Physical Review A **37**, 252-257 (1988).
4. Wm. G. Hoover and C. G. Hoover, "Hamiltonian Thermostats Fail to Promote Heat Flow", Communications in Nonlinear Science and Numerical Simulation **18**, 3365-3372 (2013).
5. G. J. Martyna, M. L. Klein, and M. Tuckerman, "Nosé-Hoover Chains : the Canonical Ensemble *via* Continuous Dynamics", The Journal of Chemical Physics **97**, 2635-2643 (1992).
6. Wm. G. Hoover and B. L. Holian, "Kinetic Moments Method for the Canonical Ensemble Distribution", Physics Letters A **211**, 253-257 (1996).
7. Wm. G. Hoover, C. G. Hoover, and J. C. Sprott, "Nonequilibrium Systems: Hard Disks and Harmonic Oscillators Near and Far from Equilibrium", Molecular Simulation **42**, 1300-1316 (2016).
8. Wm. G. Hoover and C. G. Hoover, "Singly-Thermostatted Ergodicity in Gibbs' Canonical Ensemble and the 2016 Ian Snook Prize Award", Computational Methods in Science and Technology **23**, 5-8 (2017).
9. D. Tapias, A. Bravetti, and D. Sanders, "Ergodicity of One-Dimensional Systems Coupled to the Logistic Thermostat", Computational Methods in Science and Technology **23**, 11-18 (2017).
10. Wm. G. Hoover, B. Moran, C. G. Hoover, and W. J. Evans, "Irreversibility in the Galton Board *via* Conservative Classical and Quantum Hamiltonian and Gaussian Dynamics", Physics Letters A **133**, 114-120 (1988).
11. Wm. G. Hoover and C. G. Hoover, *Time Reversibility, Computer Simulation, and/Algorithms, Chaos* (World Scientific, Singapore, 1999/2012).
12. Ch. Dellago and Wm. G. Hoover, "Finite-Precision Stationary States At and Away from Equilibrium", Physical Review E **62**, 6275-6281 (2000).
13. C. Grebogi, E. Ott, and J. A. Yorke, "Roundoff-Induced Periodicity and the Correlation Dimension of Chaotic Attractors", Physical Review A **38**, 3688-3692 (1988).
14. J. C. Sprott, Wm. G. Hoover, and C. G. Hoover, "Heat Conduction, and the Lack Thereof, in Time-Reversible Dynamical Systems: Generalized Nosé-Hoover Oscillators with a Temperature Gradient", Physical Review E **89**, 042914 (2014).
15. F. Ricci-Tersenghi, "The Solution to the Challenge in 'Time-Reversible Random Number Generators' by Wm. G. Hoover and Carol G. Hoover", arχiv 1305.1805. See also arχiv 1305.0961, Reference 16.
16. Wm. G. Hoover and C. G. Hoover, "Time-Reversible Random Number Generators : Solution of Our Challenge by Federico Ricci-Tersenghi".

17. D. J. Evans, E. G. D. Cohen, and G. P. Morriss, "Probability of Second Law Violations in Shearing Steady Flows", Physical Review Letters **71**, 2401-2404 (1993).

18. J-P. Eckmann and I. Procaccia, "Fluctuations of Dynamical Scaling Indices in Nonlinear Systems", Physical Review A **34**, 659-661 (1986).

19. Wm. N. Vance, "Unstable Periodic Orbits and Transport Properties of Nonequilibrium Steady States", Physical Review Letters **69**, 1356-1359 (1992).

20. D. J. Evans and D. J. Searles, "The Fluctuation Theorem", Advances in Physics **51**, 1529-1585 (2010).

21. Wm. G. Hoover, O. Kum, and H. A. Posch, "Time-Reversible Dissipative Ergodic Maps", Physical Review E **53**, 2123-2129 (1996).

22. J. Kumičák, "Irreversibility in a Simple Reversible Model", Physical Review E **71**, 016115 (2005), Figure 2.

23. P. K. Patra, Wm. G. Hoover, C. G. Hoover, and J. C. Sprott, "The Equivalence of Dissipation from Gibbs' Entropy Production with Phase-Volume Loss in Ergodic Heat-Conducting Oscillators", International Journal of Bifurcation and Chaos **26**, 1650089 (2016).

24. Wm. G. Hoover, H. A. Posch, B. L. Holian, M. J. Gillan, M. Mareschal, and C. Massobrio, "Dissipative Irreversibility from Nosé's Reversible Mechanics", Molecular Simulation **1**, 79-86 (1987).

25. B. Moran, Wm. G. Hoover, and S. Bestiale, "Diffusion in a Periodic Lorentz Gas", Journal of Statistical Physics **48**, 709-726 (1987).

26. N. I. Chernov, G. L. Eyink, J. L. Lebowitz, and Ya. G. Sinai, "Steady-State Electrical Conduction in the Periodic Lorentz Gas", Communications in Mathematical Physics **154**, 569-601 (1993).

27. Wm. G. Hoover and H. A. Posch, "Second-Law Irreversibility and Phase-Space Dimensionality Loss from Time-Reversible Nonequilibrium Steady-State Lyapunov Spectra", Physical Review E **49**, 1913-1920 (1994).

28. K. Aoki and D. Kusnezov, "Nonequilibrium Statistical Mechanics of Classical Lattice ϕ^4 Field Theory", Annals of Physics **295**, 50-80 (2002) = arχiv 0002160.

29. Wm. G. Hoover, H. A. Posch, K. Aoki, and D. Kusnezov, "Remarks on Non-Hamiltonian Statistical Mechanics: Lyapunov Exponents and Phase-Space Dimensionality Loss", Europhysics Letters **60**, 337-341 (2002).

30. Wm. G. Hoover, K. Aoki, C. G. Hoover, and S. V. de Groot, "Time-Reversible Deterministic Thermostats", Physica D **187**, 253-267 (2004).

31. Wm. G. Hoover and C. G. Hoover, *Simulation and Control of Chaotic Nonequilibrium Systems* (World Scientific, Singapore, 2015) , page 205.

32. D. J. Evans, Wm. G. Hoover, B. H. Failor, B. Moran, and A. J. C. Ladd, "Nonequilibrium Molecular Dynamics *via* Gauss' Principle of Least Constraint", Physical Review A, **28**, 1016-1021 (1983).

33. Wm. G. Hoover and C. G. Hoover, "What is Liquid? Lyapunov Instability Reveals Symmetry-Breaking Irreversibilities Hidden Within Hamilton's Many-Body Equations of Motion", Condensed Matter Physics **18**, 1-13 (2015) = arχiv 1405.2485.

Chapter 11

Smooth-Particle Continuum Mechanics

/ Continuum Mechanics / Eulerian and Lagrangian Frames / Continuity, Motion, and Energy Equations / Constitutive Relations / Shockwave Structure / Rayleigh-Bénard Flow / SPAM = Smooth Particle Applied Mechanics / Conservation Laws and SPAM / Rayleigh-Bénard Flows with SPAM / Equilibrating and Collapsing Columns with SPAM / Molecular Dynamics *versus* Continuum Mechanics / Summary of Lecture 11 /

11.1 Continuum Mechanics

Continuum mechanics is a discipline wholly different to molecular dynamics. Both require fast computers for the solution of differential equations. Molecular dynamics has two (q, p) ordinary differential equations for each degree of freedom. Continuum mechanics has no discrete "degrees of freedom". It is a "field theory" in which the continuum contains a density $\rho(r, t)$, a velocity $v(r, t)$, and an energy density $e(r, t)$. These fields, if known, can be inserted into "constitutive equations" to obtain other desirable fields – the pressure tensor, heat flux vector, entropy, heat capacity and so on. Rather than the force laws on which molecular dynamics is based continuum mechanics is based on constitutive relations (ideal gas, van der Waals' fluid, Grüneisen solid, ...) and conservation laws. Rather than Newton's, Lagrange's, Hamilton's, Gauss', or Nosé-Hoover's mechanics continuum mechanics evolves through the simultaneous solution of [1] the continuity equation, [2] the equation of motion, and [3] the energy equation. These are three *partial* differential equations where the evolution $(\dot{\rho}, \dot{v}, \dot{e})$ is an explicit function of not just (ρ, v, e), but also spatial derivatives or "partial derivatives" like ∇v and $\nabla \cdot Q$. The continuum evolution equations are called "partial" differential equations for this reason.

Evidently computers cannot evaluate all the details of continuous fields. To make for a manageable approach the fields need to be computed at an array of points, nearly always at hundreds to billions of points though the practical limit increases in spurts, from one "latest model" computer to the next. We will begin by deriving the evolution equations, based on the conservation of mass, momentum, and energy. We will describe solution techniques in two different coordinate frames, the "Lagrangian" one moving with the material, and the "Eulerian" one fixed in space. Then we will leave grid-based methods to develop an alternative *continuum particle (!) method*, a "smooth-particle" approach to solving the continuum equations. We will use particles to follow the flow rather than a finite-difference grid of points. This simplification is a real time saver. We believe it is the simplest possible route to solving interesting problems in continuum mechanics.

Smooth-particle methods require the use of a "weight function" which describes the influence of nearby material on the evolving properties of a smooth-particle point. As a welcome side benefit these same weight functions also provide the best possible way of averaging molecular-dynamics or Monte Carlo "point" properties to find an equivalent "field" description. Weight functions not only yield fields from points. They also simplify and automate the calculation of the gradients appearing in the righthandsides of the continuum's evolution equations.

We will examine problems in which the molecular dynamics and field-theory descriptions are isomorphic [having the *same* form]. We will point out some of the difficulties associated with modelling surfaces and boundary conditions in smooth-particle flow simulations. We will consider some demonstration problems complicated enough to reveal the advantages and shortcomings of the continuum approach but simple enough for their exploration with computer programs of only a few hundred lines. We begin our study of smooth-particle continuum mechanics with a derivation of the partial differential evolution equations for the density, velocity, and energy.

11.2   The Continuity, Motion, and Energy Equations

In a big homogeneous sample of fluid there is no problem defining density as a function of location r and time t. But in the "real world" $\rho(r,t)$ eventually gives way to a discrete particle description when the sampling volume is small enough that the fluctuations in the density are significant. We avoid that complexity here. We adopt the continuum viewpoint that our dependent variables are defined everywhere and at all times so that the space and time derivatives are well defined. The evolution equations for $(\dot{\rho}, \dot{v}, \dot{e})$ can be derived in any number of dimensions (three is the usual choice ; we generally prefer two in order to reduce the computational and the graphical time requirements). Here we will use a dimensionality of one to simplify the notation. With a good notation the one-dimensional description and the two- or three-dimensional descriptions will look exactly the same and will be equally valid.

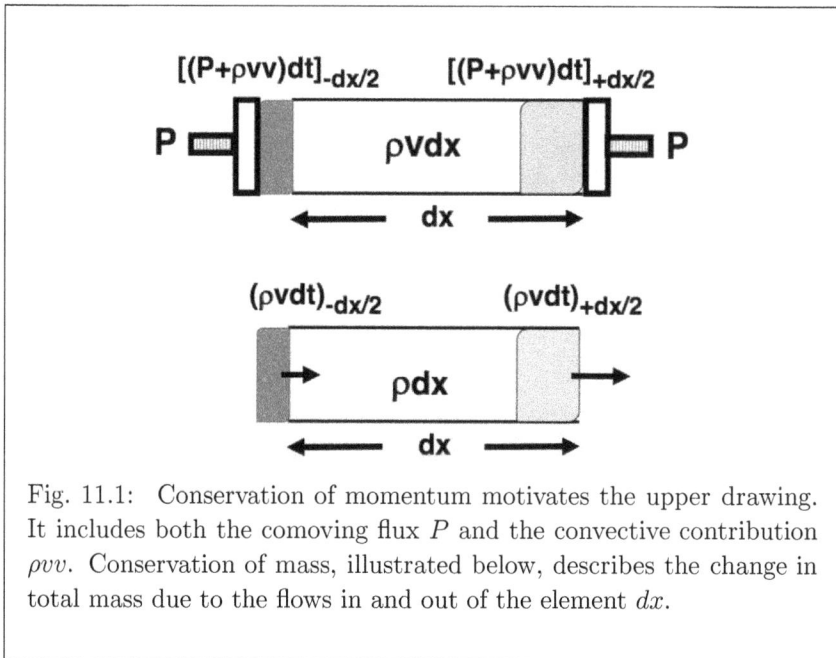

Fig. 11.1: Conservation of momentum motivates the upper drawing. It includes both the comoving flux P and the convective contribution ρvv. Conservation of mass, illustrated below, describes the change in total mass due to the flows in and out of the element dx.

Figure 11.1 shows an "Eulerian" volume element of length dx and fixed in space. During the time dt the darkly shaded material at the left, with mass $(\rho v)_{-dx/2}dt$ enters the fixed volume element dx, bringing with it the momentum $(P_{xx} + \rho v_x^2)_{-dx/2}dt$. During this same time interval the lightly shaded material, with mass $(\rho vdt)_{+dx/2}$, exits at the right, along with the momentum $(P_{xx} + \rho v_x^2)_{+dx/2}dt$. Notice that in addition to the convective flux of momentum ρvv there is an additional comoving flux of momentum *defining* the pressure tensor, just $P = P_{xx}$ in this one-dimensional case. Relating the boundary flows of mass and momentum to the total mass and momentum in dx, ρdx and ρvdx, gives the "Eulerian" continuity equation and equation of motion. Let us consider the continuity equation first :

$$(\partial \rho dx/\partial t)_x = (\partial \rho/\partial t)dx = (\rho v)_{-dx/2} - (\rho v)_{+dx/2} = -(\partial(\rho v)/\partial x)dx \longrightarrow$$

$$(\partial \rho/\partial t)_x = -\nabla \cdot (\rho v) \text{ [Eulerian Continuity Equation] }.$$

Here "Eulerian" is synonymous with a "fixed" as opposed to "comoving" grid. The grid moving with the material is called "Lagrangian". The continuity equation has a Lagrangian form which follows from its Eulerian cousin :

$$\dot{\rho} \equiv (\partial \rho/\partial t) + v \cdot \nabla \rho = -\rho \nabla \cdot v \text{ [Lagrangian Continuity Equation] }.$$

The Lagrangian continuity equation is formulated so as to follow the change in density experienced by a comoving measuring device accompanying the flow at the flow velocity v.

The Eulerian motion equation, which evolves ρv, requires the measurement of the pressure tensor as well as the velocity and density. See again the upper drawing in **Figure 11.1**. The piston to the left exerts a pressure $P_{-dx/2}$ and that on the right $P_{+dx/2}$ with $P(x)$ varying smoothly (differentiably) in between. If we consider the change in the momentum within dx there are two convective flow contributions plus the additional momentum flux [= pressure] contribution symbolized by the pistons :

$$(\partial(\rho v dx)/\partial t)_x = (\rho vv)_{-dx/2} - (\rho vv)_{+dx/2} + (P_{xx})_{-dx/2} - (P_{xx})_{+dx/2} .$$

Parts of the first three terms can be eliminating by using the product of v and the Eulerian continuity equation :

$$v(\partial \rho/\partial t) + v\nabla \cdot (\rho v) \equiv 0 .$$

What remains is the Lagrangian equation of motion :

$$\rho \dot{v} = \rho[(\partial v/\partial t) + v \cdot \nabla v] = -\nabla \cdot P \text{ [Lagrangian Motion Equation] }.$$

Comoving velocity in the *comoving* frame has no convective contribution.

Exercise : Consider the two-dimensional motion equation explicitly in order to find the contribution of the gradients of the shear stress tensor, $\sigma_{xy} \equiv -P_{xy} = -P_{yx}$ to the x and y components of the corresponding (vector) equation of motion.

The Eulerian energy equation follows a now familiar path, starting with the energy content of dx, $\rho[\, e + (v^2/2)\,]dx$ and summing the convective flow of energy, plus the work done by the pressure forces $P_{xx}v_x dt$, plus the energy transferred by heat conduction, the heat flux Q_x. Using the Eulerian continuity equation and equation of motion, the final Lagrangian energy equation has just the form one would expect from the combined First and Second Laws of Thermodynamics :

$$\rho\dot{e} = -P : \nabla v - \nabla \cdot Q \ [\text{ Lagrangian Energy Equation }]\ .$$

The pressure-volume work term contains four separate contributions in two dimensions and nine in three. A slightly more detailed derivation is given in Section 7.10 of *Computational Statistical Mechanics*, free at williamhoover.info on the web.

11.3 Constitutive Relations for Continuum Mechanics

In the momentum and energy equations we saw gradients of both the pressure tensor P and the heat-flux vector Q. But how are these to be determined ? Both require "constitutive relations" relating them to the fundamental field variables (ρ, v, e) . A fluid could be described by the ideal-gas law, $P_{eq} = \rho k T = \rho e$, or by van der Waals' equation of state, or by a simple power law, $p \propto \rho^{\gamma}$. To such equilibrium models we can add Newtonian viscosity η and Fourier's thermal conductivity κ :

$$P_{xy} \equiv P_{yx} = -\eta[\ (\partial v_x/\partial y) + (\partial v_y/\partial x)\ ]\ ;\ Q = -\kappa\nabla T\ .$$

In formulating Newtonian shear viscosity the nonequilibrium part of the pressure tensor is necessarily symmetric, in order to avoid a catastrophic rotational instability. See again Section 1.10.2 on pages 34-35.

11.4 Structure of Steady Shockwaves – Direct Integration

A strong sound wave with a large localized increase in density can be excited by a bullet, a hammer, or a jet airplane. In typical fluids the sound speed is an increasing function of density. In the moving wave the density gradient steepens as the denser fluid catches up with the less dense. Soon a near discontinuity results, a "shockwave". "Sonic booms" and thunder are the most familiar examples. The stabilizing (dissipative) effects of viscosity and heat conductivity spread the steepest wave structure out over a distance comparable to the fluid's "mean free path". The mean free path is the distance between molecular collisions, about 3×10^{-7} meters $= 3000$ Ångstroms for the sea-level atmosphere on earth. This is about one thousand times the diameter of a molecule. Under water, where the molecular free path is of order one Ångstrom, the shock width is tiny, molecular in size.

An example solution of the continuum equations[1,2] for a steady shock-wave can be found by solving two coupled differential equations. Let us find the stationary structure of such a wave based on the following constitutive assumptions for a fluid in two space dimensions :

$$e = (\rho/2) + 2T \; ; \; P = \rho e = (\rho^2/2) + 2\rho T \; ;$$
$$P_{xx} = P - (dv/dx) \; ; \; P_{yy} = P + (dv/dx) \; ; \; Q = -(dT/dx) \; .$$

The direction of travel is parallel to the x axis so that all field variables are time-independent functions of x alone.

Exercise : Show, by interpreting the pressure tensor in two dimensions as force per unit length, that the symmetric shear stress $-P_{xy}$ becomes $\pm(P_{xx} - P_{yy})/2$ when the (x, y) coordinate system is rotated $\pm 45°$.

We have assumed the value unity for both the shear viscosity and the thermal conductivity. The pressure and energy have repulsive "cold curve" contributions depending upon the density plus a thermal part linear in the temperature. The internal energy per particle e includes $2T$ just as in a solid-phase harmonic-oscillator model. A two-dimensional oscillator at temperature T has a kinetic energy of kT and an equal potential-energy contribution. We have also made the "Stokes Assumption" that the bulk viscosity vanishes (so that $P_{xx} + P_{yy} = 2P_{eq}$). The assumption dates back to the days when little was known about bulk viscosity. We use it here for simplicity's sake.

Exercise : Show that a degree of freedom with a Gaussian velocity distribution $f(v) \propto e^{-mv^2/2kT}$ has an average value of the mean-squared velocity equal to (kT/m) . In evaluating Gaussian integrals it is convenient to remember

$$\int_0^\infty 2\pi r e^{-\alpha r^2/2} dr = (2\pi/\alpha) = 4[\int_0^\infty e^{-\alpha x^2/2} dx][\int_0^\infty e^{-\alpha y^2/2} dy] \; .$$

Show that differentiation of these integrals with respect to α provides values for the integrals $\int_0^\infty e^{-\alpha r^2/2} r^n dr$ where n is a positive integer.

For simplicity we assume twofold compression from a cold state, $T = 0$ at $\rho = 1$ with velocity $+2$, to a "hot state" at temperature T and density $\rho = 2$ with a velocity $+1$. These assumptions are enough to constitute a well-posed problem. Verify that conservation of mass is satisfied and that conservation of momentum implies a hot temperature of $T = (1/8)$. This information is enough to find the steady values of the mass, momentum, and energy fluxes, 2, (9/2), and 6. The numerical solution of this problem can be simplified by assuming a time-independent solution, a solution of the wave structure in the "comoving frame" centered on the moving wave. In that frame the mass, momentum, and energy fluxes are all constant :

$$\rho v = 2 \; ; \; P_{xx} + \rho v^2 = (9/2) \; [\text{ mass and momentum }] \; ;$$

$$\rho v [\, e + (P_{xx}/\rho) + (v^2/2) \,] + Q = 6 \; [\text{ energy conservation }] \; .$$

These last two equations provide ordinary differential equations in the two variables $v(x)$ and $T(x)$:

$$P_{xx} = \rho e - (dv/dx) \; ; \; Q = -(dT/dx) \; .$$

Within the shockwave the variations of density, temperature, energy per unit mass, P_{xx}, and velocity are :

$$1 < \rho < 2 \; ; \; 0 < T < \tfrac{1}{8} \; ; \; \tfrac{1}{2} < e < \tfrac{5}{4} \; ; \; \tfrac{1}{2} < P_{xx} < \tfrac{5}{2} \; ; \; 2 > v > 1 \; .$$

A straightforward solution method proceeds by using RK4 to integrate the two ordinary differential equations for (dv/dx) and (dT/dx). The righthandsides of the two differential equations, YYP(1) and YYP(2) are calculated from the current values of v and T as follows :

```
RHO = (2/V)
PXX = 4.5D0 - RHO*V*V
E   = (RHO/2) + 2*T
P   = RHO*E
YYP(1) = P - PXX                             ! This is (DV/DX)
YYP(2) = RHO*V*(E + (PXX/RHO) + (V*V/2)) - 6 ! This is (DT/DX)
```

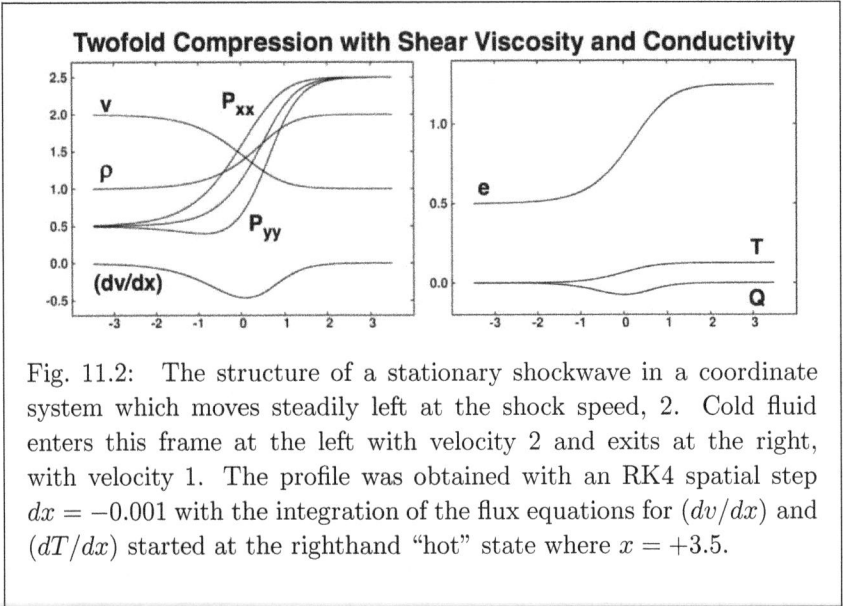

Fig. 11.2: The structure of a stationary shockwave in a coordinate system which moves steadily left at the shock speed, 2. Cold fluid enters this frame at the left with velocity 2 and exits at the right, with velocity 1. The profile was obtained with an RK4 spatial step $dx = -0.001$ with the integration of the flux equations for (dv/dx) and (dT/dx) started at the righthand "hot" state where $x = +3.5$.

Starting at the "hot" end of the shockwave at x $= +3.5$, with density 1.9999, velocity 1.0001, and temperature 0.12499 provides the smooth solution shown in **Figure 11.2**. 7000 values of the independent variable $-3.5 < x < +3.5$ are generated with 7000 steps, $dx = -0.001$.

In **Figure 11.3** we show the very similar shockwave obtained when the thermal conductivity vanishes : $\eta = 1$ and $\kappa = 0$. We have chosen to plot that simpler shockwave as a function of time. This shows, over a time range of 5, emphasizing the slower hotter region, the environment seen by a Lagrangian observer. That comoving observer enters the interval range $-3.5 < x < +3.5$ at the left at time $t = 0$ and exits at $x = +3.5$, with an elapsed time of $t = 5.0_3$. While passing through the shockwave the observer is slowed from his entrance velocity of 2 to his exit velocity of 1.

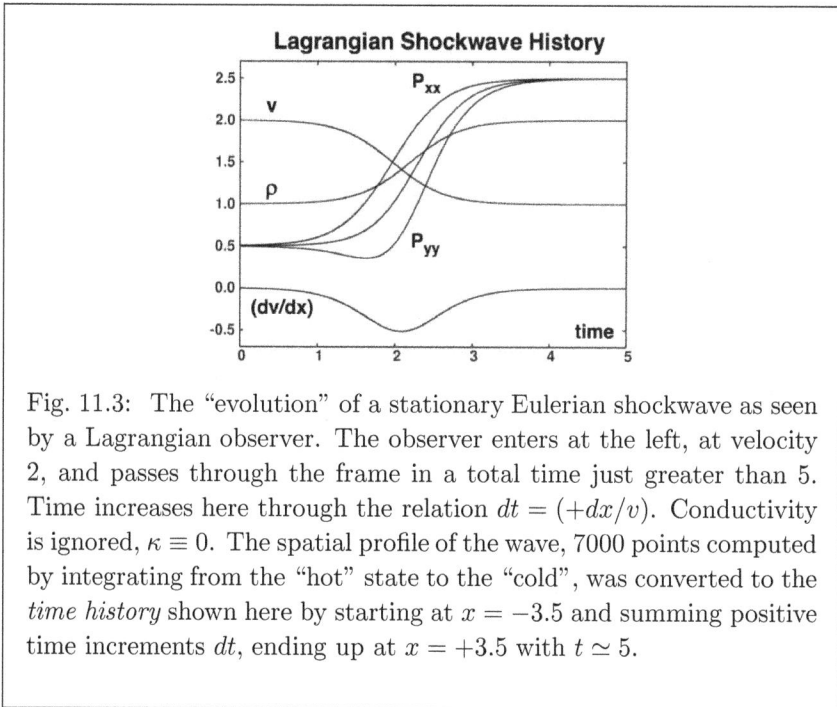

Fig. 11.3: The "evolution" of a stationary Eulerian shockwave as seen by a Lagrangian observer. The observer enters at the left, at velocity 2, and passes through the frame in a total time just greater than 5. Time increases here through the relation $dt = (+dx/v)$. Conductivity is ignored, $\kappa \equiv 0$. The spatial profile of the wave, 7000 points computed by integrating from the "hot" state to the "cold", was converted to the *time history* shown here by starting at $x = -3.5$ and summing positive time increments dt, ending up at $x = +3.5$ with $t \simeq 5$.

Evidently heat conductivity is unimportant in determining shockwave structure. In fact for the boundary conditions we have chosen to investigate there is no solution for $\eta = 0$ and $\kappa = 1$. Conductivity alone does not explain shockwaves.

Exercise : Show for $\eta = 1$ and $\kappa = 0$ that the conservation laws reduce to a single ordinary differential equation (for the same boundary conditions and constitutive assumptions as before) : $(dv/dx) = (3/v)(v - 1)(v - \frac{1}{2}) \longleftrightarrow (d\rho/dx) = (3\rho/2)(\rho - 1)(2 - \rho)$.

The two example problems solved here both relied on the steady-state assumption. Stationarity eliminates time dependence, resulting in three constant-flux equations for $(d\rho/dx)$, (dv/dx), and (dT/dx). A time-dependent problem, for instance computing the transient evolution of the shockwave from an initial condition, requires solving separate differential equations along a grid in the x direction. From the computational standpoint molecular dynamics, where the transient and the steady-state computations evolve naturally, is somewhat simpler to program, with the same computational effort for each time step. In continuum mechanics adding time dependence greatly increases the computational complexity. Let us briefly consider a simplistic approach to solving the zero-conductivity problem of **Figure 11.3** using Eulerian finite differences.

11.5   Shockwave Structure Using an Eulerian Grid

There are many approaches to grid-based solutions of continuum problems. And many of them fail, particularly when applied to problems including colliding or fragmenting materials or to problems involving steep gradients. It is particularly instructive to write a program for a problem where the longtime stationary solution is already known. It took me nearly a day's effort to write such a program, eliminate errors, improve stability, and confirm convergence. The various steps in the computation were as follows: [1] choose a grid of 1000 points with $dx = 0.01$, spanning the range of the problem; [2] initialize the values of v and ρ and e at the interior grid points (2 through 999) using interpolating guesses that closely satisfy the boundary conditions at $x = \pm 5$ where $\tanh(\pm x) = \pm 0.99991$:

$$v = (3/2) - (1/2)\tanh(x) \; ; \; \rho = (2/v) \; ; \; e = (7/8) + (3/8)\tanh(x) \; .$$

For stability it is best to evaluate the velocity and energy at the nodes. Then evaluate the density and the pressure tensor in the zones or cells between neighboring nodes. [3] use centered-difference formulæ in evaluating the derivatives needed for interior zones or nodes 2 through 999 :

$$(\partial\rho/\partial t)_i = -[\,(v\rho)_{i+1} - (v\rho)_{i-1}\,]/(2dx) \; ;$$

$$(\partial v/\partial t)_i = -v_i[\,v_{i+1} - v_{i-1}\,]/(2dx) - (Pxx_{i+1} - Pxx_{i-1})/(2\rho_i dx) \; ;$$

$$(\partial e/\partial t)_i = -v_i(e_{i+1} - e_{i-1})/(2dx) - (Pxx_i v_{i+1} - Pxx_i v_{i-1})/(2\rho_i dx) \; .$$

[4] Integrate these 3 × 998 differential equations for interior values of (ρ, v, e) using RK4 until a stationary state is reached. 25,000 timesteps with $dt = 0.0002$ was adequate. In order to avoid an odd-even instability it was expedient to "smooth" the profiles every thousand timesteps, using a two-step "artificial diffusion" process:

$$\langle\, f_i \,\rangle = (1/4)[\, f_{i+1} + 2f_i + f_{i-1}\,] \; ; \; f_i = \langle\, f_i \,\rangle \; .$$

The resulting profile was quite consistent with the stationary solution presented in **Figure 11.3**.

11.6 Rayleigh-Bénard Flow Using an Eulerian Grid

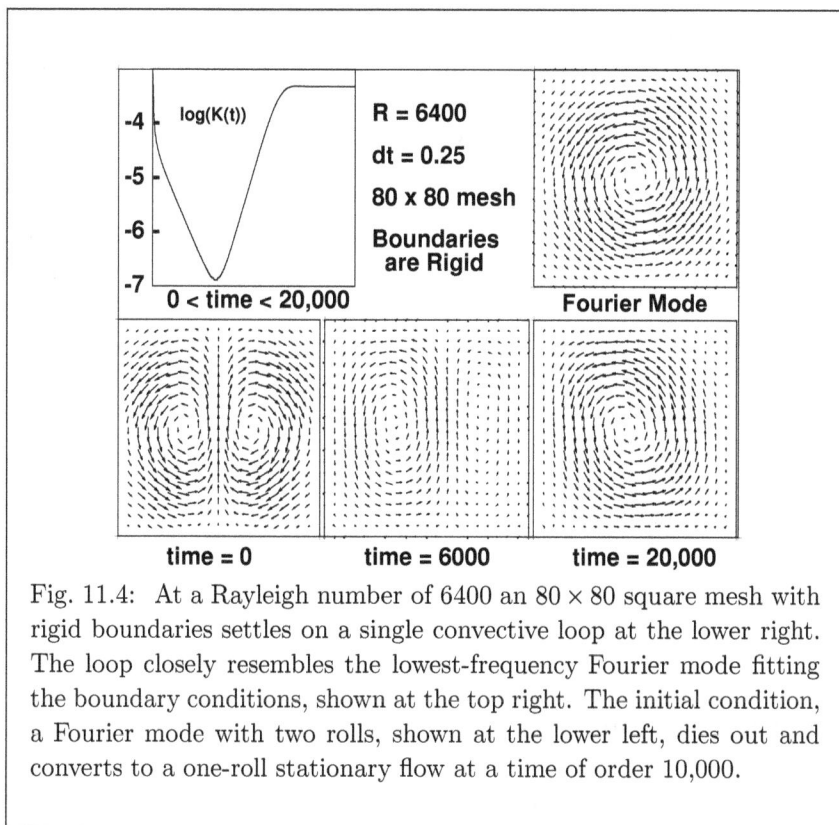

Fig. 11.4: At a Rayleigh number of 6400 an 80 × 80 square mesh with rigid boundaries settles on a single convective loop at the lower right. The loop closely resembles the lowest-frequency Fourier mode fitting the boundary conditions, shown at the top right. The initial condition, a Fourier mode with two rolls, shown at the lower left, dies out and converts to a one-roll stationary flow at a time of order 10,000.

The shockwave problem, with its known solution, is a good introduction to Eulerian continuum mechanics. Let us turn toward a two-dimensional problem, seemingly more complex, but simplified somewhat by its stationary boundaries. This is the "Rayleigh-Bénard" problem. It has an amazing variety of solutions, as we shall see.

The Rayleigh-Bénard problem treats a compressible and conducting fluid heated from below, cooled from above, and acted on by a gravitational field. In 1916 Lord Rayleigh, John William Strutt (1842-1919) was first to analyze the convective instability that enhances static Fourier heat flow by adding a steady convective flow of the type shown in **Figure 11.4**. The thermal expansion at the lower boundary, coupled with gravity, provides a mechanism for convection. Convection is limited by shear viscosity.

Figure 11.4 shows the time history of the kinetic energy starting with a two-roll initial condition. Over a time of a hundred sound traversal times the system discovers and adopts the stationary single-roll morphology shown at the bottom right. The upper right velocity field is the simplest roll pattern consistent with fixed boundaries on all four sides.

This problem geometry can provide solutions in which heat flow and viscous drag act simultaneously to generate internal chaos and turbulence even though the boundary conditions are simple and stationary. We confine our discussion here to a two-dimensional square system with vanishing boundary velocity on all four walls and with the simplest possible constitutive relations. Despite this simplicity a wide variety of solutions can result, from static linear heat flow to stationary convection[3] to wildly chaotic and unstable convective turbulence.[4] We focus on results obtained for three different flow regimes. These situations are described in greater detail in Section 4.11.2 of Reference 2, pages 144-154.

Because several properties of the heated fluid are involved in determining the flows' morphologies it is convenient to fix the gravitational field, $g \equiv (1/L)$, and the lateral boundary conditions as well as the ratio of the viscosity and heat conductivity and the top-to-bottom temperature difference. We choose the two-dimensional monatomic ideal-gas equation of state, $P = \rho e = (NkT/V)$ with a "hot" temperature of 1.5 at the base and a "cold" 0.5 at the top throughout. It is convenient and conventional to express the deviation from equilibrium for this problem in terms of the Rayleigh number $\mathcal{R}$:

$$\mathcal{R} \equiv (\partial \ln V/\partial T)_P \Delta T g L^3 /(\nu D_T) \ .$$

Here ν is kinematic viscosity (η/ρ) and D_T is thermal diffusivity $(\kappa/\rho C_V)$. Both transport coefficients have units [length^2/time] . The numerator of $\mathcal{R}$ corresponds to "lift" through thermal expansion while the denominator corresponds to "drag". Their ratio determines the flow type, if any.

We obtained the simulation results in the Figures by applying the same ideas as in the shockwave simulations of Sections 11.4 and 11.5. We do not show the static case for there is little to see : a pressure profile which decreases linearly from bottom to top, a stationary upward heat flux, and a linear temperature profile, decreasing from bottom to top. **Figure 11.4** shows the single-roll convection which replaces the static situation at a Rayleigh number near 5000. Let us increase the Rayleigh number to 160,000. We find that an interesting periodic sequence of two-, three-, and four-roll patterns has replaced the single roll. **Figure 11.5** shows a sequence of roll patterns clockwise from top left. The kinetic energy of the flow field varies by roughly a factor of five during the sequence. Then comes a precise mirror-image sequence (reflected about the vertical y axis). The period of the complete cycle, twice that of the figure, is 2592 for this 80×80-cell simulation.

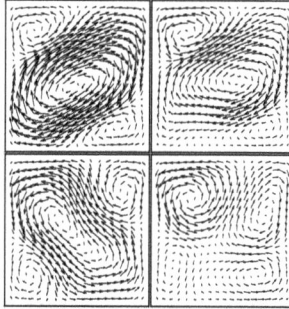

Fig. 11.5: At a Rayleigh number of 160,000 an 80×80 square mesh with rigid boundaries traces out a complicated time-periodic flow. We followed the flow for ten periods, each of which includes the four views shown here together with four more corresponding to precise left-right mirror images. This flow, though complicated, is periodic in time.

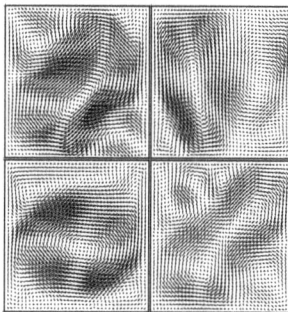

Fig. 11.6: At a Rayleigh number of 800,000 this 160×160 mesh shows the chaos which has ensued as the Rayleigh number has increased five-fold above that generating the time-periodic flow of **Figure 11.5**.

Increasing the Rayleigh number further to 800,000 gives a hint of the complexity still to come, requiring more and more zones to resolve the structure, taking more and more time as the transit time of the flow and the number of zones both increase to maintain stability and accuracy. **Figure 11.6** shows four snapshots in a chaotic 160×160-zone flow. Now the fluctuations in the kinetic energy span a factor of ten with the time dependence chaotic in appearance. To check on the possibility of chaos we sum up the separation of two simultaneous simulations, $\sqrt{\sum(\delta_\rho^2 + \delta_v^2 + \delta_e^2)}$. The time between successive peaks of kinetic energy is of order 2000. **Figure 11.7** shows the growth of the separation for a time of roughly 50 such peaks. The slope is the Lyapunov exponent, $\simeq 0.0002$, and corresponds to about two kinetic-energy oscillations per e-folding.

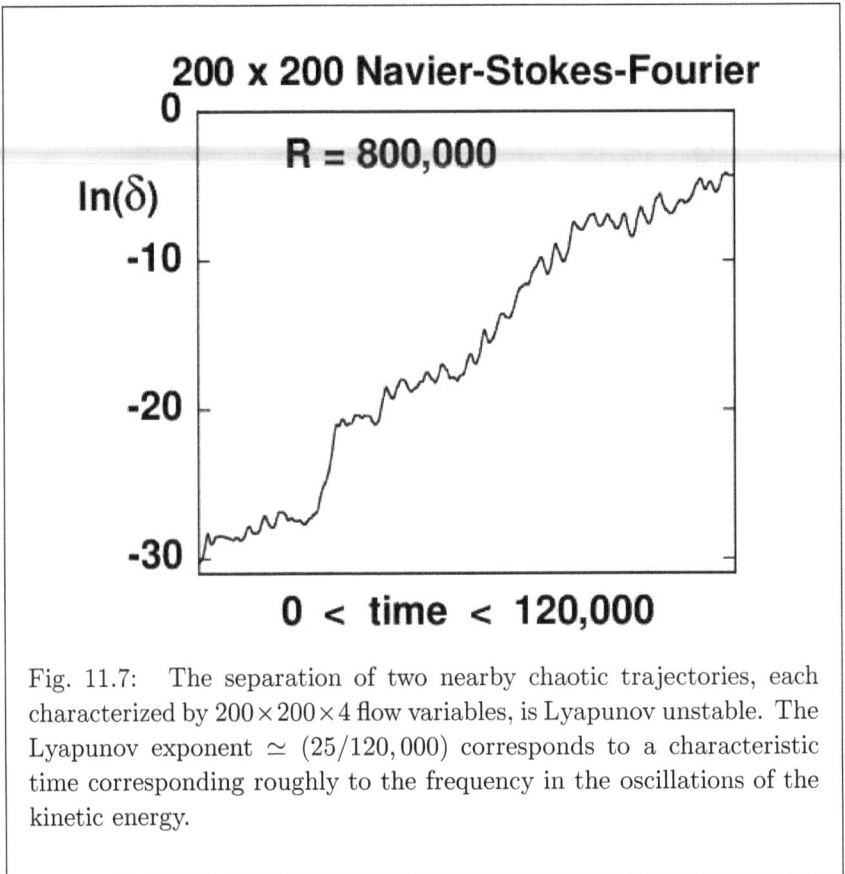

Fig. 11.7:   The separation of two nearby chaotic trajectories, each characterized by $200 \times 200 \times 4$ flow variables, is Lyapunov unstable. The Lyapunov exponent $\simeq (25/120,000)$ corresponds to a characteristic time corresponding roughly to the frequency in the oscillations of the kinetic energy.

The familiar Lorenz attractor is a caricature of the Rayleigh-Bénard problem involving just three ordinary differential equations rather than the over $200 \times 200 \times 4 \times 2 \simeq 320,000$ required for the Lyapunov analysis just described. The complexity of Rayleigh-Bénard flow is daunting. Evidently only a single transient solution can be followed, and with much greater expense, in three-dimensional simulations. In the laboratory the transition times from one flow morphology to another can take on the order of days to weeks. In addition to patience one needs to exercise common sense in deciding for how long and in what depth to study these flows.

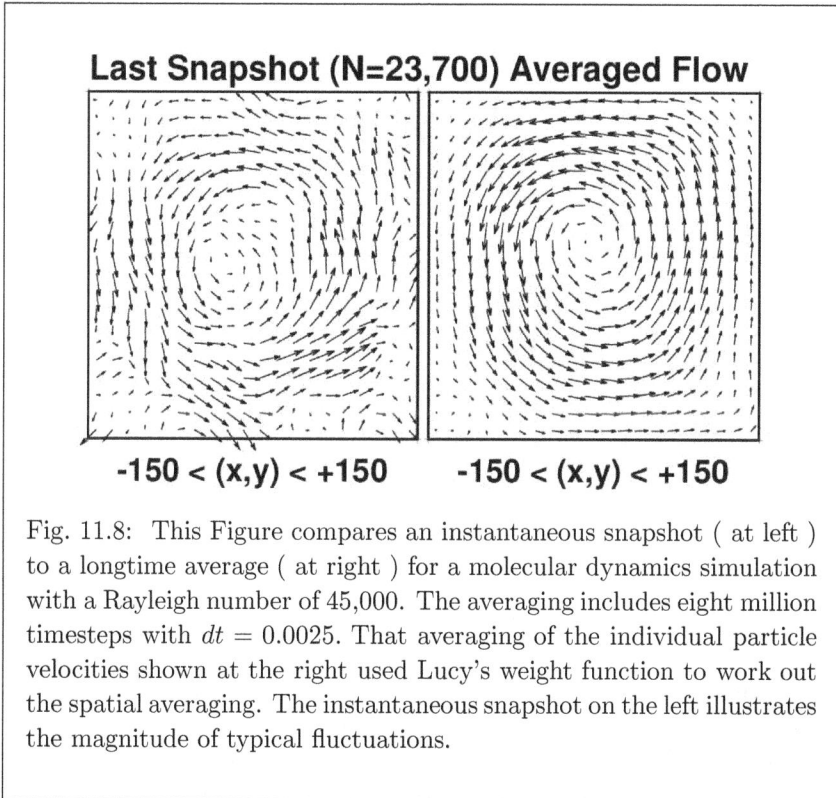

Last Snapshot (N=23,700) Averaged Flow

-150 < (x,y) < +150   -150 < (x,y) < +150

Fig. 11.8: This Figure compares an instantaneous snapshot (at left) to a longtime average (at right) for a molecular dynamics simulation with a Rayleigh number of 45,000. The averaging includes eight million timesteps with $dt = 0.0025$. That averaging of the individual particle velocities shown at the right used Lucy's weight function to work out the spatial averaging. The instantaneous snapshot on the left illustrates the magnitude of typical fluctuations.

For an even more demanding more stimulating approach to this simple model of turbulence one can follow these flows with molecular dynamics. **Figure 11.8** shows an instantaneous snapshot as well as a time average of a 23,700-particle laptop-computer simulation at a Rayleigh number of 45,000 using molecular dynamics.

Figure 11.9, from Dennis Rapaport's 2006 work, required parallel processing for treating 3,507,170 particles interacting with a short-ranged repulsive potential.[5]

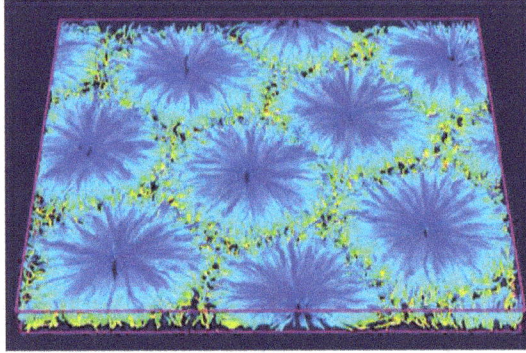

Fig. 11.9: Streamlines showing the motion and the temperature (with blue coldest and descending at the centers of the convection cells) . This three-dimensional simulation took about one month of machine time on an eight cpu cluster and is nicely described in Reference 5.

The averaging process required to convert particle data to fields like the flow field $v(x, y)$ shown in the Figure is based on "weight functions". The idea is to average the influence of particles over a neighborhood of two dozen particles or so. Density at a point r_p, for example, can be defined as the sum of weights from particles within the range $h = 3$ of $\{ r_p \}$:

$$\rho(r_p) \equiv \sum_j w(r_p - r_j) \; ; \; w(z) = (5/9\pi)(1 + 3z)(1 - z)^3 \; ; \; z \equiv (|r|/3) \; .$$

Notice that the weight function is normalized, $\int_0^\infty 2\pi r w(r) dr \equiv 1$. The nearest-neighbor spacing is about unity. Three is a typical range for a suitable weight function, ($h = 3 \rightarrow z < 1$) under these conditions. To find the local flow velocity using smooth-particle averaging the same idea is applied in summing up nearby particles' velocities :

$$v(r_p) \equiv \sum w(r_p - r_j) v_j / \sum w(r_p - r_j) \; .$$

The velocity arrows in **Figure 11.8** were constructed in just this way. It is clear that the smooth-particle averaging has several desirable properties, simplicity, flexibility, and spatial averages with two smoothly-varying derivatives, ideal for calculating the gradients of not just (ρ, v, e) but also those field variables containing gradients themselves, $\nabla \cdot Q$ and $\nabla \cdot P$.

In addition to its applications to molecular dynamics or Monte Carlo simulations, the smooth-particle idea has also proved useful in *continuum* simulations. All that is required is to replace the partial differential equations of continuum mechanics by equivalent ordinary differential equations for evolving representative smooth particles. The result is a relatively new numerical method for continuum mechanics that is no more complicated than molecular dynamics. The basic idea is primarily due to J. J. Monaghan[6] who helped introduce the method in 1977 and has developed it ever since, as his life's work. Our own perspective on the subject is described in Reference 7. We term the method "SPAM" = "**S**mooth **P**article **A**pplied **M**echanics" in order to emphasize that the technique is applicable to solids as well as to fluids other than water.

11.7   Smooth Particle Applied Mechanics

Mesh-based methods for solving the continuum equations for the evolution of (ρ, v, e) are useful so long as the flow is slow and smooth, avoiding shockwaves, and so long as fracture and fragmentation are avoided. The difficulties associated with "rezoning", changing the mesh in response to local conditions, suggest developing a mesh composed of moving particles, each of them representing a part of the fluid, with its own mass, velocity, and energy. Associating a shape function or weight function with each particle, describing its region of influence, suggests a method well-suited to deal with violent deformation and fragmentation.

The continuum motion and energy equations typically involve second derivatives of the velocity and energy fields. For this reason it is desirable that the weight functions representing particles be "smoothed" over a characteristic length h with at least two continuous derivatives at the cutoff radius h. The simplest polynomial with this property, with a smooth maximum at the origin and a very smooth cutoff at h is Lucy's quartic weight function :

$$w(r < h) = C[\, 1 - 6z^2 + 8z^3 - 3z^4\,] \text{ with } z \equiv (r/h) \ .$$

C is a normalization constant. The integral of the weight function needs to be chosen to match the mass of one particle. For unit mass particles and in two dimensions the constant $C = (5/\pi h^2)$ provides the normalization $\int_0^h 2\pi r w(r) dr \equiv 1$, as was used to construct **Figures 11.4-11.8**.

> **Exercise :** Find normalization constants for Lucy's w in one dimension, where $\int_{-h}^{+h} w\,dx = 1$ and in three where $\int_0^h 4\pi r^2 w\,dr = 1$.

We have seen the utility of the smooth-particle spatial sums of molecular dynamics data to provide the flow field $v(r,t)$ for Rayleigh-Bénard convection cells. Such sums can convert the continuum conservation laws to sets of ordinary differential equations for $(\dot\rho, \dot v, \dot e)$. In addition to freedom in choosing the range and form of the weight function there is freedom too in selecting those smooth-particle sums that best represent the field variables. It is desirable that these sums obey the same conservation relations as do the original continuum equations that they model. Let us consider the simplest instance of smooth-particle modelling, the continuity equation for the conservation of mass.

11.7.1 *The Continuity Equation is an Identity with SPAM*

Because each SPAM particle has a definite mass spatially distributed by its weight function conservation of mass is satisfied automatically. The continuity equation is the differential relation linking density and velocity. It is not obvious that this differential relation can be satisfied by smooth-particle sums. But it is possible. To see how we begin by defining the density ρ and velocity v at r by sums :

$$\rho_r \equiv \sum_j w_{rj} \ ; \ (\rho v)_r \equiv \sum_j w_{rj} v_j \ .$$

According to the continuity equation the time derivative of the density ρ is simply related to the spatial derivative of (ρv) : $(\partial\rho/\partial t)_r = -\nabla_r\cdot(\rho v)_r$. Let us see if the SPAM definitions are consistent with the continuity equation.

Without loss of generality we choose the field point r and compute the change in density there due to the motion of those particles $\{\,j\,\}$ which are close enough to influence the smooth-particle sums :

$$(\partial\rho/\partial t)_r = \sum_j (\partial w_{rj}/\partial t)_r \ = \sum_j \nabla_j w_{rj}\cdot v_j = -\sum_j \nabla_r w_{rj}\cdot v_j \ .$$

We recognize the last expression as the righthandside of the continuity equation (because $\nabla_r \cdot v_j \equiv 0$) :

$$-\nabla_r\cdot(\rho v)_r \equiv -\sum \nabla_r w_{rj}\cdot v_j \ .$$

Thus the continuity equation is reproduced precisely, independent of the choice of h and the functional form of w, making it unnecessary to integrate the $\dot{\rho}$ equation in SPAM simulations. Density can be determined precisely by evaluating a weight-function sum, $\sum_j w_{rj}$. Considerable interesting research can be carried out on corrections analogous to surface tension, where interfaces or free surfaces are involved.

11.7.2 *Exact Conservation of Momentum with SPAM*

Although there are many choices of motion equations with smooth particles, one unique choice appears specially propitious in that it conserves momentum exactly. Let us develop that choice next. We begin by rewriting the Lagrangian continuum equation of motion as the sum of two terms :

$$\rho \dot{v}_r = -\nabla_r \cdot P \longrightarrow \dot{v}_r = -\nabla_r \cdot (P/\rho)_r - (P/\rho^2)_r \cdot \nabla_r \rho .$$

Both terms on the righthand side can be written as smooth-particle sums provided that we make use of a plausible definition for the gradient of (F/ρ) where the field variable F will be either P or ρ^2 :

$$(F/\rho)_r \equiv \sum_j (F/\rho^2)_j w_{rj} \longrightarrow \nabla (F/\rho)_r \equiv \sum_j (F/\rho^2)_j \nabla_r w_{rj} .$$

The further step of evaluating the lefthandside of these equations *at* the location of Particle i then gives a set of ordinary differential equations of motion which can be used in smooth-particle simulations of continua :

$$\{\ \dot{v}_i = -\sum_j [\ (P/\rho^2)_j + (P/\rho^2)_i\ ] \cdot \nabla_i w_{ij}\ \} .$$

Notice that if we sum this equation of motion over all $\{\ i\ \}$ each of the bracketted pair terms occurs twice, once dotted into $\nabla_i w_{ij}$ and once into $\nabla_j w_{ij}$. The two cancel so that *this SPAM algorithm conserves momentum exactly.* Note also that the private velocity of the ith Particle v_i is different to the smooth-particle field variable at that same location r_i, $\sum_j v_j w_{ij} / \sum_j w_{ij}$. Judicious approximations are inherent in SPAM.

The observant reader will see that the smooth-particle equation of motion just given has a familiar form for the constitutive relation $P = (\rho^2/2)$. It is the isoenergetic microscopic equation of motion for a Lucy system, either fluid or solid. It is also the SPAM representation of a two-dimensional adiabatic and isentropic ideal gas. The SPAM model for a two-dimensional monatomic ideal gas provides equations of motion identical to the classical mechanics of a fluid with Lucy's potential function :

$$\{\ \phi_{ij} = (5/\pi h^2)(1 - 6z^2 + 8z^3 - 3z^4)\ ;\ z = (r_{ij}/h)\ \} .$$

11.7.3   *Exact Conservation of Energy with SPAM*

The continuum energy equation $\dot{e} = (1/\rho)[\,-\nabla P : \nabla v - \nabla \cdot Q\,]$ can be converted to its SPAM analog by a diligent extension of the ideas contributing to the continuity and motion equations :

$$\dot{e}_i = +\tfrac{1}{2}\sum[(P/\rho^2)_i + (P/\rho^2)_j] : v_{ij}\nabla_i w_{ij} - \sum[(Q/\rho^2)_i + (Q/\rho^2)_j] \cdot \nabla_i w_{ij} \ .$$

Here the sums are over all Particles $\{\,j\,\}$ and $v_{ij} = v_i - v_j$.

The analogy with molecular dynamics in the special inviscid nonconducting case with $P = (\rho^2/2) = \rho e$ is exact. Lucy's potential function is an interesting special case providing a second route to the "work term" in the energy equation. With pairwise-additive identical particles there are two plausible analogs for the individual particle energies. A simple analogy drawn from the motion equation is to identify the smooth-particle density with the energy. Apart from an additive constant ϕ_{ii} the smooth-particle density corresponds to the energy of particles interacting with Lucy's weight function, suggesting that the mechanical energy change in the continuum correponds to chain-rule evaluation of a Lucy's particle's density change :

$$\dot{e}_i \propto \dot{\rho}_i = (1/2)\sum_j (v_i - v_j) \cdot \nabla w_{ij} \ [\text{ Difference Form of } \dot{e}\,] \ ,$$

a shortened version of the SPAM energy equation given above.

On the other hand, if one simply divides up the total energy of a system with pairwise-additive forces the natural choice is :

$$e_i \equiv (1/2)\sum_j \phi(r_{ij}) + (1/2)(v_x^2 + v_y^2)_i \ ; \ \sum_i e_i \equiv E \ .$$

With this definition the kinetic energy change, $v_i \cdot \dot{v}_i$ makes an additional contribution to $\dot{e}_i$. The result is quite a different energy equation, with the velocities summed rather than differenced :

$$\dot{e}_i = (1/2)\sum_j (v_i + v_j) \cdot \nabla_i w_{ij} \ [\text{ Sum Form of } \dot{e}\,] \ .$$

Here $-\nabla w_{ij}$ is the repulsive force on Particle i due to its interaction with Particle j through the derivative of Lucy's weight function :

$$F(r_{ij}) = -\nabla_i w(r_{ij}) = (60/\pi)r_{ij}(1 - |r_{ij}|)^2 \ \text{ when } 0 < |r_{ij}| < 1 \ .$$

The force is a vector parallel to the line $r_i - r_j$.

The difference between these two points of view is whether or not the kinetic energy of a particle, which is crucially dependent on the choice of coordinate frame, is a part of that particle's private internal energy or not.

Without venturing a final opinion on this interesting question let us work out an example problem to verify the validity of the second point of view. Where kinetic energy is included in each Lucy particle's private energy it is the sum, rather than the difference, of particle's velocities that contributes to $\dot{e}_i$.

Figure 11.10 shows the time-dependence of the individual particle energies for a three-body Lucy fluid simulation with $h = 1$ in a 3×3 periodic box with initial conditions :

$$\{ (x, y, v_x, v_y) \} = \{ (1, 1, -1, 0), (1, -1, -1, 0), (-1, -0.5, 2, 0) \} .$$

The initial energies are purely kinetic, $(1/2), (1/2), 2$, so that $(E/N) = 1$. Summing the kinetic and potential energies for each particle gives results identical to those computed by integrating the three $\{ \dot{e} \}$ equations. The agreement is good to at least seven figures after one million RK4 timesteps with $dt = 0.001$.

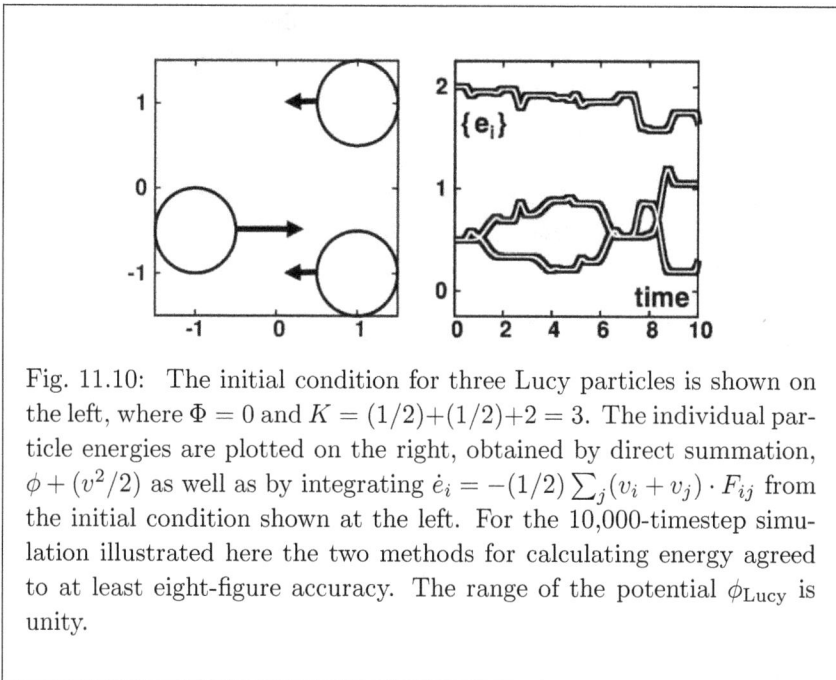

Fig. 11.10: The initial condition for three Lucy particles is shown on the left, where $\Phi = 0$ and $K = (1/2)+(1/2)+2 = 3$. The individual particle energies are plotted on the right, obtained by direct summation, $\phi + (v^2/2)$ as well as by integrating $\dot{e}_i = -(1/2) \sum_j (v_i + v_j) \cdot F_{ij}$ from the initial condition shown at the left. For the 10,000-timestep simulation illustrated here the two methods for calculating energy agreed to at least eight-figure accuracy. The range of the potential ϕ_{Lucy} is unity.

Although the choice $e_i = \phi_i + (1/2)v_i^2$ is "natural" one could equally well imagine that the velocities contribute to a fluctuating macroscopic velocity field. Under those conditions it is tempting to define a configurational temperature to replace the missing kinetic definition :

$$T_c \equiv \langle\, (\nabla \mathcal{H})^2 \,\rangle / \langle\, \nabla^2 \mathcal{H}^2 \,\rangle \ .$$

For the Lucy potential the denominator in this expression vanishes (leading to the divergence of the "temperature" and to our lack of enthusiasm for it) when $r = (h/3)$. Evidently a variety of interesting studies optimizing the smooth-particle fluid description and exploring its connections to the Lucy fluid could be carried out. A practice problem useful for developing a SPAM code is Rayleigh-Bénard flow. This illustrates the potential of the method, as we shall see from a variety of interesting results, discussed next.

11.8 Rayleigh-Bénard Flow with SPAM

Mesh-based models of three-dimensional macroscopic flows involve following dozens of variables, including the spatial gradients $\nabla\rho$ and ∇e, as well as the tensors ∇v, and P, and the divergences $\nabla \cdot P$ and $\nabla \cdot Q$. Even in the two-dimensional case there is a long list, with some variables required both in the zones and at the nodes. The programming is much simpler with SPAM where the number of particles is fixed and the key variables required for the evolution are $\{\, x, y, v_x, v_y, e \,\}$. We have seen how to compute the spatial derivatives of P and Q in the motion and energy equations. The velocity and temperature gradients needed for a Navier-Stokes description of the flow follow from definitions designed to give no momentum or energy flux between particles with the same v or T :

$$(\rho\nabla v)_i \equiv -\sum_j (v_i - v_j)\nabla_i w_{ij} \ ; \ (\rho\nabla T)_i \equiv -\sum_j (T_i - T_j)\nabla_i w_{ij} \ .$$

My Ph D student Oyeon Kum carried out a variety of Rayleigh-Bénard simulations with smooth particles and with conventional Eulerian meshes. **Figure 11.11** compares 5000-particle and 5000-zone simulations of an ideal-gas fluid with constant and equal values of the kinematic viscosity and thermal diffusivity driven by a threefold difference in temperature between the base and the top of a closed convection cell. The Rayleigh Number for these flows is 10,000, well above the approximate threshold value for convection, 1708, estimated by Chandrasekhar 50 years ago.

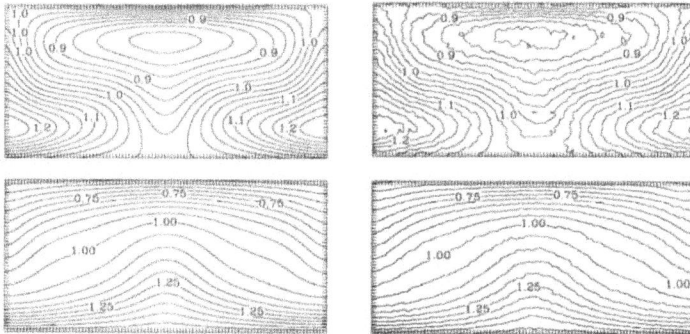

Fig. 11.11: Comparison of density (above) and temperature (below) contours for an ideal-gas flow at a Rayleigh Number of 10,000 with a mesh (to the left) and with smooth particles (to the right).

11.8.1 *Boundary Conditions with SPAM*

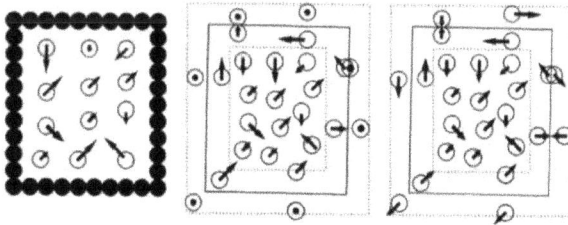

Fig. 11.12: Three implementations of rigid boundary conditions for smooth particles. In the rigid boundary case at the left the boundary is created of fixed particles with permanent values of their velocity and energy. Two other convenient boundary types use "mirror" particles assigned to any system particle that gets within the cutoff distance h of the boundary. In the middle panel the mirror particles have assigned velocities and energies characteristic of the boundary. On the right the boundary values are chosen such that the average velocity and temperature is reproduced: $v_{\text{system}} + v_{\text{mirror}} = 2v_{\text{boundary}}$, for example.

In addition to the mesh-based centered-difference versions of the continuum equations, and the various smooth-particle algorithms[8, 9] boundary conditions are a crucial component of any flow simulation. Even an "isolated system" needs boundaries, free, fixed, or periodic being the usual choices. Oyeon Kum carried out a detailed investigation of three models for rigid boundaries in SPAM simulations, the most straightforward of which is the leftmost situation in **Figure 11.12**. There sufficiently many particles are arranged on the boundary and, though motionless, these particles are included in the smooth-particle sums influencing the motion of the interior moving particles.

Three boundary types were used in Oyeon's work, which focussed on a comparison of Smooth Particle simulations with mesh-based analogs. Simulations with fixed particles at the top and bottom and with periodic boundaries at the sides were particularly straightforward to implement. A careful comparison of all three boundary types for the Rayleigh-Bénard flows of the type shown in **Figure 11.12** shows that the velocity and energy fields' dependence on the chosen boundary type is smaller than the deviation of the smooth-particle profiles from the finite-difference solution of the same problem. These differences between simulations with 5000 Lucy particles and 5000 meshpoints are of order a few percent which we view as perfectly satisfactory agreement. The simplicity of the Smooth Particle approach is a powerful argument for its adoption whenever its difficulties in dealing with surfaces can be avoided or ignored.

11.8.2 *Multiple Solutions with SPAM*

Victor Castillo carried out a careful investigation of the development and convergence of numerical methods for the grid-based solution of Rayleigh-Bénard flows. A stimulating result of his work can be found in Figure 1 of Reference 10. To place the work in context consider that many researchers, the most well-known of whom is Ed Jaynes (1922-1998), have suggested criteria for deciding which of many nonequilibrium flow solutions is the "most stable" or the "most likely". The usual criteria have involved maximizing the entropy or minimizing the entropy production. The Rayleigh-Bénard problem shows that no such simple recipe is likely to be correct.

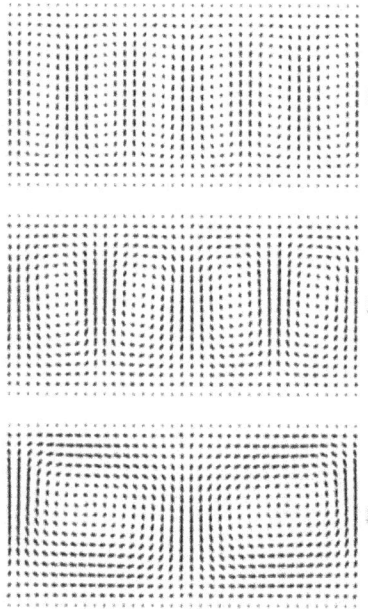

Fig. 11.13: Three different solutions of the flow equations with rigid boundaries at the top and bottom and periodic boundaries at the "sides". The six-roll solution is stable for 470 sound traversal times (after which two-roll or four-roll solutions typically emerge) while the two-roll and four-roll solutions were stable through times corresponding to 10,000 sound traversal times.

The three solutions shown in **Figure 11.13**, all for exactly the same boundary conditions at a Rayleigh Number of 40,000 show disparate properties. The six-roll solution has the least energy. The two-roll solution has the greatest entropy production. the four-roll solution has the largest growth rate when started from an infinitesimal amplitude. The fact that the two-roll and four-roll solutions are stable to machine accuracy shows that no simple recipe can distinguish their relative stability. The six-roll solution, for the same boundary conditions is stable to 470 sound traversal times suggesting that it would be fully stable for conditions close to those of the Figure. Extrapolation of finite-mesh results to the continuum limit could be carried out with three-figure accuracy in the kinetic and internal energies, the heat flux, and the growth rate from an infinitesimal mode amplitude as an initial condition. The two-roll and four-roll solutions were found to be stable in three space dimensions as well as two. Notice that any solution in two spatial dimensions is automatically a solution in three, though it is never possible to rule out a Lyapunov instability in three dimensions which has no analog in two.

The two simplest boundary conditions in continuum mechanics are "free" and "fixed". Our free-boundary models of an expanding ideal gas and our fixed-boundary Rayleigh-Bénard problems illustrate these two options. Partial containment and containment through gravity add to the variety available in continuum simulations. Let us turn to two more complicated problems. The equilibrium of a column with a rigid boundary on the bottom, periodic boundaries on the sides, and a free boundary at the top, stabilized by gravity, illustrates the three different possibilities all in the same problem. Replacing the periodic side boundaries by free ones allows for the study of collapse. We turn to these problems next.

11.9   Equilibrating and Collapsing Fluid Columns

In mid-September 2010 Carol and I participated in a delightful conference in Spain. It was a welcome chance to get together with new friends and old and to describe our research work. We had long since (2005) left the Livermore Laboratory in California behind in favor of retirement in Nevada. We took the opportunity to review our many decades of research in molecular dynamics, smooth particle applied mechanics "SPAM", and large-scale Lagrangian simulations of fluid and solid mechanics.[11] We were then, as we are now, intrigued by the close resemblance of smooth-particle mechanics and molecular dynamics. We took up two simple problems to illustrate those techniques: first, the equilibration of a column of compressible fluid in a gravitational field; second, the collapse of such a column following the release of the confining constraints perpendicular to the field direction. We used SPAM for both problems.

In all of these SPAM simulations, illustrated in **Figures 11.14-11.16**, we used a simple adiabatic equation of state to describe a fluid column in the presence of gravity :

$$P = \rho^3 - \rho^2 \longleftrightarrow e = (1/2)(\rho - 1)^2 \ .$$

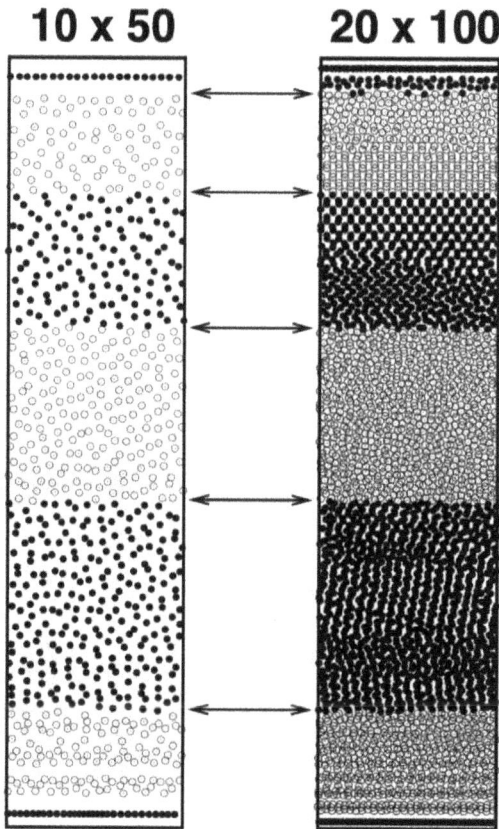

Fig. 11.14: Scale models of a smooth-particle fluid in a gravitational field. The equation of state is $P = \rho^3 - \rho^2 \longleftrightarrow e = (1/2)(\rho - 1)^2$. The gravitational field strength is $g = (1/2L)$ with $H = 5L$. The arrows indicate the five density contour values of 1.9, 1.7, 1.5, 1.3, and 1.1.

Generally it is a good plan, as implemented here, to study the "size dependence" of simulations by carrying out a series of computations. This is more informative than spending the same computational resources on just one largest-possible problem. A series of simulations gives a good idea of the reliability and convergence of the results while a single "large" simulation may not. Likewise, a single simulation in which an independent variable, such as temperature or density, varies with time is typically much more valuable than a series of separate simulations at each temperature or density.

Equilibration of a fluid column can be carried out with periodic boundaries at the sides, a powerlaw repulsion at the bottom, and a vertical gravitational field of strength g. According to continuum mechanics equilibration corresponds to a balance between the gravitational and pressure forces :

$$(dP/dy) = (dP/d\rho)(d\rho/dy) = (3\rho^2 - 2\rho)(d\rho/dy) = -\rho g \longrightarrow$$

$$g = (5/2H) \text{ and } (3\rho^2/2) - 2\rho - 2 = -gy .$$

Here we have chosen the integration constant and the gravitational field strength such that the density is 2 at the column base $y = 0$ and unity at the top of the column, $y = H$, where the pressure vanishes and the density is 1. Twofold compression at the base of a square $H \times H$ specimen containing $(8/5)H^2$ particles provides smooth pressure and density profiles.

Figure 11.14 shows excellent agreement between the density profiles obtained with 500 and with 2000 smooth particles. The plotting symbols change as the density passes through the five contour values of 1.9 to 1.1 in five equally-spaced steps of 0.2. The structures of the two scale models correspond nicely as the number of particles is quadrupled with the same pressure profile. These two simulations of equilibrated columns show that the number-dependence can be ignored for the equilibration provided a few thousand particles are used. A detailed discussion of these equilibration simulations can be found in Bill's book, *Smooth Particle Applied Mechanics*.[7] With the equilibration implemented the next step is to release the lateral boundaries in order to model a nonequilibrium collapse problem.

For the *nonequilibrium* dynamics of the collapsing column we considered two doublings of the problem scale, with widths of 20, 40, and 80 including 640, 2560, and 10,240 smooth particles. An equilibration phase like that of **Figure 11.14** was carried out first, allowing systems with original heights 32, 64, and 128 to relax and compress under the influence of gravity and a viscous damping force. By including a surface potential in these smooth-particle simulations, evaporation can be eliminated. To achieve this we used a positive potential which is near zero in the bulk fluid and which discourages density gradients :

$$\Phi_{\text{surface}} = (1/10) \sum_i (\nabla_i \rho_i)^2 .$$

The density gradient at each particle is computed from Lucy's weight function with a range of 3, $w(r) = (5/9\pi)[\ 1 - 6z^2 + 8z^3 - 3z^4\ ]$:

$$\nabla_i \rho_i = \sum_j (x_{ij}, y_{ij})(20/27\pi)[\ -(1 - z_{ij})^2\ ]\ ;\ z_{ij} \equiv (|r_{ij}|/3)\ .$$

It was also expedient to include a repulsive "core" potential to avoid the pairing of particles at short range (where the Lucy-potential repulsive force is small) :

$$\Phi_{\text{core}}(r < \sigma) = [\ 1 - (r/\sigma)^2\ ]^4 \text{ with } \sigma^2 = 0.2\ .$$

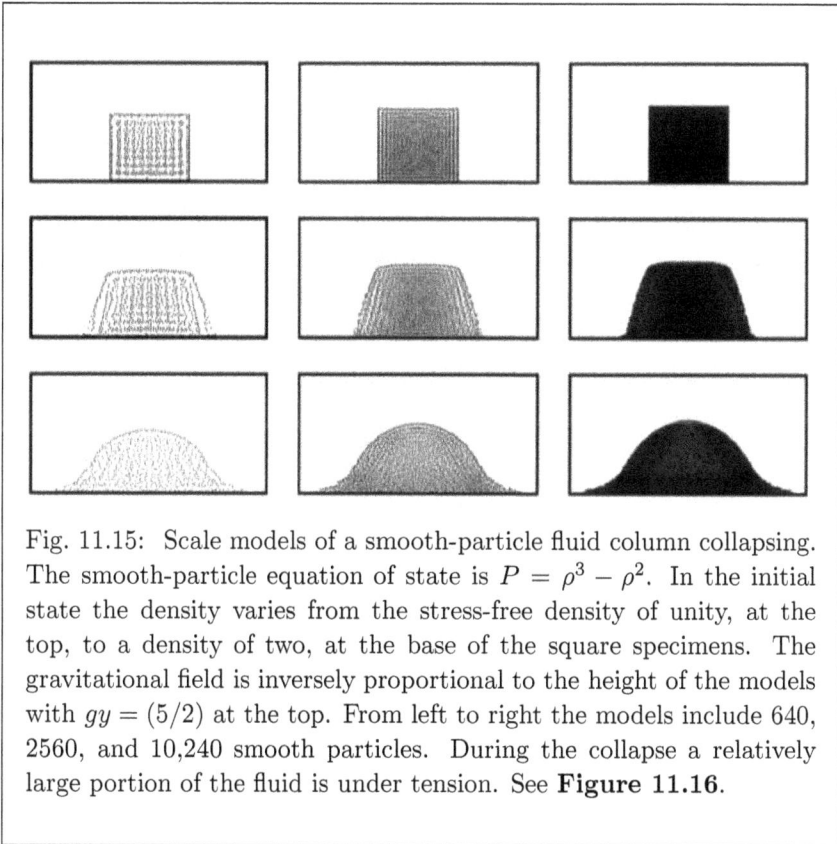

Fig. 11.15: Scale models of a smooth-particle fluid column collapsing. The smooth-particle equation of state is $P = \rho^3 - \rho^2$. In the initial state the density varies from the stress-free density of unity, at the top, to a density of two, at the base of the square specimens. The gravitational field is inversely proportional to the height of the models with $gy = (5/2)$ at the top. From left to right the models include 640, 2560, and 10,240 smooth particles. During the collapse a relatively large portion of the fluid is under tension. See **Figure 11.16**.

Figure 11.15 shows three scale models of the liquid column collapse. Only a few hundred smooth particles are required to give a good account of the deformation. With the equation of state $P = \rho^3 - \rho^2$ the sound velocity, $c = \sqrt{(dP/d\rho)}$, varies by nearly a factor of three from the top to the bottom of the initial square structures. Because the smooth-particle model follows individual particles in the two-phase region there is no need to develop and implement a separate hydrodynamic model for failure and fragmentation.

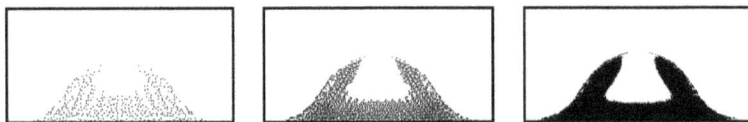

Fig. 11.16: Scale models showing the unstable tensile region during the collapse of a smooth-particle fluid column. The three models contain 640, 2560, and 10,240 particles and are shown at the corresponding times of 10, 20, and 40. The equation of state is $P = \rho^3 - \rho^2$.

Figure 11.16 shows that the interior of the collapsing structure is actually under tension and would therefore be a complicated and time-dependent mixed-phase region. Profiles and stresses in two doublings of sizes (from 640 to 2560 to 10,240 particles) showed good convergence with a tensile region covering a good fraction of the falling column.

While writing this book we have been collaborating with Karl Travis and his student Amanda Hass on the properties of two-dimensional systems with the short-ranged "84" pair potential shown at the left in **Figure 11.17** and vanishing beyond a cutoff radius of $\sqrt{2}$:

$$\phi(r < \sqrt{2})_{84} = (2 - r^2)^8 - 2(2 - r^2)^4 \longrightarrow$$

$$\phi(0) = 224 \; ; \; \phi(0.90044) = 0 \; ; \; \phi(1) = -1 \; ; \; \phi(\sqrt{2}) = 0 \; .$$

Karl and Amanda carried out phase-diagram studies based on a five-term virial series for the pressure as a (temperature-dependent) function of density. They also pursued "Gibbs'-ensemble" studies seeking to locate the gas-liquid coexistence envelope. We explored small systems, mainly with $N = 1024$, expecting to complement Karl and Amanda's work with both liquid and solid results. We were expecting to compare the SPAM column simulations with corresponding atomistic simulations using the 84 liquid.

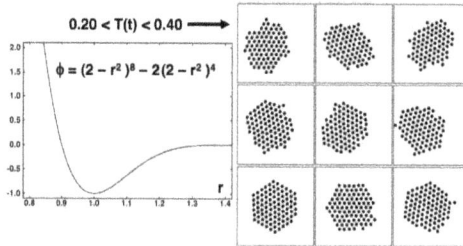

Fig. 11.17: The "realistic" "84" potential which we adopted for equilibrium studies is shown on the left. Snapshots of a 91-particle crys-
tallite over a temperature range from 0.20 (lower left snapshot has
$T = 0.22222$ with 90 particles remaining) to $T = 0.40$ (upper right
snapshot with 78 particles at $t = 18,000$). The snapshots are equally
spaced in time and temperature with temperature controlled by a Nosé-
Hoover thermostat. The initial hexagonal shape minimized the surface
energy. At these temperatures, well below any bulk melting point,
the structure undergoes shape changes through the "hexatic" sliding
of rows along the most densely packed directions with the occasional
escape of particles into the gas phase.

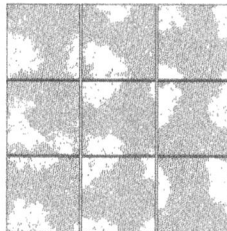

Fig. 11.18: These snapshots at a density of (5/6) are equally spaced in
time and temperature, starting out with a square lattice at $T = 0.35$.
The first snapshot (lower left) has $T = 0.35555$ and the last has T=
0.40 (upper right). There is considerable diffusion and flow but the
structure clearly shows a two-phase gas+triangular-lattice-solid system
with no trace of liquid.

It was a real surprise not to find a liquid phase for that potential ! The triangular lattice of **Figures 11.17 and 11.18** is perfectly stable at temperatures below 0.40. At higher temperatures an irregular structure appears. See **Figure 11.19**. Within the statistical fluctuations the system behaves like a harmonic solid, with a thermal energy of $\Phi = K = NkT$. It is evident that the regular triangular lattice, at a density $\sqrt{(4/3)}$ with an energy per particle of -3 is stable relative to the square lattice, which, at a density of unity, has its least energy per particle of $\Phi = -2Nk$. The evident ability of the structure to flow suggests its usefulness in modelling fluids.

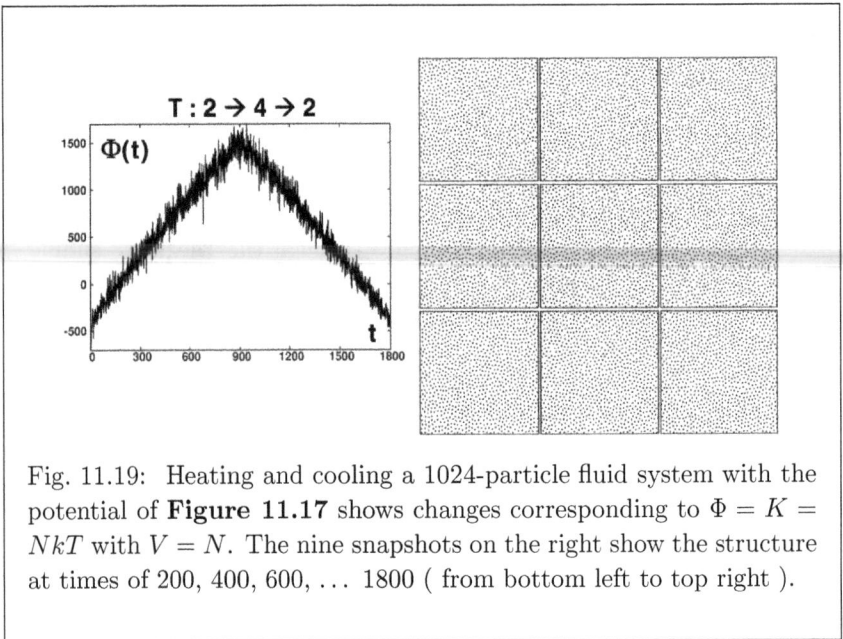

Fig. 11.19: Heating and cooling a 1024-particle fluid system with the potential of **Figure 11.17** shows changes corresponding to $\Phi = K = NkT$ with $V = N$. The nine snapshots on the right show the structure at times of 200, 400, 600, ... 1800 (from bottom left to top right).

Exercise : Write a molecular dynamics program for the 84 potential and show that the square lattice is stable at a density of $\rho = 3$ and a temperature of $T = 0.4$. Use periodic boundary conditions.

Small systems of a few thousand particles having a free surface can lose particles through evaporation or sublimation rather than melting. This mechanism can frustrate the study of collapse. A triangular-lattice crys-tallite with a free-surface boundary has some of its particles linked to no more than three neighbors. Bulk particles are linked to six, and are accordingly more closely bound. The relatively short-ranged 84 potential provides nearest-neighbor interactions with negligible second-neighbor attraction. As a result the weakly-bound surface particles provide a mechanism for sublimation rather than melting at the surface. This enhanced sublimation has been documented in three-dimensional systems and further discussed[12, 13] for the near-spherical Buckminster Fullerene C_{60} molecule.

11.10 Summary of Lecture 11

In these eleven lectures we have stressed the benefits of comparing microscopic and macroscopic models of matter. The correspondence of the two points of view is most useful and convincing for bulk matter where phase boundaries are unimportant. Surfaces present special difficulties for the smooth-particle approach to fluid dynamics because tension corresponds to purely-attractive forces. Tricks like the surface potential $\propto (\nabla \rho)^2$ and repulsive "core" potentials can mitigate this problem, as we showed in simulating the equilibration and collapse of liquid columns.

Solving Rayleigh-Bénard problems is an excellent way to begin a study of continuum mechanics. Fixed grids with static boundaries make for relatively easy programming. The variety of flow patterns in two space dimensions is a rewarding introduction to compressible fluid mechanics. It is also a problem area well-suited to smooth-particle simulations.

The atomistic approach to these same problems is particularly complicated for short-ranged attractions. To our surprise we were not only unable to demonstrate the collapse of a liquid column, we were also unable to demonstrate the existence of a liquid phase. We expected to find that the liquid exists *somewhere* in the proximity of densities and temperatures of order unity by applying gravity to a confined fluid column. Our attempts produced large grains of triangular-lattice solid but no liquid.

Why not heat a semi-infinite strip of solid slowly, using periodic boundary conditions in the x direction with free boundaries in the y ? Trying this we found that a gradual increase of the temperature in the strip gave sublimation rather than melting. The same effect, sublimation rather than melting, was displayed by the small hexagonal sample shown in **Figure 11.17**. In **Figure 11.18** the absence of the liquid phase is evident in the density and temperature regime where one expects to see two equilibrating phases, a gas stabilized by its higher entropy $\propto \ln(V)$ and a condensed phase stabilized by its binding energy.

At higher temperatures still a variety of condensed phases can coexist, particularly in three dimensions and in systems with impurities or with more than a single component. Most of the work carried out in simulating such "real-world" problems uses packaged software which required many man-years of development. In interpreting results from packages it is essential to keep in mind that the underlying models are imitations of reality and that classical mechanics itself is likewise an imitation of atomistic behavior, good in some cases, but not in others.

In these lectures we have discussed problems many of which can easily be programmed in a few minutes plus some that require a few hours or even a day of coding. The background knowledge acquired by solving such problems is particularly useful in judging the plausibility of results from packaged software. There are many examples in the literature of serious errors committed by using such software outside its logical limits. We expect to learn more from today's students as they develop their computational skills by working through a variety of interesting problems. A fringe benefit of the experience of doing the work is an enhanced ability to detect results which are either "nonsense" or "too good to be true".

References

1. Wm. G. Hoover and C. G. Hoover, *Simulation and Control of Chaotic Nonequilibrium Systems*, Section 4.7.1, page 96 (World Scientific, Singapore, 2015).
2. Wm. G. Hoover and C. G. Hoover, *Time Reversibility, Computer Simulation, Algorithms, Chaos*, Sections 6.6.1 and 6.6.2, pages 217-221 (Second Edition, World Scientific, Singapore, 2012).
3. O. Kum, Wm. G. Hoover, and H. A. Posch, "Viscous Conducting Flows with Smooth Particle Applied Mechanics", Physical Review E **52**, 4899-4908 (1995).
4. V. M. Castillo and Wm. G. Hoover, "Entropy Production and Lyapunov Instability at the Onset of Turbulent Convection", Physical Review E **58**, 7350-7354 (1998).
5. D. C. Rapaport, "Hexagonal Convection Patterns in Atomistically Simulated Fluids", Physical Review E **73**, 025301(R) (2006).
6. J. J. Monaghan, "Smoothed Particle Hydrodynamics", Annual Review of Astronomy and Astrophysics **30**, 543-574 (1992).
7. Wm. G. Hoover, *Smooth Particle Applied Mechanics – the State of the Art* (World Scientific, Singapore, 2006).
8. O. Kum. *Nonequilibrium Flows with Smooth Particle Applied Mechanics*. Ph D thesis, University of California at Davis-Livermore (July, 1995).
9. V. M. Castillo, Jr, *Cubic Spline Collocation Method for the Simulation of*

Turbulent Thermal Convection in Compressible Fluids, Ph D thesis, University of California at Davis-Livermore (January, 1999).

10. V. M. Castillo, Wm. G. Hoover, and C. G. Hoover, "Coexisting Attractors in Compressible Rayleigh-Bénard Flow", Physical Review **55**, 5546-5550 (1997).

11. Wm. G. Hoover and C. G. Hoover, "Three Lectures: NEMD, SPAM, and Shockwaves", in *Nonequilibrium Statistical Physics Today*, in the Proceedings of the 11th Granada Seminar on Computational and Statistical Physics, 13-17 September 2010, editted by P. L. Garrido, J. Marro, and F. de los Santos, American Institute of Physics Conference Proceedings **1332**, pages 23 through 55 (Melville, New York, 2011)=arχiv 1008.4947.

12. M. H. J. Hagen, E. J. Meijer, G. C. A. M. Mooij, D. Frenkel, and H. N. W. Lekkerkerker, "Does C_{60} Have a Liquid Phase ?", Nature **365**, 425 (1993).

13. Wm. G. Hoover and C. G. Hoover, "Time-Irreversibility is Hidden Within Newtonian Mechanics", Molecular Physics (submitted, 2018) = arχiv:1801.09899.

Chapter 12

Epilogue

We thoroughly enjoyed our visit to India. The stimulation, friendship, and joy of the students were the main reward. The touring and hospitality we enjoyed through Professor Bhattacharya's kindness cannot be repaid. This book is our own effort to pass on our picture of computational physics' fundamentals in a form conducive to future research and discoveries. It is an act of love. We are confident that microscopic and macroscopic models of the kinds described here will furnish intellectual stimulation and joy and will have practical consequences for the research careers of our readers.

It has been illuminating for the two of us to become aware of the unending mysteries which can be discovered, even rediscovered, and explained through computer simulation. In the process of converting our lecture notes to book form we happened on many unexpected findings, the most recent being the lack of a useful liquid phase for the "realistic" 84 potential, the usefulness of one of William Milne's predictor integrators for bit-reversible simulations of irreversible flows, and the pervasive instability of two of his integrators, one a predictor and the other not, for first-order differential equations.

We hope and expect that the simulation of simple systems will continue to surprise and stimulate students and researchers far into the unforeseeable future.

The Yale motto is our goal: *"Lux et Veritas"*.

Bill and Carol Hoover
Ruby Valley Nevada
March 2018

Index